Electrode Processes in Solid State Ionics

NATO ADVANCED STUDY INSTITUTES SERIES

Proceedings of the Advanced Study Institute Programme, which aims at the dissemination of advanced knowledge and the formation of contacts among scientists from different countries

The series is published by an international board of publishers in conjunction with NATO Scientific Affairs Division

A	Life Sciences	Plenum Publishing Corporation London and New York
B	Physics	
C	Mathematical and Physical Sciences	D. Reidel Publishing Company Dordrecht and Boston
D	Behavioral and Social Sciences	Sijthoff International Publishing Company Leiden
E	Applied Sciences	Noordhoff International Publishing Leiden

Series C – Mathematical and Physical Sciences

Volume 25 – Electrode Processes in Solid State Ionics

Electrode Processes in Solid State Ionics

Theory and Application to Energy Conversion and Storage

Proceedings of the NATO Advanced Study Institute held at Ajaccio (Corsica), 28 August–9 September 1975

edited by

M. KLEITZ

Institut National Polytechnique de Grenoble, Grenoble, France

J. DUPUY

Université Claude Bernard, Lyons, France

D. Reidel Publishing Company

Dordrecht-Holland / Boston-U.S.A.

Published in cooperation with NATO Scientific Affairs Division

Library of Congress Cataloging in Publication Data

NATO Advanced Study Institute, Ajaccio, Corsica, 1975.
Electrode processes in solid state ionics.

(NATO advanced study institutes series : Series C, Mathematical and physical sciences ; v. 25) an Includes bibliographies.
1. Electrodes Congresses. 2. Solid state chemistry Congresses. 3. Ionic crystals Congresses. I. Kleitz, M., 1937– II. Title. III. Series.
QD571.N28 1975 541'.3724 75–44113
ISBN-13: 978-94-010-1891-3 e-ISBN-13: 978-94-010-1889-0
DOI: 10.1007/978-94-010-1889-0

Published by D. Reidel Publishing Company
P.O. Box 17, Dordrecht, Holland

Sold and distributed in the U.S.A., Canada, and Mexico
by D. Reidel Publishing Company, Inc.
Lincoln Building, 160 Old Derby Street, Hingham, Mass. 02043, U.S.A.

CONTENTS

PREFACE

The idea of an "Advanced Study Institute" on the theme of electrode reactions on solid electrolytes was put forward by Dr. J. Dupuy at the meeting of the International Society for Electrochemistry in Eindhoven in September 1973. Through Dr. Dupuy, the Solid State Physics Department of Lyons University offered the Institute possibilities of accommodation in Corsica that seemed particularly tempting.

The subject matter appealed to a number of people for a variety of reasons. A great deal of development work on applications comes up against interface phenomena which appreciably reduce anticipated performances. Numerous potential applications of specific electrodes or gauges appear that would benefit from a more systematic approach. From a more fundamental viewpoint, interface phenomena on ionic crystals are the subject of independent investigations in quite distinct research fields such as solid state physics and electrochemistry. The choice of an interpretation from among the different models available is very often not a straightforward matter, and an attempt to promote a synthesis by bringing together the proponents of the various "schools" could not fail to be rewarding.

The value of these arguments was recognized by NATO which awarded a substantial grant to the Institute. Further subsidies were awarded by the Mission Interministérielle pour l'Aménagement et l'Equipement de la Corse (Inter-Ministery Body for the Development of Corsica), Thomson-CSF Company, the Physics Faculty of Claude Bernard University in Lyons, and the Institut Polytechnique in Grenoble. The National Science Foundation granted travelling scholarships to young scientists from the United States taking part in the Institute. The members of the Organizing Committee and all the participants are greatly indebted to these organizations.

The project of the "A.S.I." took shape in the course of later meetings of the European Group for Solid State Electrochemistry (Besançon February 1974, Bressanone, December 1974), which assembled most of the members of the Scientific Committee.

Dr. D.O. Raleigh took charge of liaising with the United States and was a source of sound advice for the Committee. All these persons must be warmly thanked for their enthusiastic encouragement and most efficient help.

A vast bibliographic file was used to publicize the forthcoming meeting and was implemental in contacting many young specialists who had recently obtained valuable results in research fields that are traditionally distinct from Solid State Electrochemistry. The number of applications to attend the Institute far exceeded acceptable limits and almost half these requests had to be regretfully turned down.

This work is a collection of the teaching papers presented over the twelve days of study at the Institute. The courses were followed by fruitful discussions and work sessions on the following themes :

. electrode polarization and mechanisms
. electrode materials
. electrolyte reduction and decomposition
. electrochemical methods
. electrode reactions on semiconductors
. typical applications

During these work sessions original results were presented. Unfortunately, for economic reasons the number of pages of this book is stringently limited. Consequently, it is to be regretted that all this second facet of the Institute is not covered in the proceedings, and in particular the summaries of the discussions which followed the teaching papers and which would have formed a very valuable complement.

In this field where development is so rapid and the number of people concerned so much greater than the actual participants in the Institute, it was felt that a prompt publication of the proceedings was essential. Consequently, some material defects are unavoidable and we trust that the reader will take them in good part.

To conclude, I would like to thank those who, although their names do not appear in any official capacity, made an essential contribution to the success of the Institute :

- the late Pascal Rossini, Mayor of the City of Ajaccio, and Jacques de Rocca Serra, Directeur de la Mission Régionale

for the sightseeing outings they organized and offered to the participants and which delighted us all

- Madame Cavalier and Monsieur Fouletier

for the considerable work they accomplished in the preliminary stages of the organization.

Grenoble
October 1975

Michel Kleitz

LIST OF PARTICIPANTS

Directors :

M. KLEITZ, I.N.P. Grenoble, France
C.H.S. DUPUY Co-Director, Université Claude Bernard, Lyons,France
J. DUPUY Co-Director, Université Claude Bernard, Lyons, France

Scientific Committee :

R.D. ARMSTRONG, University of Newcastle-Upon-Tyne, England
R.M.DELL, UKEA-Harwell, England
C. DEPORTES, Institut National Polytechnique, Grenoble, France
D.H. COLLINS, Admiralty Engineering Laboratory, England
H. GERISCHER, Fritz-Haber-Institut der Max Plank Gesellschaft, Berlin, Germany
S. PIZZINI, University of Milano, Italy
B.C.H. STEELE, Imperial College, London, England
D.O. RALEIGH, Science Center, Rockwell Int. California, U.S.A.
H. RICKERT, University Dortmund, Germany
H. TANNENBERGER, Institut Battelle, Geneva, Switzerland

Organization Committee :

C. CHASSAGNE, Secretary General, Université Claude Bernard, Lyons
M. DUCLOT, Treasurer, I.N.P. Grenoble, France

Secretarial and Technical staff :

P. BELL, Translator Interpreter, Bureau de traduction, Grenoble
M. BONNAT, I.N.P. Grenoble, France.

<u>Australia</u> :

L.B. HARRIS - School of Physics, University of New South Wales, P.O. Box 1, Kensington NSW 2033.

<u>Canada</u> :

T.H. ETSELL - Department of Metallurgy and Materials Science, University of Toronto, 184 College Street, Toronto M5S 1A4.

M. GAUTHIER - Institut de Recherche de l'Hydro-Québec, 1800 montée Ste Julie, Varennes, Québec.

<u>Denmark</u> :

S.A. MARKLAND - Vicking Chemicals, P.O. Box 226, 4000 Roskilde

<u>France</u> :

M. ARMAND - Laboratoire de Cinétique Electrochimique Minérale, ENSEEG, Domaine Universitaire, 38401 St Martin d'Hères.

J. BERT - Dept. Physique des Matériaux, Université Claude Bernard 43 Bd du 11 Novembre, 69100 Villeurbanne.

G. CHASSAGNE - Dept. Physique des Matériaux, Université Claude Bernard, 43 Bd du 11 Novembre, 69100 Villeurbanne.

B. CLAUDEL - Institut de Recherches sur la Catalyse, 39 Bd du 11 Novembre 1918, 69100 Villeurbanne.

M. COULON - Laboratoire d'Adsorption et Réaction de gaz sur solides, ENSEEG, Domaine Universitaire, 38401 St Martin d'Hères.

M. CROSET - Thomson CSF, Laboratoire Central de Recherche, BP 10 Corbeville/Orsay 91401.

M. DUCLOT - Laboratoire de Cinétique Electrochimique Minérale, ENSEEG , Domaine Universitaire, 38401 St Martin d'Hères.

C.H.S. DUPUY - Dept. Physique des Matériaux, Université Claude Bernard, 43 Bd du 11 Novembre, 69100 Villeurbanne.

J. DUPUY - Dept. Physique des Matériaux, Université Claude Bernard, 43 Bd du 11 Novembre, 69100 Villeurbanne.

P. FABRY - Laboratoire de Cinétique Electrochimique Minérale, ENSEEG, Domaine Universitaire, 38401 St Martin d'Hères.

J. FOULETIER - Laboratoire de Cinétique Electrochimique Minérale, ENSEEG, Domaine Universitaire, 38401 St Martin d'Hères.

C.G. GABRIELLI - Physique des Liquides et Electrochimie, Université Pierre et Marie Curie, Tour 22, 4 Place Jussieu, 75230 Paris Cedex 05.

M. GOUET - U.E.R. Sciences, Département de Physique, Université de Paris Val de Marne, 94000 Créteil.

M.H. GUENIN - Institut de Recherches sur la Catalyse, 39 Bd du 11 Novembre 1918, 69100 Villeurbanne.

M. KLEITZ - Laboratoire de Cinétique Electrochimique Minérale, ENSEEG, Domaine Universitaire, 38401 St Martin d'Hères.

G. LETISSE - U.E.R. Sciences, Université de Paris Val de Marne, 94000 Créteil.

G. ROBERT - Laboratoire d'Electrochimie, Université de Besançon, Faculté des Sciences, La Bouloie, Route de Gray, 25030 Besançon Cedex.

E. SCHOULER - Laboratoire de Cinétique Electrochimique Minérale, ENSEEG, Domaine Universitaire, 38401 St Martin d'Hères.

J. SIEJKA - Groupe de Physique des Solides de l'ENS, Tour 23, 2 Place Jussieu, 75005 Paris.

J. VEDEL - Laboratoire d'Electrochimie Analytique et Appliquée, 11, rue Pierre et Marie Curie, 75231 Paris Cedex 05.

Germany :

H. DOEBRICH - Battelle Institut e.V. 6000 Frankfurt am Main 90, Am Römerhof 35, Postfach 900 160.

P. BINDRA - Fritz Haber Institut der Max Planck Gesellschaft, Faradayweg 4-6, 1 Berlin 33.

L. CONTRERAS-VENEGAS - Universität Dortmund, Lehrstuhl für Physikalische Chemie, 46 Dortmund 50, Postfach 50 05 00.

G. EICHINGER - Technische Universität 8 München 2, Arcisstrasse 21, Postfach 20 24 20.

H. GERISCHER - Fritz Haber Institut der Max Planck Gesellschaft, D1 Berlin 3/Dahlem, Faradayweg 4-6.

J. GROBRECHT - Fritz-Haber-Institut der Max Planck Gesellschaft, D.1000 Berlin 33, Faradayweg 4-6.

B. HARTMANN - University of Dortmund, Lehrstuhl für Physikalische Chemie, 46 Dortmund Hombruch, Postfach 500.

G. HOLZÄPFEL - Universität Dortmund, Lehrstuhl für Physikalische Chemie, 46 Dortmund 50, Postfach 50 05 00.

P.J. MESCHTER - Max-Planck Institut für Eisenforschung, 4000 Düsseldorf, Max Planck Strasse 1, Postfach 14 01 40.

B. PETTINGER - Fritz-Haber-Institut der Max Planck Gesellschaft, Faradayweg 4-6, 1 Berlin 33.

Great Britain :

R.D. ARMSTRONG - The University of Newcastle-Upon-Tyne, School of Chemistry, Newcastle-Upon-Tyne NE1 7RU.

N. BONANOS - Applied Sciences, Stage II, University of Sussex, Falmer Brighton, Sussex BN1 9QT.

D.H. COLLINS - Admiralty Engineering Laboratory, West Drayton, Middx, UB7 9BZ.

R.M. DELL - Atomic Energy Research Establishement, Applied Chemistry, Division, Building 220.23 Harwell, Didcot OX11 ORA Oxon.

T. DICKINSON - Electrochemistry Research Laboratories, University of Newcastle-Upon-Tyne, Newcastle-Upon-Tyne NE1 7RU.

G.J. DUDLEY - Dept. Metallurgy, Materials Science, Imperial College, Prince Consort Rd, London SW7.
M. GREEN - Imperial College of Science and Technology, Department of Electrical Engineering, Exhibition Road, London SW7, 2BT U.K.
A.H. HARKER - Theoretical Physics Division, AERE Harwell OX11 ORA.
D.A. HOOPER - Building 552, Materials Development Division, AERE Harwell, DIDCOT. Oxon.
J.M. POLLOCK - School of Chemistry, Leicester Polytechnic, Leicester.
B.C.H. STEELE - Imperial College of Science and Technology, Dept. of Metallurgy, Royal School of Mines, Prince Consort Road, London SW7 2BP.
J.E. STRUTT - Applied Sciences, Stage II, University of Sussex, Falmer, Brighton, Sussex BN1 9OT.
K. TAYLOR - Electrochemistry Research Laboratories, University of Newcastle-Upon-Tyne, Newcastle-Upon-Tyne NE1 7RU.

Israël :

M. STEINBERG - The Hebrew University of Jerusalem, Department of Inorganic and Analytical Chemistry, Jerusalem.

Italy :

G.B. BARBI - Euratom, Casella Postale 1, Ispra, Varèse.
C. CALANDRA - Istituto Di Fisica, Via Universita 4, 41100 Modena.
F. COLOMBO - Universita di Milano, Istituto di Elettrochimica e Metallurgia, Via Venezian 21, Milano.
A. GENNARO - Institute of Physical Chemistry, University of Padua, Via Loredan 2, 35100 Padua.
C.L. MARI - Istituto di Elettrochimica dell'Universita di Milano, Via Venezian 21, 20133 Milano.
A. MASSERINI - Istituto di Elettrochimica del Politecnico, Piazza Leonardo da Vinci 32, 20133 Milano.
L. OSCAR - Universita di Milano, Istituto di Elettrochimica e Metallurgia, Via Venezian 21, Milano 20133.
S. PIZZINI - Universita di Milano, Istituto di Elettrochimica e Metallurgia, Via Venezian 21, 20133 Milano.

The Netherlands :

R.W. BONNE - Solid State Department, State University of Utrecht, Sorbonnelaan 4, Utrecht, De Uithof.
D. den ENGELSEN - Philips Research Lab. Eindhoven.
L. HEYNE - Philips Research Laboratories, Eindhoven.
J. SCHOONMAN - Solid State Department, State University of Utrecht, Sorbennelaan 4, Utrecht, De Uithof.

Norway :

S. STUEFLOTTEN - University of Trondheim, Norwegian Institute of Technology, Division of Physical Electronics, N 7034, Trondheim NTH.

Romania :

V.M. LAZARESCU - Centre de Chimie Physique, Rue Galati 31, Bucarest.
S. TANASESCU - Centre de Chimie Physique, Rue Galati 31, Bucarest.
N. TOTIR - Centre de Chimie Physique, Rue Galati 31, Bucarest.

Spain :

F. JAQUE - Departamento de Fisica, Universidad Autonoma de Madrid, Cantoblanco, Madrid.
M.T. MONTOJO - Centro de Investigacion y Desarrollo de la Armada, Arturo Soria 289, Madrid 33.

Sweden :

B. HEED - Chalmers University of Technology Fack S.402-20 Göteborg 5.
K. SCHROEDER - Chalmers University of Technology Fack S-402 20 Göteborg 5.

Switzerland :

M. VOINOV - Battelle, Centre de Recherche de Genève, 7 Route de Drize, 1227 Carouge, Genève.

U.S.A :

J.M. BLAKELY - Dept. of Materials Science and Eng.College of Engineering, Cornell University, Bard Hall, Ithaca, New York 14853.
T.R. BRUMLEVE - Dept. of Chemistry, University of North Carolina, Chapel Hill, North Carolina 27 514.
R.L. DOTY - Medtronic Inc. 3055 Old Highway 8, Minneapolis, Minnesota 55418.
G.C. FARRINGTON - General Electric Corporate Research & Development Center, Building K-1, Room 3B2, P.O. Box 8, Schenectady NY 12301.
D.L. FOSTER - Clemson University, Clemson, South Carolina 29631.
H.K. HENISCH - The Pennsylvania State University, University Park, Pennsylvania 16802.
R.A. HUGGINS - Center for Materials Research, Stanford University, Stanford California 94305.
T.G. JOW - Department of Materials Science Tech. Institute, Northwestern University, 2145 Sheridan Road, Evanston, Illinois 60201.
F.A. KROGER - Department Materials Science, Univ. Southern California, University Park, Los Angelès, Calif. 90007.
J.R. MACDONALD - Department of Physics and Astronomy, The University of North Carolina, At Chapel Hill, North Carolina 27514.
P. MARK - Department of Electrical Engineering, Brackett Hall, Engineering Quadrangle, Princeton, New Jersey 08540.
I.D. RAISTRICK - Center for Materials Research, Stanford University, Stanford California 94305.

D.O. RALEIGH - Science Center, Rockwell International, P.O. Box 1085,1049 Camino Dos Rios, Thousand Oaks, California 91360.

R.D. SHANNON - Central Research Dept. Experimental Station, E.I. du Pont de Nemours, Wilmington, DEl 19898.

R.J. SOJKA - 132 Larsen Hall, Dept. of Electrical Engineering, University of Florida, Gainesville Flo. 32601.

J.B. WAGNER - Materials Research Center, Technological Institute, Northwestern University, Evanston, Illinois 60201.

D.H. WHITMORE - Dept. of Materials Science and Engineering, Northwestern University, Evanston, Illinois 60201.

DESCRIPTIONS OF JUNCTIONS BETWEEN ELECTRONIC AND SOLID IONIC CONDUCTORS

M. Kleitz, P. Fabry, E. Schouler

Laboratoire de Cinétique Electrochimique Minérale
Associé au C.N.R.S. (E.R.A. N° 384)
E.N.S. d'Electrochimie et d'Electrométallurgie de
GRENOBLE - Domaine Universitaire - BP 44 -
38401 SAINT MARTIN D'HERES - FRANCE

From a simple experimental viewpoint an electrode can be defined as a location in a chain of electric conductors where the conductivity changes in nature from an electronic to an ionic conductivity. Figure 1 shows a schematic electrode where the ionic conductivity results from the mobility of A^+ ions. The transition can occur either as a discontinuity at a phase boundary or even as a gradual change in a single phase material[1]. The junctions between electronic and solid ionic conductors belong to this group. The originality in their studies arises mainly from the plurality of the descriptions of solid ionic conductors. They are sometimes considered as electrolytes and treated within the scope of the electrochemical formalism. In many respects they can also be considered as semi-conductors.

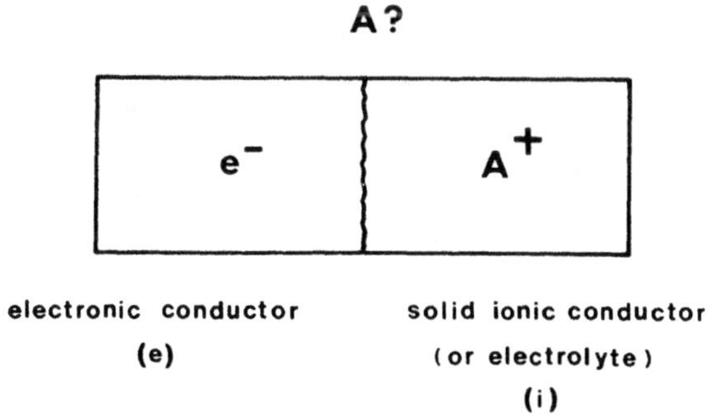

Fig. 1- Sketch of an electrochemical junction.

M. Kleitz and J. Dupuy (eds.), Electrode Processes in Solid State Ionics, 1–17.

With reference to traditional electrolytes, the peculiarity of solid electrolytes results from their additional electronic conductivity. Other particularities may be mentioned : the solidity of the electrolyte suppresses any convection, the ions are not solvated. Some solid electrolytes have a very simple crystalline structure where the energy profile of the path followed by the ions is known. This would make the situation ideally simple for electrode process studies. Unfortunately, this is frequently counteracted by the great complexity of the electrode surface which is generally formed by a contact between two solids. This point has already been stressed by Raleigh[2].

With reference to traditional semi-conductors, the peculiarity of these materials results essentially from the rather high mobility of some ions. This frequently means that some electronic traps are mobile and play a dominant part, instead of the majority electronic carriers, in determining the electrostatic potential profile in the material. It also means that any current passing through the material may have chemical effects. When a current is flowing through the electrode sketched in figure 1, in such a direaction that A^+ions migrate from the right to the left, at the electrode, the electrons e^- combine with the ions to form neutral A particles. This may result in particular properties according to the nature of the reservoir where these neutral particles are accommodated. If they are retained as adsorbed particles at the phase boundary, where they create surface states, the local electronic distribution and Fermi level position may be functions of the current which has passed through the electrode. In the same way, a current passing through a solid electrolyte may modify the local concentrations of electronic traps in the bulk of the material. In fact, the solid regarded as a semi-conductor is frequently heterogeneous.

The existence of these different descriptions has led to different interpretations of the electric and chemical phenomena observed at electrodes, and, a few years ago, when starting research in the field, the crucial point was the choice of a "school". If the investigator opted for the physicist school, the voltage U of a cell involving an ionic conductor, for instance, would be described as a variation of the electrostatic potential within the crystal (Fig.2a). This would result in particular from space charges in the vicinity of the electrodes. If the investigator opted for the Wagner theory, the voltage would be viewed as proportional to the variation of the electron electrochemical potential $\tilde{\mu}_e$ through the ionic crystal (Fig.2b). This variation would be due to changes in composition resulting from the equilibria with the phases in contact at the electrodes. If the investigator opted for the electrochemist family, the voltage would be viewed again as proportional to a variation of the electron electrochemical potential (Fig.2c) but, in this case, it would only be locally determined at the electrodes by the reactions taking place there.

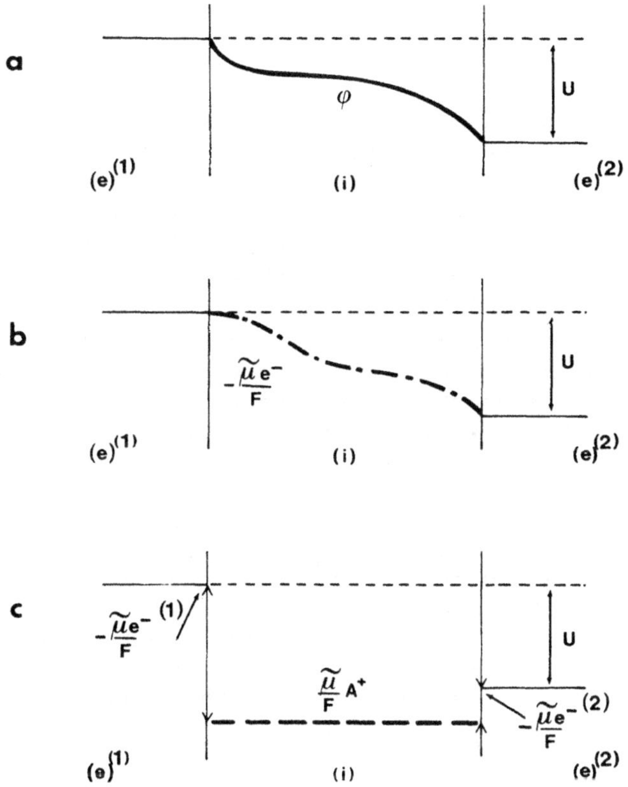

Fig. 2- Determination of a cell voltage according to different schools (see text).

The frequently observed E(I) characteristic :

$$I = I_o \left(\exp \frac{FE}{RT} - 1\right) \qquad (1)$$

is another good example. In the literature, its writing varies in accordance with the different terminologies employed. (For the sake of consistency all the formulae are written here with the electrochemical parameters R, the gas constant and F the Faraday constant). It frequently involves specific correcting terms. It is interpreted within the framework of the Wagner theory[3,4] as describing the influence of the polarization potential on the small additional electronic conductivity of the electrolyte. The parameter I_o is written :

$$I_o = \frac{RT}{LF} \sigma_+ \qquad (2)$$

where σ_+ is the electronic conductivity of the material in equilibrium with the counter electrode (and L the sample length). An electrochemist[5,6] would rather write this characteristic :

$$E = \frac{RT}{F} \ln \frac{I_o - I}{I_o} \tag{3}$$

and consider it as evidence of the existence of a limiting supply rate for one electrode reactant (the ion A^+ or the neutral A particle in our schema). A physicist[7] would immediately recognize in it the characteristic of a Schottky barrier. Obviously additional information can help to substantiate one of these options. However, it must be admitted that, in the past, the option was frequently a matter of faith.

A number of such dramatic differences can also be quoted in the experimental field. With the cell C / AgBr/ Ag generally studied within the framework of the Wagner theory, it is commonly admitted that after a while, under open circuit conditions, the silver activity is equal to one through the whole AgBr crystal if the gas surrounding the carbon electrode is inert. With the cell $Ar\text{-}O_2$, $Pt/ZrO_2\text{-}Y_2O_3/Pt$, Air, mainly used for thermodynamic measurements, it is also commonly admitted that whatever the partial pressure of the oxygen traces in argon, they are measured by the cell. This means that however "inert" the argon, the platinum electrode is in equilibrium with it. In one case, the homogeneization trend within the electrolyte is regarded as sufficient to dominate the behaviour of the cell, in the other, it is ignored.

Nowadays, the scientific schools are not so well delineated, but, for a newcomer the situation is still somewhat confusing. One way to disentangle this situation could be found in a systematic search for the dominant electrode elementary process and reactants. The dominant electrode elementary process is simply the one with the larger exchange current. The dominant elementary reactants are those for which changes in concentration directly result in noticeable changes in the electrode properties. The possibility and the interest of considering certain constituents of a system as dominant in the determination of its electrical and chemical properties has been emphasized for instance by Roberts and Schmidlin[8] for the electronic distribution in large band gap semiconductors. However obvious such a reference to the concept of dominant process and reactants may be in our case, more insights could certainly be gained by promoting it. For instance, it could be an easy way to correct an excessive reference to the ideal equilibrium conditions which is sometimes misleading. The cells based on solid oxide electrolytes provide a good illustration of the point. Probably due to the fact that they have been mainly used for thermodynamic measurements, it is usual to characterize the chemical systems which form the electrodes by the oxygen pressures which are, or would be, in equilibrium with them, consequently the interfacial process is described as an oxygen electrode reaction (written in Kröger's notations)

$$1/2\ O_2 + 2e^- + V_o^{\cdot\cdot} \leftrightarrows O_o \qquad (4)$$

The experimental studies of such a reaction performed with pure oxygen have shown, as expected, that its overall rate decreases as the oxygen pressure decreases. When dealing with an electrode where the oxygen pressure is fixed by CO-CO_2 mixtures, the observed electrode reaction rates are much higher than those extrapolated from the previous results. As long as no further data are known on the reaction kinetics it would be more reasonable not to look for peculiar variations of reaction-(4)-parameters but to interpret the discrepancy by the fact that another reaction has taken over the role of dominant process. The reaction which has been proposed :

$$CO_2 + 2e^- + V_o^{\cdot\cdot} \longrightarrow CO + O_o \qquad (5)$$

which does appear to occur in a specific electrode-potential range, supposes the possibility of direct exchange of oxygen between the electrolyte and the carbon monoxide and dioxide molecules. This phenomenon has already been observed in studies of adsorption and catalysis on CeO_2 based materials[10]. Unfortunately, results from similar studies on thoria based materials are rather contradictory[11]. With an electrode formed with a copper-cuprous oxide mixture Cu-Cu_2O[12] the overall reaction rate is also much higher than that expected for reaction (4). The dominant process, in this case, could be a simple exchange of oxide ions.

$$O^{2-} + V_o^{\bullet\bullet} \leftrightarrows O_o \qquad (6)$$

A similar reaction was studied by Fabry[13] at the interface between two stabilized zirconia crystals and was observed to occur without an intermediate step of neutral oxygen formation.

Possible dominant electrode processes - ion transfer

Let us now consider in some detail the simplest possible dominant processes and, first, the situation where the electronic conductor also plays the part of main reservoir for the A particles (Fig.1). Many experimental examples can be quoted :

S-NaS_n / β-Alumina[14]
Hg-(Na) / β-Alumina[15]
Intercalated compound (Na) / β-Alumina[16]
Ag/AgCl[17]
$Ag_{liq.}$ (O_2) / ZrO_2-Y_2O_3[18]
Fe-FeO / ThO_2-Y_2O_3[19]

In the last two situations oxygen and O^{2-} anions are involved. A priori, the electronic conductor which contains e^- electrons and A particles can also be viewed as containing A^+ions. Consequently, we could consider the simple exchange of A^+ions :

$$A^+_{(e)} \rightleftarrows A^+_{(i)} \tag{7}$$

as a possible dominant process for the electric charge transfer from one conductor to the other. An example has just been mentioned with the exchange of O^{2-} ions. This description will certainly appear oversimplified to many electrochemists. A simple process is however more likely to prevail here than with aqueous electrolytes ; no solvation occurs in solid electrolytes and similar bindings are more likely to link the particle to its surrounding on both sides of the phase boundary. So far this process has not been held in great favour among theoreticians. However, the recent development of new electrode materials such as intercalated compounds[16] encourages its recognition as a genuine electrode process.

For such an electrode the equilibrium condition is simply expressed by the equality of the relevant electrochemical potentials :

$$\tilde{\mu}_{A^+}^{(e)} = \tilde{\mu}_{A^+}^{(i)} \tag{8}$$

With the usual splitting of an electrochemical potential into a chemical potential μ_{A^+} and an electrostatic energy $F\phi$

$$\tilde{\mu}_{A^+} = \mu_{A^+} + F\phi \tag{9}$$

we find from this equilibrium condition, that the electrostatic potential difference between the bulks of the two phases is fixed by their chemical composition :

$$\phi^{(e)} - \phi^{(i)} = -\frac{1}{F}\,(\mu_{A^+}^{(e)} - \mu_{A^+}^{(i)}) \tag{10}$$

The distribution of the A^+ ions in the electrolyte double layer and the electrostatic potential profile between the limits $\phi^{(e)}$ and $\phi^{(i)}$ can be in principle calculated,for instance according to the method proposed by Blakely[20] and co-workers.

With regard to the behavior of such an electrode under current, the exchange reaction can be characterized by a forward and a backward reaction rate. Under equilibrium, these are proportional to a value which can be called, as is usual in electrochemistry, the exchange current i_o. If the deviation :

$$\Delta\tilde{\mu}_{A^+} = \tilde{\mu}_{A^+}^{(i)} - \tilde{\mu}_{A^+}^{(e)} \tag{11}$$

from the equilibrium conditions is small, i.e. if the overall current I of A^+ ions is small with respect to i_o, it can be deduced, according to the usual considerations, that I and $\Delta\tilde{\mu}_{A+}$ are linearly related as :

$$I = a.\Delta\tilde{\mu}_{A+} \qquad (12)$$

A faradic resistance r_F :

$$r_F = \frac{1}{aF} \qquad (13)$$

can then be defined to characterize the electric properties of the phase boundary. To take into account the electric double layer, various equivalent circuits can be proposed. In fact in many situations of this type the exchange current i_o is rather high. The double layer capacitance frequently appears shortcircuited by a rather low faradic resistance. In this field, most of the experiments were performed with silver ion conductors[17,21-23].

Ideally blocked electrode

Let us now consider the situation where the A particles and all other electrolyte constituents can be supposed insoluble in the electronic conductor. Two cases are possible : either no electric charge carrier at all can be exchanged at the phase boundary or only electronic charge carriers can be exchanged. The first case is typically that which has been dealt with by the physicists[24,25]. It is usually referred to as an ideally blocked electrode. The dominant electric link between the two phases is then, only the electrostatic reciprocal influence. In fact, in this approach, it was at first frequently assumed in the relevant mathematical developments that any change ΔU of the electronic conductor potential induced an equal variation $\Delta\phi$ of the electrostatic potential at the surface of the ionic conductor[24].

Electron transfer

In the second case the electrode reaction can be written simply as an exchange of electrons (or electron holes) between the electronic conductor and the surface of the ionic conductor :

$$e^-_{(e)} \leftrightarrows e^-_{(i)} \qquad (14)$$

The equilibrium condition is hence expressed by the equality of their electrochemical potentials :

$$\tilde{\mu}_{e^-}^{(e)} = \tilde{\mu}_{e^-}^{(i)} \qquad (15)$$

This leads to the following expression of the electrode potential E :

$$E = -\frac{\mu_{e^-}^{(i)}}{F} + \phi \qquad (16)$$

where $\mu_{e^-}^{(i)}$ measures the local position, on the surface of the electrolyte, of the Fermi level with respect to the band edges. According to the terminology by Roberts and Schmidlin[8] the question now is : which electronic states play the role of the dominant level and hence appear as the dominant electrode reactant? Using a simple macroscopic band scheme, the vicinity of the electrolyte surface can be described as shown in figure 3.

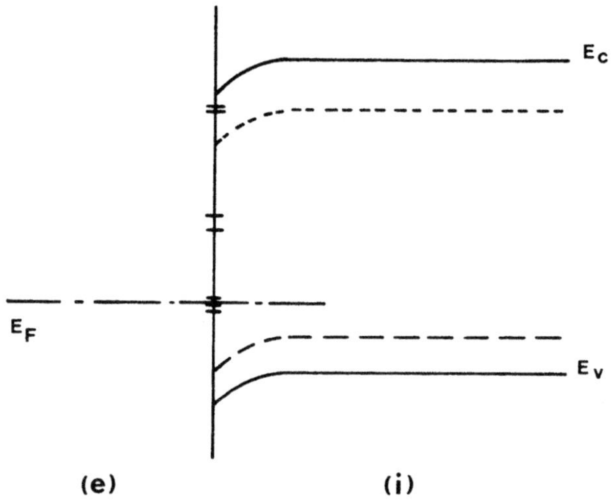

Fig.3- Band scheme of an electrochemical junction.

The bulk is characterized by a band gap and discrete electronic levels due to point defects, some of them being mobile ; the surface, by specific,intrinsic and extrinsic, electronic levels called surface states. In semiconductor physics, numerous experimental results have conclusively shown that the surface states frequently have a dominant influence in locally fixing the position of the Fermi level[26]. With solid ionic conductors, at the present time, this is still questionable. However, two extreme possible situations can be reasonably delineated.

One is typically represented by materials having a rather narrow band gap, or a noticeable additional electronic conductivity, or a variable departure from stoichiometry. Here bulk levels are likely to be dominant. This is typically the case which is treated within the framework of the Wagner theory[4] under the denomination of ion-blocking electrode. With regard to the phase boundaries, it must be mentioned, at this point, that the concept of blocked or blocking electrodes used in this theory frequently differs from that used by the physicists. In the latter, the phase boundary can be viewed as a capacitor on the plates of which charges accumulate. The reciprocal influence between them is determined by the densities of electric charges irrespective of their electronic or ionic nature and their chemical properties. With a typical ion-blocking electrode of the Wagner theory we can say that the capacitor model still holds only for the ionic carriers. This capacitor is shortcircuited by a circuit through which electrons can flow (Fig. 4).

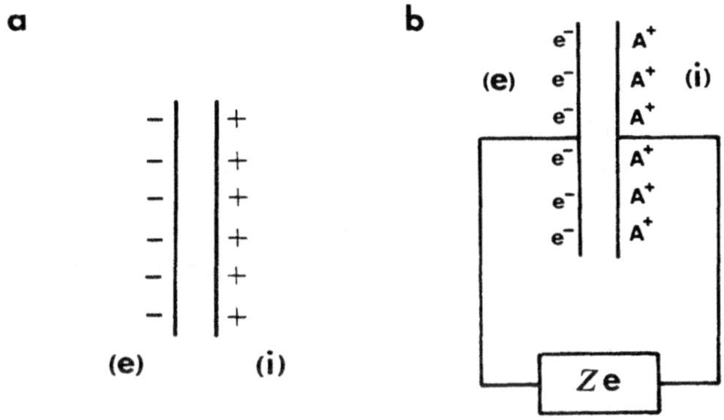

Fig.4- a) Sketches of an ideally blocked electrode.
b) a blocked electrode in the Wagner theory.

Equilibrium conditions are then determined by the electric characteristics of the capacitor and by those of the electronic short circuit. Among other things, these reflect the chemical properties of the phases in contact. Any charge of the capacitor may involve a change in the electronic circuit characteristics. This has been studied by Weiss[27], Joshi and Wagner[28].

The other possible extreme situation is that of materials presenting a large band gap, such as alkali halides, a very small electronic conductivity and a fixed stoichiometry. In such materials, the Fermi level is necessarily far away, on an RT scale, from any band edge and from any concentrated discrete levels. Under these conditions it is reasonable to assume that these levels have a weak buffer effect on the Fermi level position and that some intrinsic or extrinsic surface states are likely to take over the role of dominant level. Although the presentation may appear unusual, the situation is rather traditional. It is quite close to that of a simple electrode in an aqueous electrolyte where the electron electrochemical potential is supposed to be locally determined at the electrode. It is also somewhat similar to that of an adsorption with partial transfer envisaged by Mitoff and Charles[29]. This electrochemical point of view has been mainly developed by the Karpachev's group[30] and Hladik[31]. Using the semiconductor physics presentation in parallel has the great advantage, for a solid state electrochemist, of opening up a very promising field. Referring to the results already obtained on semiconductors, it seems that, because of the great simplicity of some ionic crystals. their surface state levels can be theoretically calculated[9,32-35]. In these cases we can then expect a theoretical determination of the relative position of the electrode potentials. This is obviously very promising. One question that comes to mind is : what could be the influence of the electronic conductor in contact with the surface, on the surface state levels, and would it be calculable ?

Localized electrodes

Let us consider the last situation in rather more detail, and first, schematically describe the electrolyte inserted in an electrochemical cell. For the sake of simplicity, its electrical conductivity will be supposed to mainly result from the mobility of only one type of A^+ ions. This property can be expressed by the traditional relationship :

$$I = - C\tilde{U}F \text{ Grad } \tilde{\mu}_{A^+} \qquad (17)$$

where I is the current density, C the mobile A^+ ions concentration, $\tilde{U}$ their electrochemical mobility, F their electric charge.

With the usual splitting of an electrochemical potential (9) and the expression of a chemical potential :

$$\mu_{A^+} = \mu^\circ_{A^+} + RT \ln \gamma C \qquad (18)$$

where γ is an activity coefficient, function of C, we have :

$$I = - C\tilde{U}FRT \text{ Grad}(\ln\gamma C) - C\tilde{U}F^2 \text{ Grad}\phi \qquad (19)$$

If the electrolyte is operated in its electrolytic domain, or, more precisely, in a stability redox domain[1] where C is a constant, this simplifies to :

$$I = - C\tilde{U}F^2 \text{ Grad}\phi \qquad (20)$$

where the coefficient $C\tilde{U}F^2$ which is in fact the electrolyte conductivity, is a constant. This is the simple Ohm law. It shows that the variations of ϕ are simply equal to the ohmic drop. In particular, under open circuit conditions when the current density is nil, Grad ϕ is nil. In a macroscopic band scheme of the electrolyte, the band edges are flat.

As emphasized in an earlier paper[36] if we refer to the authors who have dealt with the double layers in solid ionic compounds, the electrostatic potential variations in such layers are traditionally supposed to be only determined by the ion distribution. The electronic carriers are always supposed to have a negligible influence. This implies that any change in their concentration, i.e. any change in the local position of the Fermi level, also has a negligible influence. This conclusion has also been reached by Weiss[27]. Such an assumption is even more likely to be correct when dealing with materials having a high ionic conductivity such as β-alumina. In these materials, the double layer is certainly very narrow and almost non-existant.

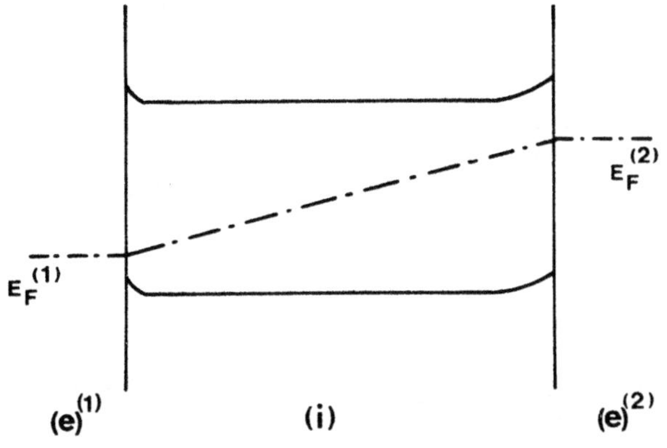

Fig.5- Band scheme of a solid state electrochemical cell.

We arrive at the unusual conclusion that the macroscopic band profile is no longer determined by the electronic distribution as in traditional semiconductors but by the properties of the mobile ionic carriers, which, in fact, are the majority carriers. Then the electrolyte inserted in an electrochemical cell, under open circuit conditions, can be simply represented as a rigid reference frame (Fig.5). The Fermi level is principally determined at the surfaces by the local distribution of electrons among surface states. According to eq. (16) the cell voltage is simply related to the difference between these Fermi levels at the electrodes :

$$E_1^2 = - \frac{I}{F} (\mu_{e^-}^{(2)} - \mu_{e^-}^{(1)}) + (\phi^{(2)} - \phi^{(1)}) \qquad (21)$$

where $(\phi^{(2)} - \phi^{(1)})$ is a constant which only depends on the ions. Under these conditions, any change of the cell voltage can be a direct measurement of the shift of the Fermi level with respect to the band edges at one electrode, if it is fixed at the other.

The Fermi level gradient shown in Fig.5 results in a microelectronic current which is compensated by an equivalent microionic current. This is the so-called electrochemical semipermeability phenomenon[37].

For the sake of example, let us suppose that the cell O_2, Pt/stabilized zirconia/Pt, O_2 behaves according to this simple model. Then the fact that the voltage simply depends on the oxygen pressure in the surrounding gas implies that at least one set of the dominant surface states are due to adsorbed oxygen species. Under such conditions, the time lag of the cell voltage after a change in oxygen pressure in the gas phase is determined by the oxygen sorption kinetics. The readjustment of the Fermi level in the bulk of the material should have no important influence. We have in fact observed under appropriate conditions a much shorter time lag for the cell voltage than for the oxygen semipermeability flux which depends on the Fermi level distribution in the bulk.

It is worth recalling for the benefit of those who might not be familiar with it, that with our assumption the electrode potential variations ΔE are simply correlated by the Nernst law to the variations of the adsorbed-oxygen-atom-chemical-activity A_O :

$$\Delta E = \frac{RT}{2F} \ln \left(1 + \frac{\Delta A_O}{A_O}\right) \qquad (22)$$

This is a valuable piece of information. A nice problem for the theoreticians would be to calculate the Fermi level variations from the microscopic elementary properties of the adsorbed oxygen species and deduce from them the Nernst law.

Let us now consider the influence of a current flowing through the cell. This current is carried in the electrolyte by the O^{2-} ions. According to our first sketch of an electrode, it produces or consumes a certain quantity of oxygen right on the surface of the electrolyte. This quantity q is related to the current I by the Faraday law :

$$q = \frac{I}{2F} \qquad (23)$$

(q is expressed in terms of oxygen-gram-atoms). When the oxygen exchange rate between the adsorbed layer and the surrounding phase is low, this results in an accumulation or depletion of neutral adsorbed oxygen species, i.e. of empty surface states and therefore in a shift of the Fermi level. As previously noted[36], this could even occur without any change in the local electrostatic potential (apart from the ohmic drop). Numerous experimental results have already shown that the magnitude of such a shift can be very large[38]. In this way the Fermi level can reach bulk trap levels where electrons are injected. This phenomenon is usually called electrochemical coloration[1].

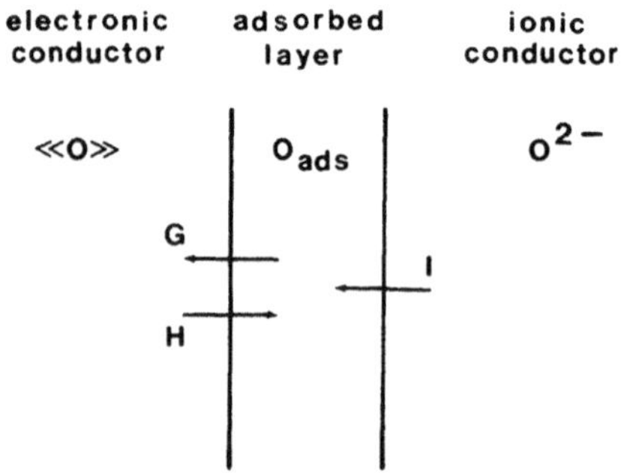

Fig.6- Localized electrode on an oxide solid electrolyte.

Such a Fermi level shift can result in effects which can easily be confused with capacitive effects[36]. These effects are called by the electrochemists, pseudo-capacitive effects. In solid state electrochemistry they are sometimes also described as an adsorption with partial transfer. To illustrate the point, let us consider the schematic situation depicted in Fig.6. It is familiar to electrochemists. Without entering into the details of the relevant surface levels and densities, we can say, from a thermodynamical viewpoint, that the electrode measures the chemical activity of O atoms in a perturbed layer located at the phase boundary. The electrode potential variations are related to the O-chemical-activity-variations by formula (22). Extending the hypotheses, here, the O atoms are supposed to be soluble in the electronic conductor but the direct exchange of oxide ions is regarded as impossible.

The experiment described is simply performed by passing a small a.c. current $I_o e^{i\omega t}$. Under steady state conditions, the quantity Γ of O adsorbed atoms per unit surface is :

$$\Gamma = \Gamma_o + \Delta\Gamma \tag{24}$$

where Γ_o is a constant value equal to the equilibrium condition value and $\Delta\Gamma$ a small variation :

$$\Delta\Gamma = \gamma e^{i(\omega t+\alpha)} \tag{25}$$

In the same way, the concentration C of O atoms dissolved in the electronic conductor near the phase boundary can be written as :

$$C = C_o + c e^{i(\omega t+\alpha')} \tag{26}$$

If H(C) and G(Γ) are the adsorption and desorption rates of oxygen from and to the electronic conductor, under the chosen conditions, they are equal to :

$$H = H_o + hce^{i(\omega t+\alpha')} \tag{27}$$

$$G = G_o + g\gamma e^{i(\omega t+\alpha)} \tag{28}$$

where : $$h = \left(\frac{\partial H}{\partial C}\right)_{C_o} \quad \text{and} \quad g = \left(\frac{\partial G}{\partial \Gamma}\right)_{\Gamma_o} \tag{29}$$

Within this formalism the O-atom-balance in the absorbed layer is:

$$\frac{d}{dt}\left(\gamma e^{i(\omega t+\alpha)}\right) = \frac{I_o e^{i\omega t}}{2F} + H_o + hce^{i(\omega t+\alpha')} - G_o - g\gamma e^{i(\omega t+\alpha)} \tag{30}$$

The difference $(H_o - G_o)$ which is the difference between the rate of adsorption and desorption under equilibrium is nil. So, from this equation we deduce :

$$\frac{\gamma}{I_o} e^{i\alpha}(g+i\omega) = \frac{1}{2F}\left(1 + \frac{2Fhc}{I_o} e^{i\alpha'}\right) \tag{31}$$

This formula can be used to calculate the expression of the electrode impedance as follows. The electrode potential has the form:

$$E = E^{\circ} + \frac{RT}{2F} \ln A(\Gamma) \tag{32}$$

Under the selected conditions, its variations ΔE which are supposed to be small are equal to :

$$\Delta E = \frac{RT}{2F}\frac{a}{A_o} \gamma e^{i(\omega t+\alpha)} \tag{33}$$

where : $$A_o = A(\Gamma_o) \quad \text{and} \quad a = \left(\frac{\partial A}{\partial \Gamma}\right)_{\Gamma_o} \tag{34}$$

and consequently the differential impedance of the electrode is :

$$Z = \frac{\Delta E}{I} = \frac{RTa\ \gamma}{2FA_o I_o} e^{i\alpha} \tag{35}$$

In the simple limiting case where the concentration C in the electronic conductor can be considered as a constant (c = 0) we deduce from the expression of I and eq. (35) that the impedance Z varies as a function of the frequency as :

$$Z = \frac{R_\eta}{1 + i\frac{\omega}{\omega_o}} \tag{36}$$

with $$R_\eta = \frac{RT}{4F^2}\frac{a}{A_o g} \quad \text{and} \quad \omega_o = g \tag{37}$$

This is the traditional semicircle complex impedance equation. At the present time research work is in progress in our laboratory on solid silver-zirconia electrodes to see whether such a model can be adequate and whether the second terms which appear in eq. (31) can account for the slant of the semi-circle impedance.

Concerning surface states, other comparisons may be of interest. In a simple picture of an NaCl crystal, for instance, the 3 s orbitals of the Na particles overlap to form the conduction band in the bulk of the material. On the surface they form intrinsic surface states presumably located at a lower level. Within the framework of this simple scheme, forming a neutral sodium particle on the surface is just entrapping an electron in a relevant surface state. Entrapping lots of electrons, i.e. forming lots of neutral sodium particles, certainly implies a shift of the Fermi level close to the corresponding surface-state-level. Therefore it is reasonable to assume that the Fermi level is located very close to the relevant surface-state-level when a pertinent massive metal, sodium in our case, is in equilibrium with the electrolyte. This would allow us to compare experimental results typically obtained in solid state electrochemistry with theoretical calculations performed by the physicists. According to the few papers dealing with calculations of intrinsic surface states on ionic crystals, it seems largely accepted that the more ionic the bonding in the crystal, the closer the intrinsic surface states are to the corresponding band edge. Within our scheme, this roughly means that the metal in equilibrium with the electrolyte induces a larger electronic conductivity. The variations observed experimentally seem to lie in this direction. Silver in equilibrium with silver halides does not induce noticeable electronic conductivities and neither does copper in contact with cuprous chloride. On the other hand, alkali metals generally induce large electronic conductivities in the alkali halides with which they are in equilibrium. When non-metals are concerned, unfortunately, the experimental results do not look so nicely distributed. Obviously a detailed investigation should take into account the energy required to form massive metal from the elementary adsorbed neutral particles. If we refer to the results by Raleigh[39] on the formation of silver on graphite, it seems that the relevant overpotential can be rather small.

Another point which could be worth considering is the effect of the trend towards homogeneization resulting from the additional electronic conductivity. If, in some points of the surface, the Fermi level is fixed by the experimental conditions close to the intrinsic surface states corresponding to the formation of neutral metal particles, it may tend to reach the same position in the vicinity of these points and metal particles may appear in this vicinity ; the higher the additional electronic conductivity and the higher the electrode overpotential, the stronger this trend. This may occur, for instance, if the electrolyte exhibits a specific surface-electronic-conductivity. This applies to free surfaces and also to grain boundaries and dislocations. The importance of such a phenomenon in dendrite growth may be worth considering. Of course, with an electrode involving as a dominant process a direct exchange of ions as previously sketched, the constraints are different. The overpotential, is not necessarily synonymous with an "overshift" of the Fermi level and of a reinforced trend to form neutral metal particles. This is certainly why electrodes involving such a direct ion exchange are successful in technological devices in avoiding dendrite formation.

References

1. P. FABRY, M. KLEITZ, This Book
2. D.O. RALEIGH, Electroanal. Chem. 6 87 (1972)
3. M. HEBB, J. Chem. Phys. 20 185 (1952)
4. J.B. WAGNER, This Book
5. M. KLEITZ, J. BESSON, C. DEPORTES, Proceed. "2ièmes Journées Int. Etudes Piles à Combustible, Publ. SERAI-COMASI p.354 Bruxelles (1967)
6. T. ETSELL, S.N.FLENGAS, 137th Meeting Electrochem. Soc. Extended Abst. n° 329 (1970)

7. E.H. RHODERICK, J. Phys. D 3 1153 (1970)
8. G.G. ROBERTS, F.W. SCHMIDLIN, Phys. Rev. 180 785 (1969)
9. S.G. ZAVT, Y.Ya.SAKS, Soviet Physics Solid State 14 2502 (1973)
10. M. GUENIN, Ann. Chim.(Paris) 8 147 (1973)
11. M. BREYSSE, Ann. Chim.(Paris) 2 367 (1967)
12. W.L. WORRELL, J.L. ISKOE, Fast Ion Transport In Solids p.513, W. Van Gool Edit. North Holland (1973)
13. P. FABRY, Thesis Grenoble (to be published)
14. R.M. DELL, This Book.
15. R.D. ARMSTRONG, T. DICKINSON, J. TURNER, J. Electrochemical Soc. 118 1135 (1971)
16. M. ARMAND, Fast Ion Transport in Solids p.665, W Van Gool Edit. North Holland (1973)
17. J. BERT, J.L. PICOT, J. DUPUY, Phys. Status Solidi A 19 119 (1973)
18. A.T. FILYAEV, S.V. KARPACHEV, I.D. REMEZ, Elektrokhimiya 6 283 (1970)
19. C.C. RICCARDI, Thesis London (1971)
20. J.M. BLAKELY, This Book
21. L. CONTRERAS, H. RICKERT, Fast Ion Transport In Solids, W.Van Gool Edit. p.523 North Holland (1973)
22. T. TAKAHASHI, O. YAMAMOTO, Electrochemica Acta 11 911 (1966)
23. R.D. ARMSTRONG, T. DICKINSON, H.R. THIRSK, R. WHITFIELD, J. Electroanal. Chem. Interfacial Electrochem. 29 301 (1971)
24. G. JAFFEE, Ann. Physik 16 217 and 249 (1933)
25. J.R. MACDONALD, This Book
26. S. KURTIN, T.C. Mc GILL, C.A. MEAD, Physical Rev. Letters 22 1433 (1969)
27. K. WEISS, Ber. Bunsenges Phys. Chem. 76 379 (1972)
28. A.V. JOSHI, Fast Ion Transport In Solids p.173 W Van Gool Edit. North Holland (1973)
J.B. WAGNER, Fast Ion Transport In Solids p.489 W Van Gool Edit. North Holland (1973)
29. S.P. MITOFF, R.J. CHARLES, J. Appl. Phys. 43 927 (1972)
30. S.V. KARPACHEV, 24th I.S.E. Meeting Eindhoven Extended Abst. p.100 (1973)
31. J. HLADIK, Electrochemica Acta 15 483 (1970)
32. C.M. BERTONI, C. CALANDRA, Phys. Status Solidi B50 527 (1972)
33. E. LOUIS, F. YNDURAIN, Phys. Status Solidi B57 175 (1973)
34. R. St.C. SMART, P.J. JENNINGS, Trans. Faraday Soc. 67 1193 (1971)
35. P. MARK, This Book
36. M. KLEITZ, P. FABRY, E. SCHOULER, Fast Ion Transport In Solids, p.439, W Van Gool Edit. North Holland (1973)
37. J. FOULETIER, P. FABRY, M. KLEITZ, J. Electrochem. Soc. (to be published)
38. P. FABRY, M. KLEITZ, C. DEPORTES, J. Solid State Chemistry 6 230 (1973)
39. D.O. RALEIGH, J. Phys. Chem. 71 1785 (1967)

DISCUSSION

HEYNE : As you stated a lot of confusion exists due to different points of view from which solid electrolyte problems are approached. I should like to emphasize that such confusions could be considerably reduced if thermodynamic arguments and model considerations would always be clearly separated. For instance splitting up of a components chemical potential into either an electrochemical potential plus an electrostatic potential (usual in normal electrochemistry) or into an ion electrochemical pot. plus an electron chem. pot. (= Fermi level), is completely arbitrary (and unnecessary) from a purely thermodynamic point of view. As soon as we split in one way or another we must be aware, and clearly state, that we use a certain model such as for instance the band picture of a semi-conductor, or the defect structure of a solid electrolyte.

KLEITZ : As far as ideal equilibrium conditions are concerned, I agree that the choice of such or such a description is arbitrary and may appear unnecessary. When dealing with experimental measurements, these conditions unfortunately are seldom reached. Even under the so-called equilibrium conditions, the investigated system is generally submitted to various perturbing effects. The resulting deviations are determined by the kinetics of the possible reactions. Then the choice of a model is no longer a formal problem. It must evidence for instance the fast reactions which are likely to remain equilibrated and the slow ones which can be limiting.

STRUCTURE AND ENERGETICS OF THE IONIC SURFACE*

Peter Mark

Department of Electrical Engineering
Princeton University
Princeton, New Jersey 08540

INTRODUCTION

This chapter is a brief review of recent advances in the structural and electronic properties of the surfaces of semiconductor materials with partial bonding ionicity and of how these influence chemical activity at surfaces. As all the material to be covered here can be found in the published literature (see reference 1 for a complete listing of pertinent literature), this review will concentrate on summarizing the broad conclusions that have been drawn from recent experimental, theoretical and computational work. The following items are covered: 1) Surface structure of clean non-polar and polar surfaces as revealed experimentally from low energy electron diffraction (LEED) symmetry and intensity data and interpreted by dynamical electron diffraction analyses. It is shown that the non-polar surfaces of the tetrahedrally coordinated materials (zincblende and wurtzite structure) are terminated to within 0.1 Å of ideal lattice positions of the terminating ions. Alternatively, LEED symmetry studies alone indicate that the polar faces are never ideally terminated. They exhibit either a 6-fold rather than the expected 3-fold rotational terminating symmetry (as if some portions of the terminating plane are wurtzite connected while others are zincblende connected), reconstruction (that is, atoms displaced over large distances parallel to the surface in the terminating plane so as to produce a new unit mesh in the surface plane) or facets whose

* Supported by The Office of Naval Research (N00014-75-C-0394), National Science Foundation (GK 38575) and The Army Research Office (DAHC 04-74-G0133).

M. Kleitz and J. Dupuy (eds.), Electrode Processes in Solid State Ionics, 19–44. *All Rights Reserved.*

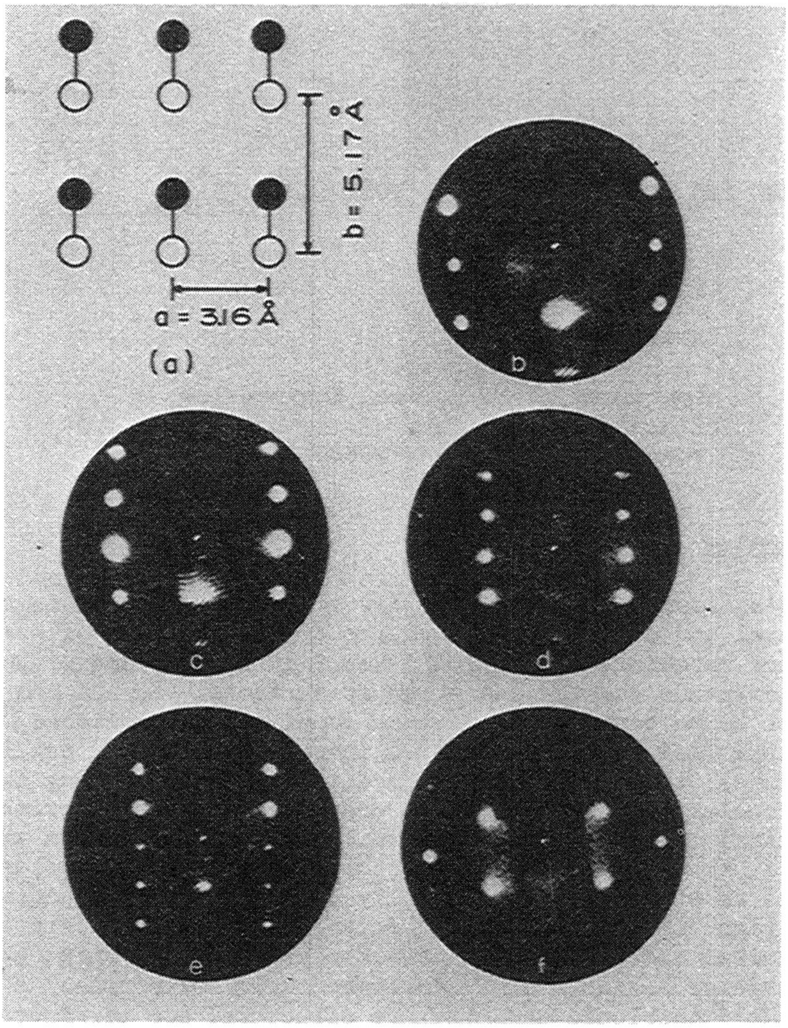

Fig. 1

(a) Arrangement of surface atoms of undistorted ZnO (10$\bar{1}$0) surface. Polar axis ([0001] azimuth) is directed upward; the [11$\bar{2}$0] azimuth is horizontal. The surface unit cell is defined. (b) to (f) LEED patterns recorded at 68.5 V, 102 V, 144 V, 158 V and 193 V, respectively. The photographs are oriented to confirm with (a). From reference 2.

faces are stable non-polar surfaces. Frequently these structures occur simultaneously over the area sampled by the primary electron beam (about 1 mm^2). Which structure is actually observed depends both on the material and on the prior processing history. 2) Surface electronic structure of non-polar surfaces (surface states or surface energy bands) as revealed experimentally by photoemission and electron energy loss spectroscopy. Here it is shown that the non-polar surfaces consistently exhibit a density of states that locates a surface conduction band in the upper half of the bulk band gap (InAs appears to be the only exception) and a surface valence band slightly below the bulk valence band edge with a surface band-gap separating the surface bands. It is shown that these recent experimental determinations are consistent with the results of earlier proposed models based on pure bonding ionicity and with more recent surface band structure calculations based on the tight binding approach. The surface electronic structure of polar faces has also been examined experimentally but as the structure of these surfaces is as yet unknown, no significant theoretical work on the electronic structure of these surfaces is available. 3) Surface chemical activity as revealed experimentally by optical ellipsometry, LEED intensity degradation, electron energy loss spectroscopy (EELS) and Auger electron spectroscopy (AES). The principal conclusion here is that for well ordered surfaces chemical activity appears to derive entirely from the covalent contribution to the bonding. That is, the surfaces become less active with increasing bonding ionicity (e.g., increasing Pauling electronegativity difference ΔX of the lattice constituents). Second, for a given material (i.e., given ΔX) it appears that the disordered surface is about 10^3 times as reactive as the same surface in an ordered state and that this order-disorder effect appears also to derive from the ionic contribution to the bonding.

SURFACE STRUCTURE

1. Non-Polar Surfaces

There is a considerable amount of literature that demonstrates the ideal character of the termination of clean, non-polar surfaces prepared by the sequence of mechanical and chemical polishing followed by ultrahigh vacuum processing consisting of argon ion beam bombardment and thermal annealing.[1-4] Only a typical example, the ZnO ($10\bar{1}0$) surface, will be given here to illustrate the techniques and results.[2] The LEED symmetry diagrams of this surface are illustrated in Fig. 1 for five different primary electron beam voltages. The dimensions of the surface unit mesh can be obtained by measuring the angular dependence of the principal first order

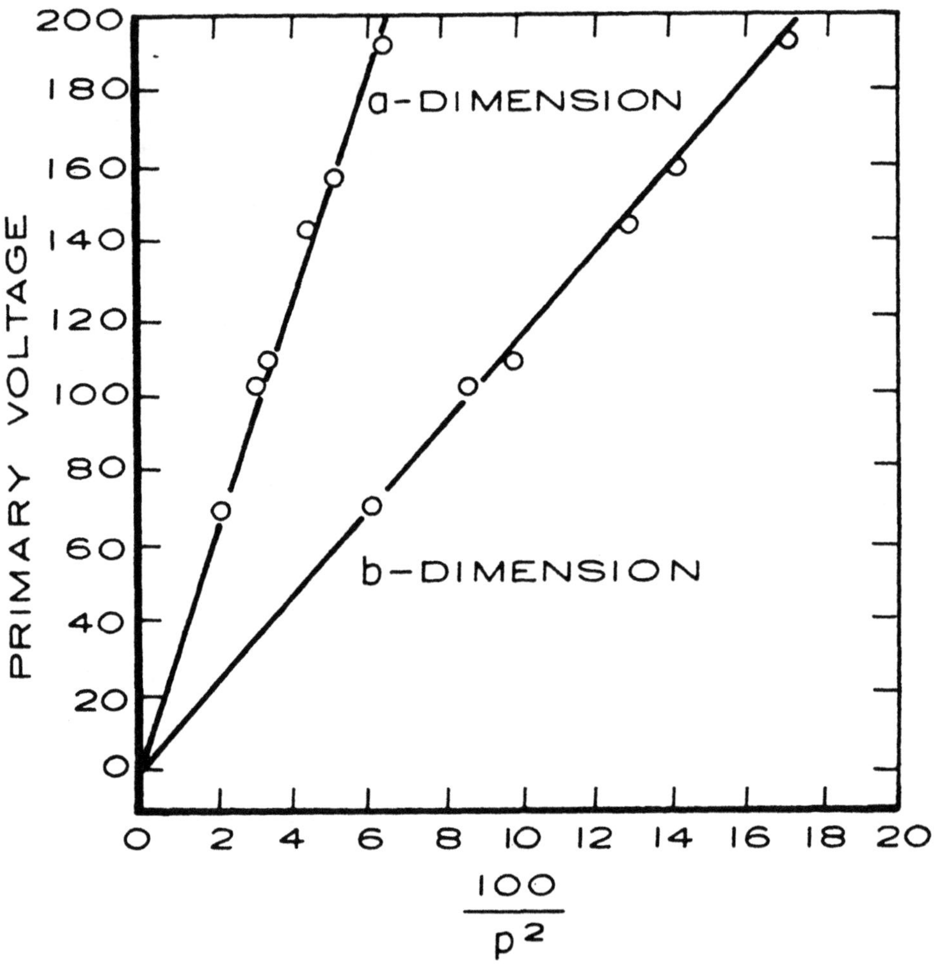

Fig. 2 Primary beam voltage versus $(100)/p^2$.

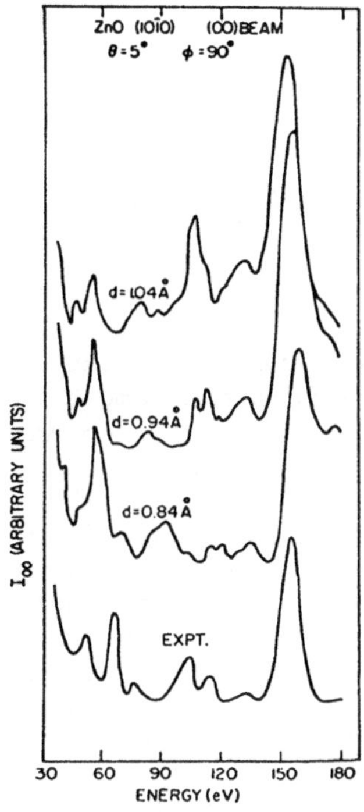

Fig. 3

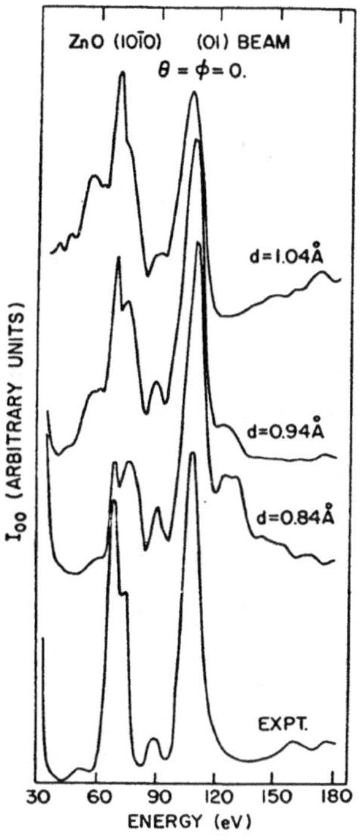

Fig. 4

3. Comparison of the calculated and observed intensities of the (00) beam of electrons diffracted from ZnO (10$\bar{1}$0). The incident beam parameters (θ,φ) and the upper layer spacing (d) are included in the figure. The azimuthal angle φ is measured from the a axis of the unit mesh (see Fig. 1). The phase shifts were obtained from overlapping ionic charge densities, T = 1°K, λ_{ee} = 12 Å and V_o (inner potential) = 10 eV. From reference 4.

4. Same as Fig. 3 for the (01) diffracted beam. From reference 4.

non-specular beams as a function of the primary beam energy. This follows from the fact that the parallel component of the electron momentum is conserved to within a reciprocal lattice vector and for the first order non-specular beams, the reciprocal lattice vectors are the unit vectors of the two-dimensional reciprocal lattice of the surface plane.[5,6] The relationship between the diffraction angle θ_{hk} (where h,k label the two-dimensional order of the diffracted beam) and the primary electron energy is

$$E = \left[\frac{2\pi^2\hbar^2 r^2}{m\,R^2(hk)}\right]\left[\frac{1}{p(hk)}\right]^2 \; ; \; p(hk) = r\sin\theta_{hk} \qquad (1)$$

where m is the free electron mass, $\hbar$ is Planck's constant divided by 2π, r is the radial distance from the sample surface to the electron detector and R(hk) is either a or b depending on whether (hk) is (01) or (10) [the direct lattice surface unit mesh dimensions a and b are defined in Fig. 1]. The slope of Eq. (1) clearly yields a and b. The data obtained from ZnO $(10\bar{1}0)$ is shown in Fig. 2.[1] Two straight lines are obtained for the two principal orientations a and b. The unit mesh dimensions obtained from this data are: a = 3.21 Å; b = 5.14 Å. The accepted nominal values are: a = 3.16 Å and b = 5.17 Å. This agreement is typical for LEED optics using a fluorescent screen to display the symmetry of the diffraction pattern.

The location of the lattice atoms in the surface region is conventionally obtained from a comparison of the experimental LEED intensities for the various order diffraction beams with intensities computed from a dynamical scattering analysis that embodies the lattice atom locations as parameters.[5,6] The details of such a dynamical scattering analysis are too complex to be reproduced in detail. There is a considerable theoretical literature on this subject, principally applied to metal surfaces and adsorbed overlayer on metal surfaces.[5,6] Recently, dynamical scattering analyses have been successfully applied to non-polar compound semiconductor surfaces.[4] Figures 3 and 4 illustrate a comparison between experimental and analytically computed LEED intensity curves for two diffraction beams of the Zn $(10\bar{1}0)$ surface.[4] The bottom trace of each figure is the experimental curve while the remaining traces are computed via a dynamical diffraction analysis with the spacing of the last atomic layer from the underlying lattice as a parameter. Ideal termination corresponds to a layer spacing d = 0.94 Å. A qualitative comparison from this figure suggests that computations using the ideal spacing generates LEED intensity curves that agree best with experiment. At most, the deviation is ± 0.1 Å. A more extensive comparison between experimental and computed

Fig. 5

LEED pattern of InAs (111) 2 x 2 recorded at 44 V, showing three-fold rotational symmetry. From reference 9.

LEED intensity curves involving data and computations at off-normal incidence for all the principal non-specular beams leads to the same result. Similar analyses have also been completed for the GaAs (110) and ZnO (11$\bar{2}$0) surfaces with similar results: best agreement between measured and computed LEED intensity curves is achieved when the computations are performed with the surface ideally terminated.

2. Polar Surfaces

Polar surfaces are those that terminate ideally in only one sign of lattice ion and consequently they are electrostatically unstable. This instability arises from the fact that a dipole moment always builds up along the polar axis of the lattice that would make the crystal behave as an electret if the lattice were ideally terminated.[7] As such behavior has never been observed, the lattice apparently always provides local excess charge distribution in the vicinity of the surface that cancels this dipole moment. As a volume dipole moment can always be cancelled by a surface charge distribution, one naturally looks for evidence of the latter on the polar terminating surfaces. This surface charge can be in the form of electronic charge in surface states, adsorbed impurity ions, doping profiles in the vicinity of the surface, a rearrangement of lattice atoms in the vicinity of the surface (facet formation) or a rearrangement of the atoms in the surface plane (reconstruction). For tetrahedrally coordinated surfaces a 1/4-monolayer sheet charge is required on electrostatic grounds.[7] For termination of a clean surface in an ultrahigh vacuum, it is most likely that this surface charge is supplied by a rearrangement of the lattice ions through facet formation or reconstruction.

It is an established experimental fact that a polar surface never terminates with a LEED symmetry suggestive of ideal termination.[1] In some cases, only integral order LEED beams are observed but in all these cases there is a 6-fold rotational symmetry rather than the expected 3-fold rotational symmetry. This suggests that the surface terminates in patches some of which are zincblende-connected while the others are wurtzite-connected. This merely means that the final double layer of these two different regions are rotated 180° with respect to each other relative to the underlying lattice. This situation is observed primarily with ZnO.[8,8a] Most of the other compound semiconductor polar surfaces terminate through reconstruction after low temperature annealing and ultimately through facet formation after high temperature annealing. An example of a reconstructed surface is given in Fig. 5 for InAs (111).[9] The figure shows a 2 x 2 reconstruction indicative of a surface unit mesh having twice the unit dimensions of the ideal unit cell.

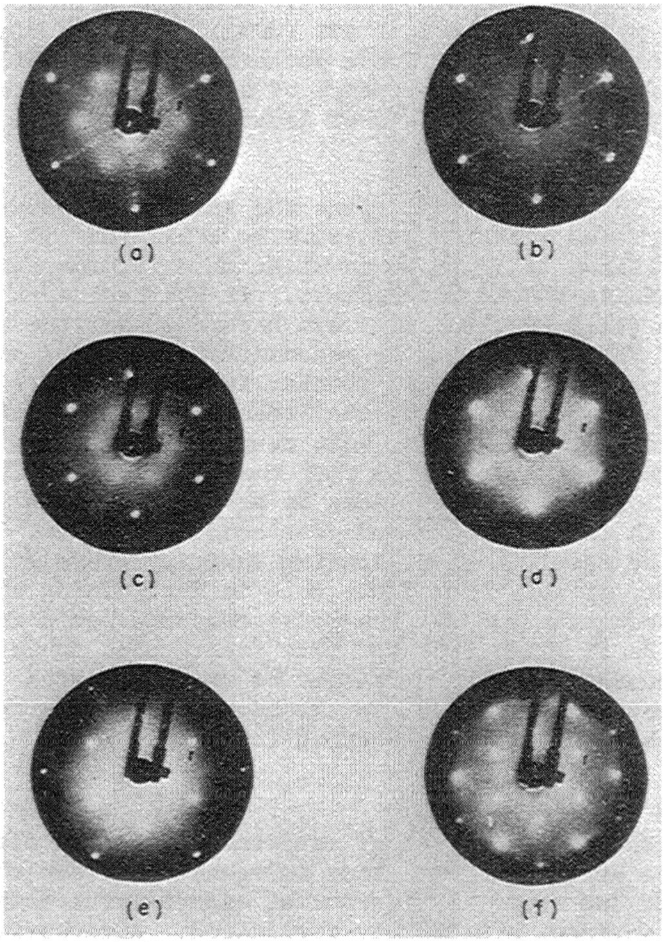

Fig. 6

LEED patterns of CdS $(000\bar{1})$-S. (a) through (f) were recorded at 30 V, 35 V, 46 V, 55 V, 74 V, and 121 V, respectively. Faint $\sqrt{3} \times \sqrt{3}$ -30° reconstruction (a and c) and facet streaks are evident in some of the patterns. Note the six-fold rotational symmetry. From reference 10.

This can easily be achieved with an ordered removal of a 1/4-monolayer of terminating atoms.[7] The pattern also shows the expected 3-fold rotational symmetry. A more complicated diffraction pattern is that of CdS ($000\bar{1}$), shown in Fig. 6 for six different voltages.[10] These patterns show a 6-fold symmetry. Also, there is evidence of a $\sqrt{3} \times \sqrt{3}$ -30° reconstruction (panels a and c) and of facet formation (the streaks in panels b and d).

It is interesting to compare the surface structure of these polar faces with the structurally equivalent Si (111) face. As silicon is a purely covalent material, no effects of bonding ionicity should be apparent. It has been established that the Si (111) surface is always reconstructed and never faceted.[11] The details of the reconstruction depend on how the surface is prepared, that is, whether it is cleaved (2 x 1 reconstruction) or ordered by the bombard-anneal technique (7 x 7 reconstruction) and to what temperature it is annealed.[12] Such comparisons have revealed that the observed Si (111) surface reconstruction for clean surfaces is never the same as that observed with the polar compound semiconductor faces.[12] It may, therefore, be concluded that bonding ionicity plays at least an important, if not a dominant, role in determining the observed reconstructions of the polar compound semiconductor surfaces. As Si (111) does not form facets when annealed to the highest temperatures, the existence of facets on polar compound semiconductor faces annealed to the highest possible temperatures can be attributed to bonding ionicity.

3. Summary

The non-polar surfaces of compound semiconductors display a remarkable structural stability independent of how they are prepared and there is substantial experimental evidence that these surfaces are ideally terminated. Alternatively, the polar surfaces demonstrate a strong structural instability and experimental evidence suggests that they are in no case ideally terminated. These structural properties of both classes of surfaces can be traced to bonding ionicity.

SURFACE ELECTRONIC STRUCTURE

The following discussion is limited to the non-polar surfaces because reliable structure information is available only for these surfaces.

The surface electronic structure, or surface band structure, of a finite lattice is defined by the characteristic property of their eigenfunctions that these be attenuated in the surface normal direction. Thus they are a subset of the entire band structure. There are formidable difficulties in characterizing the surface band structure both experimentally and theoretically. The major experimental difficulty stems from the fact that the surface band structure must be observed in the background of the bulk band structure, the latter being that set of eigenvalues whose wavefunctions are unattenuated in all directions. The conventional probes, photon and electrons, used to stimulate electronic transitions and thereby characterize the lattice band structure both have finite absorption lengths and hence always provide a mixture of surface and bulk information. It is the task of the experimentalist to design his experiments so that the probe maximizes surface sensitivity. This is generally done either by choosing electrons in the energy range of about 20 to 300 eV over which their inelastic scattering length λ_{ee} is less than 10 Å and is slowly varying with energy,[5] or by choosing photons as the probe but electrons in the same energy range as the emerging information carrying signal. The former technique is the basis of electron energy loss spectroscopy (EELS)[13-15] while the latter underlies the techniques of photoemission spectroscopy (PES).[16-20]

On the theoretical side, there are two fundamental difficulties. First, and most basic, the fundamental translational periodicity of the lattice is destroyed by rendering it finite; hence the Bloch functions are no longer accurate eigenfunctions although they still serve as very good approximations for basis functions in the expansion of eigenfunctions. Second, the structure of the lattice is not accurately known in the vicinity of the surface and even in those cases where it is known best, as for the non-polar compound semiconductor surfaces, good pseudo-potentials are still not available because the interactive process of defining good pseudo-potentials through comparison of computed band structures with experimental band structures is just now beginning. Nevertheless, despite these difficulties, the results of both experimental and computational efforts for non-polar compound semiconductor surfaces are beginning to emerge and to reveal a consistent picture for the surface band structure.

1. Experimental Technique and Results

It is necessary to probe both filled and empty surface states to obtain a complete sampling of the surface band structure. Filled surface states are most conveniently probed with conventional PES.[16,17,19] The process is illustrated in Fig. 7. Photons of a constant energy $h\nu_1$ are directed at the surface and excite electrons from occupied into empty states. If the electron energy is less than about 300 eV (and greater than about 20 eV), $\lambda_{ee} \lesssim 10$ Å and only electrons very close to the

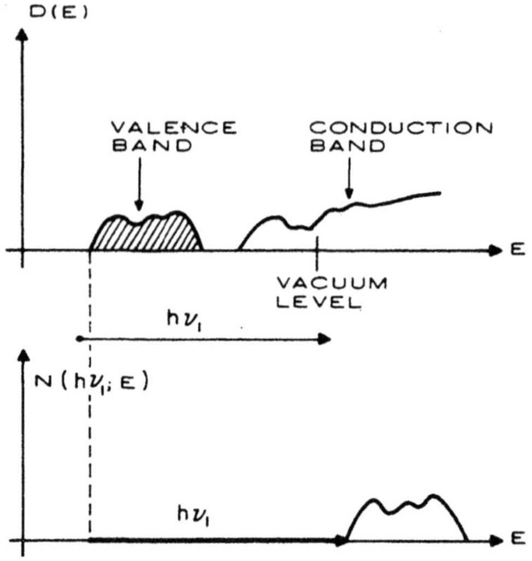

Fig. 7 Schematic representation of photoemission spectroscopy.

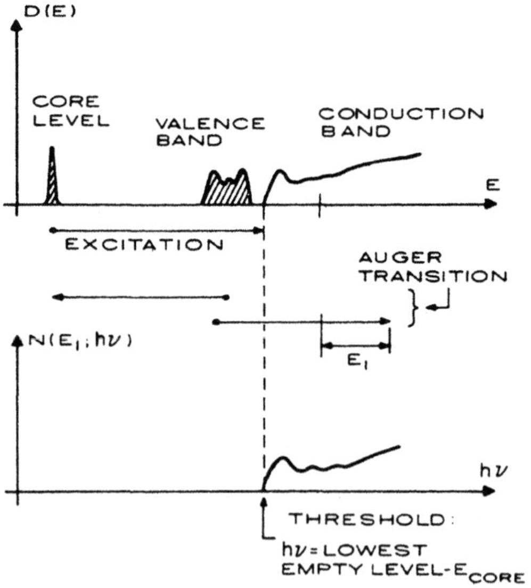

Fig. 8 Schematic representation of partial yield photoemission spectroscopy.

surface, that is, having participated in transitions in which surface states may be playing a relatively significant role, will escape from the solid and be detected. The total electron distribution $N(h\nu_1;E)$ is proportional to the product of the initial and final state densities. However, if $h\nu_1$ is large enough ($\gtrsim$ 50 eV) the conduction band density of states is reasonably structure-free and $N(h\nu_1;E)$ is a good qualitative reproduction of the valence band density of states. Because λ_{ee} is so small, a sizable contribution from the surface valence band is expected. If the surface valence band protrudes above the bulk valence band edge, the threshold for photoemission, that is, the lowest energy value of $N(h\nu_1;E)$ is reduced. If the surface valence band lies completely within the energy range of the bulk valence band, the surface band is singled out by adsorbing an impurity onto the surface and noting the change in $N(h\nu_1;E)$. It is then conventionally assumed that the adsorption sensitive parts of $N(h\nu_1;E)$ derive from surface bands.

Empty surface bands can also be probed by a variant of the PES method recently developed by Eastman[18,21] and Spicer[20] and their collaborators and called by Eastman partial yield photoemission spectroscopy (PYPES). It can be understood with the aid of Fig. 8. In the PYPES technique, the surface is illuminated with photons of varying energy and the photoelectron detector is set to record emitted electrons of a constant energy E_1. The surface is then probed with increasing photon energies starting, in principle, with zero energy. As the energy increases, electronic transitions between the delocalized states of the valence band and the conduction band are first induced with no accompanying electron emission. As the energy is further increased, transitions between the valence band and higher conduction band states are induced giving rise to the type of photoemission described in the previous paragraph. When the incident photon energy reaches about 50 - 100 eV, excitation from the localized atomic core states of the lattice into empty conduction band states, both bulk and surface are initiated. The core state hole left behind by this excitation transition can now be neutralized by downward transitions from the valence band giving rise to the conventional Auger process, namely the resonance transfer of the energy released by the core hole neutralization to another valence band electron and the subsequent emission of that electron from the solid. It is this electron, whose energy is characteristic of the solid, that is detected in the PYPES method. Clearly, the flux of these Auger electrons will not commence until the incident photon energy is large enough to excite a core state electron into an empty lattice state. Hence this minimum energy is the threshold for the appearance of energy distribution $N(E_1;h\nu)$ and for greater incident photon energies, $N(E_1;h\nu)$ will be approximately given by the

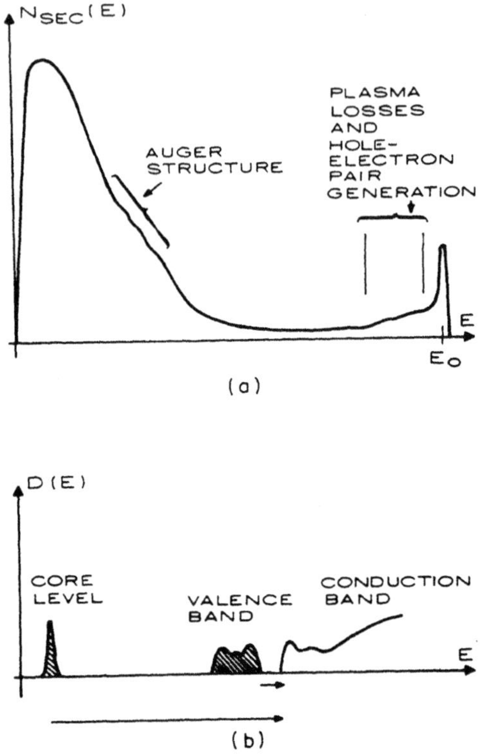

Fig. 9

(a) Schematic curve of total secondary emission electron distribution. Primary energy: E_o. The structure owing to Auger secondary electrons and to inelastic losses to collective plasma excitations and discrete electron-hole generation is shown. (b) Density of states curve showing transitions involving electron-hole pair generation. Short arrow: transitions from the valence band; long arrow: transitions from the core level.

empty density of states because the initial density of states comprises the relatively narrow core states. The selection of empty surface states from the empty bulk conduction band density of states now proceeds similarly to that employed in PES for filled states. If the empty surface bands extend below the bulk conduction band edge, the threshold of $N(E_1;h\nu)$ is lowered. If the empty surface states lie entirely within the bulk conduction band, then surface adsorption must be employed in the manner already discussed.

A second way of observing empty surface states make use of EELS.[13-15] Here a mono-energetic electron beam serves as the probe and the emerging particles are secondary electrons that have lost energy (i.e., that have been inelastically scattered) to interband transitions. The method is understood with reference to Fig. 9. Part a of this figure shows schematically the total secondary electron distribution $N_{sec}(E)$ of a typical solid. The various component of $N_{sec}(E)$ are shown in the figure; only the structure immediately to the low energy side of the elastically scattered peak at the primary energy E_o is pertinent to this discussion. This structure derives from two principal sources: collective plasma excitation of the valence electrons and localized interband transitions (hole-electron pair generation). The latter transitions are illustrated by the short arrow on the density of states diagram of Fig. 9b. Clearly, transitions involving surface bands will also be among the losses observed in the spectrum. The long arrow corresponds to transitions induced from the core states. In practice,[14] $dN_{sec}(E_o - E)/dE$ vs. $E_o - E$ is plotted, as is conventional in AES, and plasma losses are identified and "subtracted away." The remaining structure is then associated with interband transitions and from these the transitions involving surface bands are singled out with the aid of surface adsorption as described above. None of these experimental techniques are simple and their current energy resolution is poor, of the order of 0.2 - 0.5 eV which constitutes a substantial portion of the bandgap of some of the more important semiconductors.

These techniques have been applied to ultrahigh vacuum cleaved surfaces of several III-V materials (GaSb, GaAs, GaP, InSb and InAs) and it has been possible to establish some features of the surface band structure common to all these materials.[21] The volume and surface densities of states $D_v(E)$ and $D_s(E)$, respectively, typical of compound semiconductors are illustrated in Fig. 10a. It is evident from PES, PYPES and EELS that compound semiconductors have a surface band structure with a bandgap separating a low-lying filled valence surface band from a higher-lying empty surface conduction band.[21] It is further observed that the surface valence band maximum consistently lies below the bulk valence band maximum and that

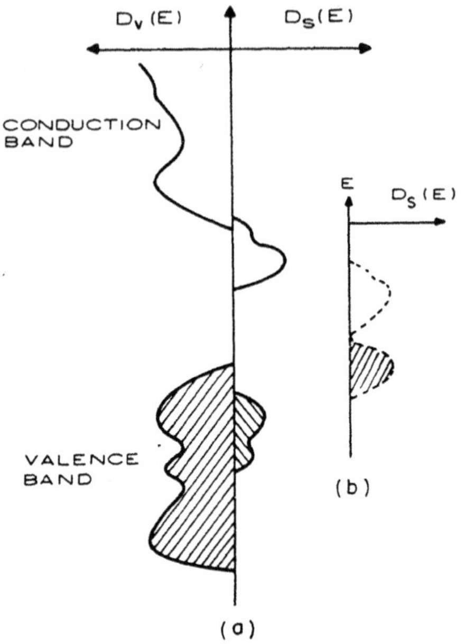

Fig. 10 Schematic density of states distribution of volume and surface bands $D_V(E)$ and $D_S(E)$, respectively, for compound semiconductor non-polar surfaces as determined from experiment. (b) Schematic surface band density of states for elemental semiconductors on the same energy scale as (a).

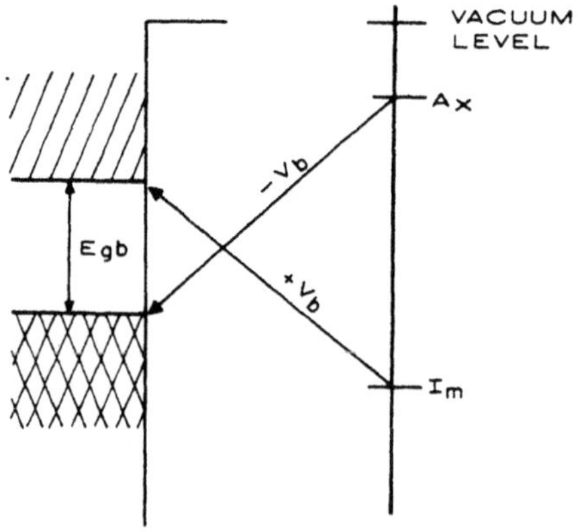

Fig. 11 Schematic representation of the generation of purely ionic band structure. A_X: free-space anion electron affinity; I_m: free-space cation ionization energy; V_b: infinite lattice Madelung energy; E_{gb}: bulk bandgap.

the surface conduction band minimum lies approximately at the midgap energy of the bulk band gap with the possible exception of InAs where the entire surface conduction band appears to lie above the volume conduction band minimum. It is interesting to note here that similar measurements have also been reported with ultrahigh vacuum cleaved elemental semiconductor surfaces (Ge and Si).[15,21] Here it has been found that there is no surface bandgap; instead, there is a cusp in the surface density of states situated in the lower half of the bandgap as illustrated by 10b. The surface bandgap observed with the compound materials consequently finds its origin with bonding ionicity.

2. Theoretical Concepts

The principal feature of the surface band structure of the compound semiconductors that can be traced to bonding ionicity is the surface bandgap. The existence of such a surface bandgap was predicted 10 years ago on the basis of pure bonding ionicity[22,23] and more recently, tight binding band structure calculations taking explicit account of bonding ionicity have reproduced the experimental surface band densities of states with remarkable accuracy.[24,25]

The surface bandgap derives from bonding ionicity arguments much in the same way as the bulk band gap. In the bulk lattice, the free-space electron affinity A_x and the ionization energy I_m levels of the anionic (X) and cationic (M) constituents, respectively, are inverted by the Madelung potential, V_b, that is, by the electrostatic potential of all the ions of the lattice evaluated at the site of either the M or the X ion:

$$V_b = \frac{CZe^2}{4\pi\epsilon_o a_o} \tag{2}$$

Here e is the electronic charge magnitude, Z is the valence, ε_o is the free space static electric permittivity, a_o is the nearest neighbor lattice constant and C is the bulk Madelung constant computed from the lattice sum of an infinite extent. The bulk bandgap E_{gb} is then given by

$$E_{gb} = 2V_b - (I_m - A_x) \tag{3}$$

as illustrated by Fig. 11. The Madelung potential, V_s, at a site in the surface plane of an ideally terminated non-polar surface, is always less than V_b because the associated lattice sum C_s is always less than C. Representative values of C, C_s

and the ratio $\gamma_s = C_s/C$ are listed in Table I for several non-polar surface indices of several lattice structures. Hence, by extension of Eq. (3), the surface bandgap E_{gs} is

$$E_{gs} = 2V_s - (I_m - A_x) < E_{gb} \tag{4}$$

also illustrated by Fig. 11. Because of the symmetry between M and X sites in this rigid lattice, point ion approximation, the reduction of the bandgap at the surface is symmetric; the bulk and surface midgap energies coincide. However, when it is realized that the anions are generally more polarizable than the cations, it is possible that the charge center of the former are displaced slightly more inward toward the bulk than the charge centers of the latter. This would have the effect of introducing an asymmetry into the Madelung potential evaluated at surface M- and X-sites. Specifically, V_{sX} will be increased because it must be evaluated more closely to an M-site one lattice layer further in while V_{sM} will decrease slightly because the X-ions in the surface plane on either side will have moved their charge centers a bit further away. This asymmetry has the effect of lowering both the surface conduction and valence band edges and so will lower the surface midgap energy relative to the bulk midgap energy as has been observed by experimentation.

Recently, a tight binding calculation of the surface band structure of GaAs has been reported (at this summer school by Dr. C. Calandra[25]) that reproduces the observed density of states quite accurately. The calculation used sp-hydridized orbitals to construct the band structure in lattice layers parallel to the surface and then considers the interaction of the lattice layers in the tight binding sense. This computation is patterned after that of Levine and Freeman[24] who used s- and p-orbitals located on neighboring M- and X-ions to simulate ionic binding rather than the sp-hybridized orbitals used by Calandra. This is not the place to go into the details. Suffice it to say, quantum mechanical band structure calculations are beginning to appear in the literature that reproduce the major features of the observed surface band density of states and it is this initial agreement between experiment and computation that is the beginning of the iterative process of obtaining an accurate description of surface energy bands.

3. Summary

From both recent experimental and computational studies the surface electronic structure of non-polar compound semiconductors comprises a filled valence band separated from an empty conduction band by a surface bandgap. The surface

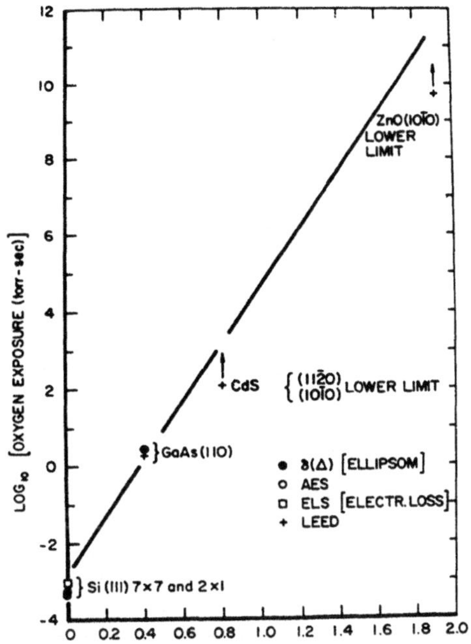

Fig. 12

Log$_{10}$ (oxygen exposure) needed to produce 50% change in measured surface parameter versus Pauling electronegativity difference ΔX. ●: change in ellipsometric parameter Δ; o: oxygen AES signal; □: oxygen EELS signal; +: LEED intensity attenuation. From reference 26.

TABLE 1

Lattice Type	C	Surface Index	C_s	$\gamma_s = \frac{C_s}{C}$	C_a	$\gamma_a = \frac{C_a}{C}$
NaCl	1.748	(100)	1.682	0.962	0.066	0.038
NaCl	1.748	(110)	1.503	0.86	0.245	0.140
NaCl	1.748	(211)	1.049	0.60	0.699	0.40
NaCl	1.748	(210)	1.346	0.77	0.402	0.23
CsCl	1.763	(110)	1.587	0.90	0.176	0.10
Wurtzite	1.641	(11$\bar{2}$0)	1.506	0.918	0.135	0.082
Wurtzite	1.641	(10$\bar{1}$0)	1.459	0.889	0.182	0.111
Zincblende	1.638	(110)	1.491	0.910	0.147	0.090

valence band appears to lie entirely in the energy range of the bulk valence band while the surface conduction band appears to extend down into the upper half of the bulk bandgap (with the possible exception of InAs). These major features appear to be a direct consequence of bonding ionicity.

SURFACE CHEMICAL ACTIVITY

1. Ordered Surfaces

It has been observed that there is a relationship between the rate of oxygen uptake (adsorption or oxidation) of an ordered non-polar compound semiconductor surface and the Pauling electronegativity difference ΔX of the semiconductor lattice constituents.[26] This relationship includes the cleavage surfaces of Si for which $\Delta X = 0$. The dependence is shown in Fig. 12 which shows the logarithm of the oxygen exposure necessary to obtain an irreversible 50% change in several measured surface parameters (see figure caption) at room temperature vs. ΔX. Note that the most active surface is Si (111) and that chemical activity decreases drastically as ΔX increases. The following explanation is offered for this effect.

For rather ionic semiconductors, it appears that oxygen uptake is predominantly by electron transfer adsorption in which the electronic charge participating in the adsorption bond comprises electrons from the semiconductor conduction band. The principal evidence for this is the frequent observation that the kinetics for this type of gas-solid interaction are governed by an Elovich rate law typical of a reaction whose rate is limited by the diffusion of electronic majority carriers through a Schottky-type barrier.[27,28] Semiconductivity modulation has also been observed to accompany this type of gas-solid interaction.

To provide stability for adsorbed ions, their electronic state must lie inside the semiconductor bulk bandgap by several kT. The relative inertness to such ion formation on ordered, non-polar semiconductor surfaces with substantial bonding ionicity can be understood by recognizing that a substantial shift in energy (usually greater than 3 eV) is required to displace the electronic energy level of the adsorbate ion from its free-space value (in this case, the electron affinity of oxygen) into the energy range spanned by typical III-V and II-VI semiconductor bandgaps. In the limit of pure ionic bonding, this shift can be attributed to the Madelung energy available at an adsorption site.[29] This energy can be written formally as

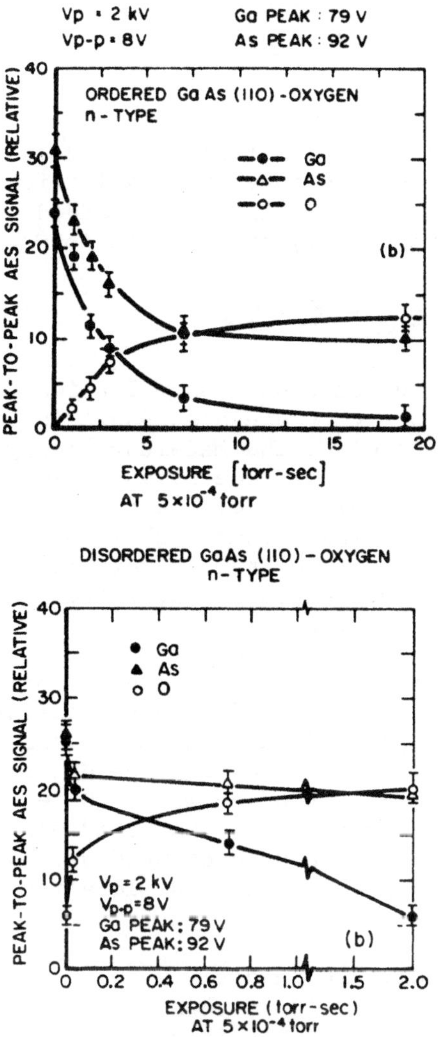

Fig. 13

(a) Oxygen up-take at room temperature of an ordered GaAs (110) surface, n-type material, as determined by change in AES signal of Ga, As and oxygen. (b) same as (a) after the surface had been disordered by argon ion beam bombardment. From reference 1.

$$V_a = \frac{C_a Z e^2}{4\pi\epsilon_o a_o} \tag{5}$$

where C_a is the Madelung constant computed at a site just outside the ideal flat surface where a lattice constituent would be if another lattice layer were added. It has been shown that C_a is given rigorously by the relation[30]

$$C_a = C - C_s \tag{6}$$

for all non-polar surfaces of binary lattices. Some values of C_a and the ratio $\gamma_a = C_a/C$ are also shown in Table 1. As C_a is at most 10% - 12% of C, the Madelung energy available at a cite on an ideal ordered surface is about 10% or less of the bulk Madelung energy which is of the order of 10 eV to 15 eV, depending on the material. Thus a shift of no more than 1.5 eV can be expected in the free-space adsorbate electron affinity and this is usually not enough to move the free-space level into the adsorbent bandgap.

For the opposite limit of no bonding ionicity, gas-solid interactions at the surfaces of purely covalent elemental semiconductors are predominantly chemical reactions associated with local chemical bonds involving semiconductor valence states. The oxidation of Si is a good example. Here it is not possible to define a one-electron state associated with the adsorbate. Rather, a new state in the valence band of the semiconductor has been created with a more or less localized wavefunction to represent the surface bond.[26]

The gradual decrease in activity with increasing bonding ionicity evident in Fig. 12 is thus ascribed to the decreasing residual bonding covalency retained by the semiconductor as ΔX increases. In the limit of ideal bonding ionicity, there would be a discontinuity in this curve, starting from some finite value at $\Delta X = 0$ determined by covalent interactions to essentially infinity to represent the total inertness of an ideal ionic material.

2. Disordered Surfaces

It has been observed that when an initially ordered non-polar semiconductor surface is disordered, its chemical activity increases substantially.[1,31] Evidence for this is given in Fig. 13 which shows changes in the AES signals of Ga, As and oxygen vs. oxygen exposure for an ordered GaAs (110) surface (a) and when the same surface has been disordered by argon ion sputtering (b).[1] The phenomenon can be understood as follows.

The principal effect of the argon ion beam is to introduce disorder in the surface plane. This disorder can be viewed as a collection of surface defects on an atomic scale, such as steps, edges and corners. Each of these surface defects has the effect of increasing the coordination number of a neighboring site outside the lattice. That is, where a flat surface might yield sites where an adsorbed oxygen ion can have only one nearest neighbor lattice cation, a surface rich in monatomic steps provides many sites where an adsorbed oxygen ion may have two or three nearest neighbor lattice cations. The Madelung constant at such a site is clearly larger than at an adsorption site on an ideal flat surface. Hence, the Madelung stabilization energy will also be greater at a step site than at a site on a flat surface. The computation of Madelung constants at adsorption sites corresponding to various surface defects on an atomic scale have been computed and can be found in the literature.[1,32] It need only be mentioned here that Madelung constants up to values greater than 50% of the bulk Madelung constant can arise from simple step-edge-corner configurations and that is enough to shift the free-space adsorbate energy level into the adsorbent band-gap for virtually any compound semiconductor adsorbent.

3. Summary

It has been observed that the chemical activity for simple gas-solid interaction on ordered non-polar compound semiconductor surfaces is always less than on ordered elemental (purely covalent) semiconductor surfaces and that the activity decreases with increasing bonding ionicity. It has also been observed that the activity of a given compound semiconductor non-polar surface is drastically increased when that surface is kinetically disordered by argon ion bombardment in vacuum. Both of these phenomena are ascribed to bonding ionicity via the lattice Madelung energy availalbe at an adsorption site outside the lattice.

SYNOPSIS

It has been demonstrated through resort to published literature that bonding ionicity plays a dominant role in several important surface properties of all binary semiconductors with partial bonding ionicity. The following are included: 1) The strong structural stability of non-polar surfaces. 2) The equally strong structural instability of polar surfaces. 3) The characteristic surface electronic structure observed with non-polar cleavage planes comprising a surface valence band whose maximum is below the bulk valence band maximum and a surface conduction band whose minimum is below the bulk conduction band minimum. 4) The relative inertness of ordered non-polar compound semiconductor surfaces to simple gas-solid chemical interactions. 5) The strong increase in chemical activity when an initially ordered non-polar compound semiconductor surface is disordered on an atomic scale.

ACKNOWLEDGEMENTS

Collaboration with Dr. J. D. Levine of RCA Laboratories, Princeton, N. J. and Dr. C. B. Duke of Xerox Research Laboratories, Webster, N. Y. during various phases of this work, is gratefully acknowledged.

REFERENCES

1. P. Mark, S. C. Chang, W. F. Creighton and B. W. Lee, CRC Critical Reviews of Solid State Sciences 5, No. 2, p. 189 (1975).

2. S. C. Chang and P. Mark, Surface Science 45, 721 (1974).

3. S. C. Chang and P. Mark, J. Vac. Sci. Technol. 12, 624 (1975).

4. A. R. Lubinsky, C. B. Duke, S. C. Chang, B. W. Lee and P. Mark, J. Vac. Sci. Technol. 13. xxx (1976).

5. C. B. Duke, in DYNAMIC ASPECTS OF SURFACE PHYSICS; Proceedings of the International School of Physics "Enrico Fermi," Course LVIII, (Editrice Compositori, Bologna, Italy, 1974), F. O. Goodman, Ed.

6. J. B. Pendry, LOW ENERGY ELECTRON DIFFRACTION, (Academic Press, New York, 1974).

7. R. W. Nosker, P. Mark and J. D. Levine, Surface Science 19, 291 (1970).

8. D. Kohl, M. Henzler and G. Heiland, Surface Science 41, 403 (1974).

8a. Y. Margoninski and R. E. Kirby, J. Phys. C: Solid State Phys. 8, 1516 (1975).

9. J. T. Grant and T. W. Haas, Surface Science 26, 669 (1971).

10. S. C. Chang and P. Mark, J. Vac. Sci. Technol. 12, 629 (1975).

11. J. J. Lander and J. Morrison, J. Chem. Phys. 37, 729 (1962).

12. P. Mark and W. R. Bottoms, Prog. Solid State Chem. 6, 17 (1971).

13. H. Ibach, K. Horn, R. Dorn and H. Lüth, Surface Science 38, 433 (1973).

14. R. Ludeke and L. Esaki, Phys. Rev. Letts. 33, 653 (1974).

15. J. E. Rowe, Solid State Commun. 15, 1505 (1974).

16. D. E. Eastman and W. D. Grobman, Phys. Rev. Letts. 28, 1378 (1972).

17. D. E. Eastman and J. L. Freeouf, Phys. Rev. Letts. 33, 1601 (1974).

18. D. E. Eastman and J. L. Freeout, Phys. Rev. Letts. 34, 1624 (1975).

19. L. F. Wagner and W. E. Spicer, Phys. Rev. Letts. 28, 1381 (1972).

20. P. E. Gregory, W. E. Spicer, S. Ciraci and W. A. Harrison, Appl. Phys. Letts. 25, 511 (1974).

21. J. L. Freeouf and D. E. Eastman, CRC Critical Reviews in Solid State Science 5, No. 3, p. 245 (1975).

22. J. D. Levine and P. Mark, Phys. Rev. 144, 751 (1966).

23. J. D. Levine, J. Vac. Sci. Technol. 6, 549 (1969).

24. J. D. Levine and S. Freeman, Phys. Rev. (B) 2, 3255 (1970).

25. C. Calandra, this NATO Summer School.

26. P. Mark and W. F. Creighton, Appl. Phys. Letts. 27, 400 (1975).

27. P. B. Weisz, J. Chem. Phys. 20, 1483 (1952); 21, 1531 (1953).

28. R. Williams, J. Phys. Chem. Sol. 23, 1057 (1962).

29. P. Mark, J. Phys. Chem. Sol. 29, 689 (1968).

30. R. W. Nosker and P. Mark, Surface Science 20, 421 (1970)

31. S. Baidyaroy and P. Mark, Thin Solid Films 12, 129 (1972).

32. I. Stranski, Z. Phys. Chem. 136, 259 (1929); Z. Electrochem. 36, 25 (193).

DISCUSSION

BLAKELY : How important do you feel it is to take account of contributions to the potential from the electronic dipole moments of the surface ions ?

MARK : Dipole energies are of the order of 10 % of bulk Madelung energies and about equal to adsorption site Madelung energies on ideal flat surfaces. However, such surfaces of ionic character are quite inert from experimental studies so that neither electrostatic nor polarization energies appear to be sufficient to generate bound-electron-transfer states.

RALEIGH : If a charge carrier is to leave a material into an adjacent phase, must it leave via a surface state, or may it also leave via a bulk state ?

MARK : In principle it can do either depending on the kinetic properties of the surface state. If the surface state is a bound state, its capture cross section for the carrier in question determines whether or not the carrier traverses the surface by "hopping" through the surface state.

GERISCHER : The vacant surface states can be detected in the partial yield spectroscopy because the matrix element for an electronic transition from the core states to these surfaces states is large since they are concentrated at the same atom. What is the chance to observe an optical transition between occupied and vacant surface states ?

MARK : My estimate is very little, owing to the delocalized character of the surface-valence-band-states in the orientation parallel to the surface. There is no evidence to my knowledge of such transitions although it could turn out that some materials might have very narrow surface valence bands in which case such transitions might be more likely.

ADSORPTION AND CATALYSIS ON IONIC SOLIDS

Bernard CLAUDEL

Institut de Recherches sur la Catalyse, and
Institut National des Sciences Appliquées de Lyon
Villeurbanne, France

Adsorption and catalysis represent a very broad domain, which in the following will be restricted to its most important aspects, in relation with ionic solid properties, and divided into three parts :

A. General definitions and formal considerations on adsorption and catalysis ;
B. Relationships between electronic properties of the bulk and surface properties;
C. Collective properties of the solid and active surface centers.

A. GENERALITIES ON ADSORPTION AND CATALYSIS (Ref. 1, p.97 and ff.)

I. DEFINITIONS

The term "catalysis" includes any process other than energy input (heat, light) which accelerates a given chemical reaction without modifying the material balance. A catalyst is a chemical species or a mixture of chemical species able of such an acceleration. By definition, it is not involved in the material balance, and thus, it does not alter the overall free enthalpy change ΔG. In particular, it cannot make a reaction occur which is thermodynamically impossible, and it does not shift the equilibrium when there is one. The role of a catalyst is then purely a kinetic one, and thus is to be related to the fundamental definitions of reaction kinetics.

M. Kleitz and J. Dupuy (eds.), Electrode Processes in Solid State Ionics, 45–81.

1. Fundamental definitions on kinetics and catalysis

A reaction consumes the reactants and yields the products. Let dn_i be the number of moles of one of these constituents which disappear, or appear, during the time interval dt. The reaction rate can be defined by the ratio

$$\frac{dn_i}{dt}$$

with the reservation of a proportionality constant chosen so that:

a. this rate is always positive;
b. it does not depend on the considered constituent.

It is easy to show that the definition :

$$(1) \qquad r = \frac{1}{\nu_i} \frac{dn_i}{dt}$$

fulfils both of these conditions, if ν_i is the <u>algebraic</u> stoichiometric coefficient, such as ν_i is positive when the constituent is a product, and ν_i is negative when the constituent is a reactant (2). Such a reaction rate is expressed in moles of the constituent i consumed or produced per unit time. It is customary, especially when the reaction medium is held at a constant volume V, to refer to this volume, by introducing a <u>new</u> reaction rate :

$$(2) \qquad r_V = \frac{r}{V}$$

If V is constant,

$$(3) \qquad r_V = \frac{1}{\nu_i} \frac{d\left(\frac{n_i}{V}\right)}{dt} = \frac{1}{\nu_i} \frac{dC_i}{dt}$$

r_V is thus expressed in moles per unit time and per unit volume

If a contact catalysis is to be considered, i.e. if a reaction between fluid constituents is catalyzed by a solid of a given extent, the reaction rate can be referred to this extent, for instance to the catalyst mass m :

$$(4) \qquad r_m = \frac{r}{m}$$

which can be called, after GERMAIN (3), specific activity. Or the catalyst extent can be taken as its surface area a_S. Thus ,

$$(5) \qquad r_S = \frac{r}{a_S}$$

This new rate can be called "intrinsic activity" of the catalyst (3).

2. Kinetic parameters .

These are the variables which affect the reaction rate. We shall distinguish among them the amounts of the constituents and the temperature .

It is customary to express the former by the concentrations in the fluid phase C_i, or the partial pressures in the gas phase P_i, and to separate the variables, so that

(6) $$r = k\ f(C_i)$$

where k is a function of temperature alone, and therefore, does not depend any longer on time. For this reason, k is called (wrongly) the "rate constant" . Generally,

(7) $$k = A \exp\left(-\frac{E}{\mathcal{R}T}\right)$$

where A is a proper constant and E the apparent activation energy.

It is customary (but not always adequate, especially in the case of heterogeneous catalysis) to write f under the form :

(8) $$f = \prod_i C_i^{q_i}$$

where the product $\prod$ is to be taken over <u>all</u> the components. The exponent q_i (integral or fractional, positive, negative or zero) is the order of the reaction with respect to component i. $\sum_i q_i$ is called the overall reaction order, and the form (8) of the function f is the power rate law. In heterogeneous catalysis, this form is far from being general, and, when it applies, it can be a mere deceptive appearance, as we shall see later .

3. Reaction mechanism. Elementary steps .

The aim of chemical kinetics is not only to measure reaction rates, but also to justify their expressions by a so-called mechanism. A mechanism is a set of elementary steps into which the overall reaction decomposes. Each step involves given intermediates. Putting forward a mechanism amounts to making an hypothesis on the elementary steps and on the intermediates. Heterogeneous catalysis is no long range interaction, and thus implies a kind of chemical combination between the reaction constituents and the solid surface. Such a combination is an elementary step and its investigation is a prerequisite in any study of catalysis .

II. ADSORPTION

1. Physisorption and chemisorption

Adsorption designates the fixation of fluid molecules (or particles) on a solid surface, without there being any penetration of these particles into the solid bulk. Desorption is the reverse phenomenon. Two types of adsorption are to be distinguished, according as the interaction between the adsorbent and the adsorbate is of physical or chemical nature. In the former case, the adsorption is called physisorption (or physical adsorption), in the latter, it is called chemisorption (or chemical adsorption) . Table I below summarizes the main features by which these two types can be distinguished , although no sharp distinction can be made and intermediate cases exist .

TABLE I
CLASSICAL DISTINCTIONS BETWEEN PHYSISORPTION AND CHEMISORPTION

	PHYSISORPTION	CHEMISORPTION
Bonds	VAN DER WAALS type	Chemical
Energies	A few kcal/mole (order of magnitude of liquefaction heat)	Dozens of kcal/mole (Order of magnitude of heat of a chemical reaction)
Modification of electronic properties of the adsorbate	Minimal	May be detectable by suitable physical means
Kinetics	No activation energy involved	Usually involves non-zero activation energy
Selectivity	Minimal	Selective
Number of adsorbed layers	Multilayer adsorption	Monolayer adsorption

Although physisorption is a major tool in texture studies (i.e. all the studies concerned with surface area, pore radii, capillary condensation, etc...) it will not be considered here.

2. Chemisorption

a) <u>Definitions</u> . In a chemisorption process, an adsorbed molecule is linked by valence bonds to a particular location of the surface, which is called an adsorption site. Such a process may be non-dissociative when the molecule is adsorbed without fragmentation, or dissociative when the molecule undergoes a dissociation into two or more adsorbed fragments. The sites are uniform when there is no variation of energy parameters when the adsorption goes on. In the opposite case, the sites are non-uniform.

b) Chemisorption kinetics and equilibria . The energy diagram relative to the inverse elementary steps of adsorption and desorption is represented by fig. 1 .
As a consequence of the principle of microscopic reversibility,

$$(9) \qquad \Delta H = E_{ads} - E_{des}$$

Let us first consider the case of a non-dissociative adsorption :

$$A + s \rightleftarrows As$$

The adsorption rate will be given by :

$$(10) \qquad r_{ads} = a\, P\, (1 - \theta)\, \exp\left(- \frac{E_{ads}}{\mathcal{R}T}\right)$$

where P is the partial pressure of the adsorbate , θ the surface coverage (i.e. the ratio of the adsorbed amount to the monolayer capacity) . Likewise, the desorption law is expressed by :

$$(11) \qquad r_{des} = b\, \theta\, \exp\left(- \frac{E_{des}}{\mathcal{R}T}\right)$$

Formulas (10 and (11) are of general validity. If it is further assumed that E_{ads} and E_{des} are independent of the surface coverage (uniform sites) , it is dealt with the LANGMUIR model. By equating r_{ads} and r_{des} the equilibrium law (also called isotherm) is obtained :

$$(12) \qquad \frac{\theta_e}{1 - \theta_e} = \lambda P$$

$$\text{with (13)} \qquad \lambda = \frac{a}{b} \exp\left(- \frac{E_{ads} - E_{des}}{\mathcal{R}T}\right) = \frac{a}{b} \exp\left(- \frac{\Delta H}{\mathcal{R}T}\right)$$

This LANGMUIR isotherm is still very popular in the interpretation of adsorption and catalysis experiments, although it is obvious that the underlying hypotheses are rather restrictive .

It is possible to get rid of the most stringent (and most unlikely) of these hypotheses by assuming that the surface is not uniform, i.e. that E_{ads}, E_{des} and ΔH vary during the adsorption. The non uniformity is called " intrinsic" when these variations are due solely to the nature of the adsorbent. "Induced" non-uniformity arises when it is the presence of an adsorbed particle on one site which leads to a variation in the energy characteristics of a neighbouring site. It is possible to show (Ref. 1, p. 136 and ff.) that in the former case, a FREUNDLICH adsorption isotherm is obtained :

$$(14) \qquad \theta_e = c\, P^m \quad (m < 1)$$

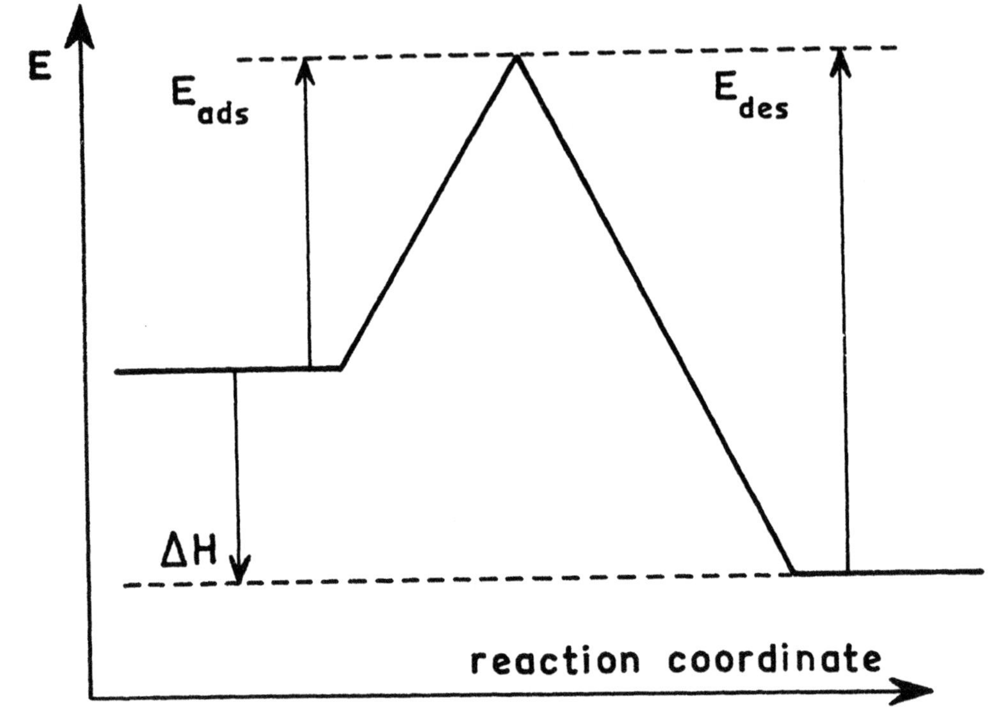

Fig.1 : Energy diagram for adsorption and desorption

In the latter case, as we shall see in the following part, the simplest assumption which can be made is that E_{ads} and E_{des} vary proportionally to θ :

(15) $E_{ads} = E_{o_{ads}} + \alpha \mathcal{R} T \theta$ ($E_{o_{ads}}$ and α positive constants)

(16) $E_{des} = E_{o_{des}} - \beta \mathcal{R} T \theta$ ($E_{o_{des}}$ and β positive constants)

Putting the value (15) of E_{ads} into the expression (10) gives :

(17) $r_{ads} = a\ P\ (1 - \theta)\ \exp\left(- \frac{E_{o_{ads}}}{\mathcal{R} T}\right) \exp(-\alpha\theta)$

At the beginning of the adsorption, θ is negligible with respect to 1, and

(18) $r_{ads} = r_{o_{ads}} \exp(-\alpha\theta)$

This rate expression has been popularized in western countries as the ELOVICH equation, although ROGINSKII and ZELDOVICH seem to have been the first to introduce it . An analogous law holds for desorption. Interestingly, equation (9) is always valid so that the chemisorption heat ΔH is also a linear function of θ :

$$(19) \quad \Delta H = \Delta H_o + \gamma \mathcal{R} T\theta \ , \ \text{with} \ \gamma = \alpha + \beta$$

An isotherm can be derived (Ref. 1, p. 140 and ff.), and shown to be expressed, in its simplified form, as :

$$(20) \quad \theta_e = \frac{1}{\gamma} \ \ell n \ \lambda_o P \quad \text{(TEMKIN isotherm)}$$

If now chemisorption is supposed to be dissociative,

$$A_n + ns \rightleftarrows n\,As$$

LANGMUIR model leads to the following isotherm :

$$(21) \quad \theta_e = \frac{(\lambda P)^{1/n}}{1 + (\lambda P)^{1/n}}$$

For non dissociative chemisorption of i adsorbates $A_1 \ldots A_i$ competing for the same sites, LANGMUIR isotherm takes the form :

$$(22) \quad \theta_{e,i} = \frac{\lambda_i P_i}{1 + \sum \lambda_i P_i}$$

c) <u>Comparison with experiment</u>. The greatest care must be taken in the interpretation of experimental results according to one of the preceding models. The experimental accuracy is usually not great enough in order to allow to discriminate between the different equations. In other words, many expressions can represent equally well a given set of experimental data. A good example of the non discriminating character of the kinetic laws and isotherms is provided by CO chemisorption with thoria surface (4). Satisfactory straight lines are obtained when θ is plotted as a function of log t as well for adsorption as for desorption. This corresponds to the "ELOVICH" rate law, whichs leads to the TEMKIN isotherm (20). Indeed, plotting θ_e as a function of log P shows that the linear relationship between these two variables is fairly well obeyed. But satisfactory straight lines are obtained as well when the adsorption kinetics is drawn in the representation log θ vs log t, thus showing that :

$$(23) \quad \log \theta = \frac{1}{n} \log t$$

with a value of n close to 6 at each working temperature. The power rate law inferred by equation (23) is often referred to as BANGHAM rate law (5). Likewise, when the isotherm is represented by plotting log θ_e as a function of log P, a good straight line is obtained, pointing out the apparent validity of FREUNDLICH law (14). It is worth noting that at low concentrations (or partial pressures), all the isotherms for a non-dissociative adsorption reduce to the linear relationship :

$$(24) \quad \theta_e = \lambda P$$

which is sometimes called HENRY's law .

As a conclusion, from these experimental results alone, it is impossible to decide in favour of either adsorption law. So, adsorption studies have to resort to other methods than purely kinetic ones in order to unravel the intricacies of an adsorbate-adsorbent system. In the case in point, microcalorimetry allows to exclude at least one model, since chemisorption heat is a linear function not of the adsorbed volume, as it should be if the ELOVICH-TEMKIN model were obeyed, but of its logarithm .

III. CATALYSIS

Finding a mechanism for a catalysis is alike finding a mechanism for a complex reaction, and suffers from the same drawback , i.e. the fact that many mechanisms may fit the same rate expression .

The most general hypothesis which can be made is the steady-state hypothesis (noted "S.S.H." below), which states that the net rate of reactive intermediate production is nil. Let us examplify this rule with a non-conventional mechanism proposed in Ref. 2, p.62 . It is assumed that the water gas shift catalysis

$$CO_g + H_2O_g \rightleftarrows CO_{2\,g} + H_{2\,g}$$

occurs through the two step sequence :

$$(1 - 1') \qquad H_2O_g + s \underset{k'_1}{\overset{k_1}{\rightleftarrows}} H_{2\,g} + Os$$

$$(2 - 2') \qquad Os + CO \underset{k'_2}{\overset{k_2}{\rightleftarrows}} CO_{2\,g} + s$$

The reaction rate is :

$$(25) \quad r = k_1(1 - \theta)\,(H_2O) - k'_1\,\theta(H_2)$$

where the symbols between round brackets designate the concentrations. Now, the S.S.H. expressed for the intermediate s gives :

$$(26) \quad k_1(1-\theta)(H_2O) + k'_2(CO_2)(1-\theta) = k'_1\theta(H_2) + k_2\theta(CO)$$

It can easily be checked that the same equation is derived from the S.S.H. for the intermediate Os. Hence ,

$$(27) \qquad \theta = \frac{k_1(H_2O) + k_2'(CO_2)}{k_1(H_2O) + k_1'(H_2) + k_2(CO) + k_2'(CO_2)}$$

and thus

$$(28) \qquad r = \frac{k_1k_2(H_2O)(CO) - k_1'k_2'(H_2)(CO_2)}{k_1(H_2O) + k_1'(H_2) + k_2(CO) + k_2'(CO_2)}$$

In particular, at equilibrium, $r = 0$, and the "mass action" law is found :

$$\frac{(H_2)(CO_2)}{(H_2O)(CO)} = K = \frac{k_1k_2}{k_1'k'_2}$$

The general rate equation (28) can be further simplified if some intermediary rates are neglected in comparison with others. For instance, if r_2 and r'_2 are much smaller than r_1 and r_1' ,

$$(29) \qquad r = \frac{k_1k_2(H_2O)(CO) - k_1'k_2'(H_2)(CO_2)}{k_1(H_2O) + k_1'(H_2)}$$

This expression (29) can be deduced from the "quasi-equilibrium" 1 - 1' (Fig.2)

$$(30) \qquad \frac{\theta\ (H_2)}{(1 - \theta)\ (H_2O)} = \frac{k_1}{k_1'}$$

$$(31) \qquad r = k_2\theta(CO) - k_2'(1 - \theta)(CO_2)$$

whence (29) is easily derived. In this simplified case, 1 and 1' are so fast as to be at "quasi-equilibrium" whereas 2 is the slowest step of the forward reaction, and 2' the slowest step of the backward reaction. Either of these slowest steps is "rate determining", i.e. the rate of the overall reaction (in either direction) adjusts itself to the rate of this rate-determining step. It must be borne in mind that the assumption of a rate determining step is less general than the S.S.H., and that the same rate expression may be obtained by making different assumptions on the nature of the rate determining step . Thus, ambiguity is the rule in formal kinetics .

As a matter of fact, there are numerous examples of such ambiguities in more conventional treatments of catalytic mechanisms. A " first-order" ambiguity lies in the analytical expression of

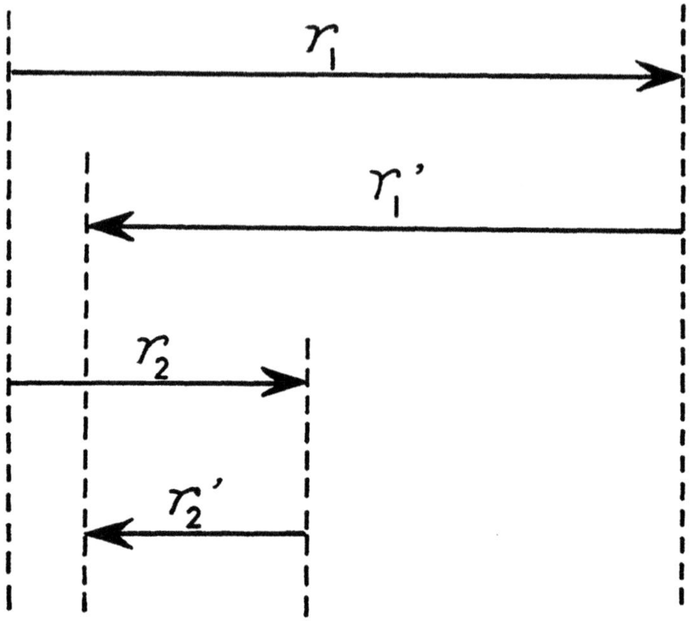

Fig. 2 : Diagrammatic representation of the reaction rates of a complex mechanism (Ref. 2, p.83)

the experimental data. It is customary to look for a rate equation of type (8), in which appear the partial orders q_i. For instance, in the case of carbon monoxide oxidation on the same thoria surface as above, with an excess of oxygen, the value of the order with respect to CO is found to be 0.4 (6) (Fig. 3)

But such a representation can be a mere camouflage. If it is assumed that the rate depends on CO concentration according to a LANGMUIR type expression :

(32) $$r = \frac{k\,P_{CO}}{1 + \lambda P_{CO}}$$

straight lines as good as those of fig. 3 can be obtained (Fig.4) Moreover, a rate expression proportional to the logarithm of P_{CO} is obeyed as well (Fig. 5) .

Here comes the "second order" ambiguity, which lies in the choice of the mechanism. The above results can be expressed as :

(33) $$r = k\,\theta_{CO}$$

This expression corresponds to a mechanism where the surface reaction is the rate determining step,the CO adsorption being at quasi-

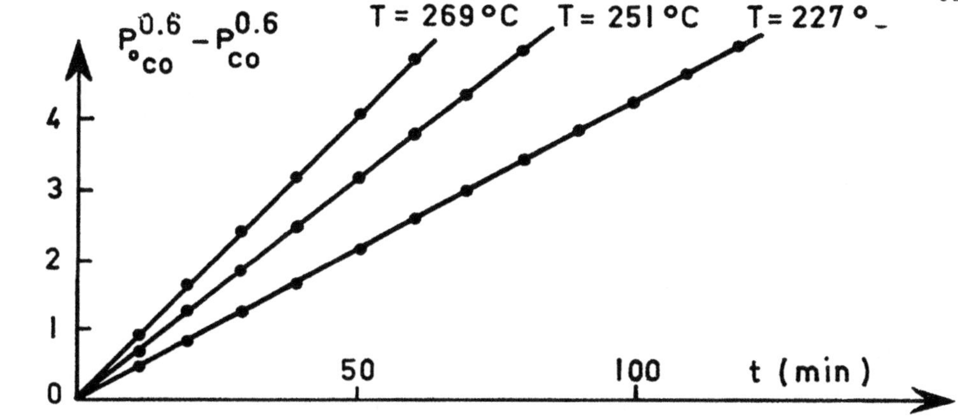

Fig. 3 : CO oxidation on ThO_2 . Order with respect to CO (After (6))

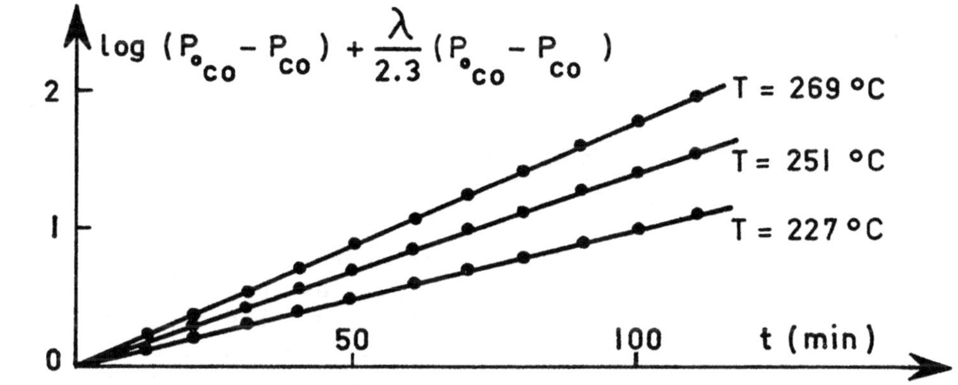

Fig. 4 : CO oxidation on ThO_2. Rate expression (32) (After (6))

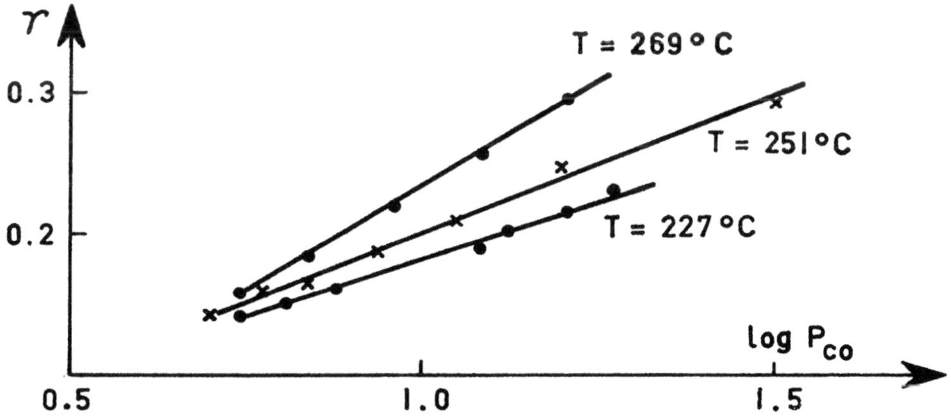

Fig. 5 : CO oxidation on ThO_2. Rate expression $r = k' \log P_{CO}$ (After (6))

equilibrium. This quasi-equilibrium can be expressed by a LANGMUIR or TEMKIN isotherm. The latter may be considered as more consistent with the precedeing adsorption data if it is admitted (perhaps questionably) that the adsorption conditions of an unique component still hold in the presence of other ones .

But let us now assume that mechanism is a sequence of irreversible steps :

(1) $CO + s \longrightarrow COs$

(2) $O_2 + 2s' \longrightarrow 2\ Os'$

(3) $COs + Os' \longrightarrow CO_2s + s'$

(4) $CO_2s \longrightarrow CO_2 + s$

with $(s) + (COs) = N_S$, const. and $(s') + (Os') = N'_S$, const.

It can be assumed that when $P_{O_2} \gg P_{CO}$, (Os') = const. , and :

$$r = \frac{k_1 k_3 N_S (Os')\, P_{CO}}{k_1 P_{CO} + k_3 (Os')}$$

which is identical with (32) but with another meaning of the parameters. Thus, the same experimental rate expression can be derived from two different mechanisms. This is a basic hindrance, although it can in some cases be overcome by having recourse to computers, which can give the degree of confidence which can be attached to the postulated mechanism .

B. RELATIONS BETWEEN SURFACE AND BULK PROPERTIES OF THE CATALYTIC SOLID (7)

In order to escape the mere formalism which has prevailed in the preceding chapter, it is necessary to develop a model of the interaction between the fluid species and the solid. In the following, we shall consider the collective model known as the band model. This is not to say that this model is the only one. As a matter of fact, after having known a considerable success in the fifties, its shortcomings have later been pointed out, and it has met with an undeserved unpopularity. It remains, however, that this model has stimulated a great deal of researches, especially on ionic solids, and it seems that a more balanced opinion now prevails, which attempts to reconcile it with the findings gained on the behaviour of localized surface centers. This attempt will be briefly accounted for in the third chapter. Suffice it to say, at present, that high temperature fuel cells with semiconducting electrodes of high electronic conductivity (8) offer a direct evidence of the validity of the charge transfer model. The latter rests on the figure 6, valid when the surface charge is negative, and where the dipole component has been neglected.

I. EARLIER CHARGE TRANSFER THEORIES ("Boundary layer theories"- B.L.T. - Randschichttheorien)

1. Derivation of the model.

In the earlier theories (9) (10) (11) (12), the emphasis is put upon the calculation of the potential barrier V_S by assuming that every adsorbed particle picks up a charge equal, in absolute value, to the value e of the elementary charge.

The basic considerations are the following : the electronic density n in a plane of abscissa x can be expressed :

$$(34) \qquad n(x) = N_c \exp\left(-\frac{E_c - E_F}{kT}\right)$$

N_c being the density of states in the conduction band. Putting :

$$(35) \qquad u = \frac{eV}{kT}$$

(by convention, $u > 0$ if the adparticle is positively charged, $u < 0$ in the opposite case) one obtains :

$$(36) \qquad n(x) = n_b \exp u$$

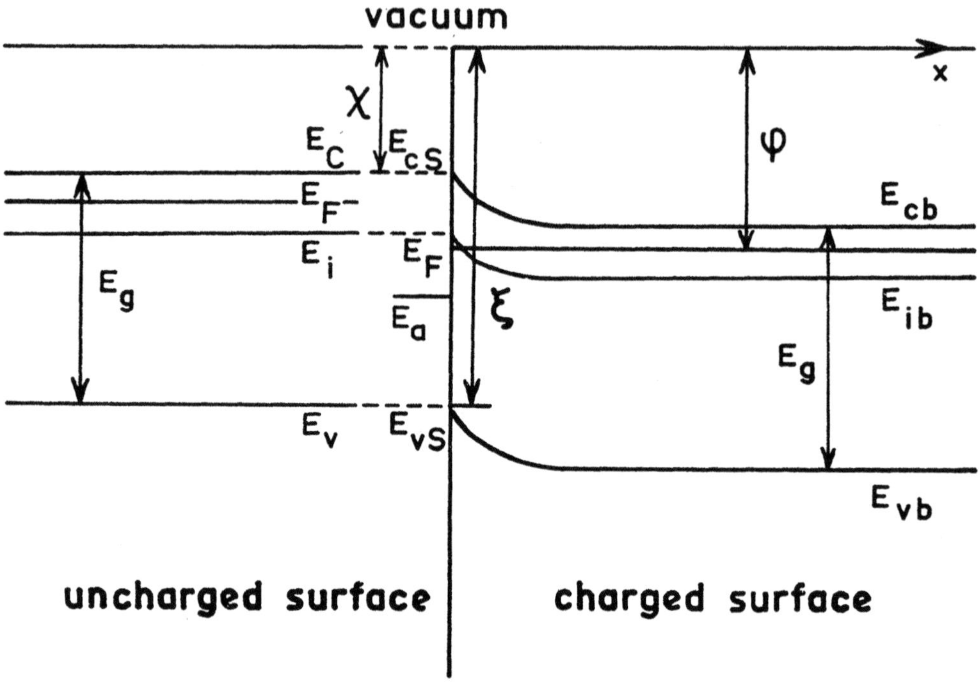

Fig. 6 : Band model of a neutral semiconductor (left) and of a semiconductor bearing a negative surface charge (right) .

(Legend of fig.6)

E_F, FERMI level ; E_c, energy of the bottom of conduction band ; E_v, energy of the top of valence band; E_g, band gap; E_a, surface acceptor level; φ,work function; χ, electron affinity; ξ, ionization energy

n_b being the electronic density in the bulk. Likewise, the hole density is :

$$(37) \qquad p(x) = p_b \exp(-u)$$

The POISSON equation is now taken into account :

$$(38) \qquad \frac{d^2V}{dx^2} = \frac{1}{\varepsilon_o \varepsilon_r} \rho(x)$$

where $\rho(x)$ is the volume charge density at abscissa x
ε_o the free space permittivity
ε_r the dielectric constant. (In the following, we shall put $\varepsilon = \varepsilon_o \varepsilon_r$)

$\rho(x)$ can be expressed, for a semiconductor containing N_a acceptor ions per unit volume and N_d donor ions per unit volume, as :

$$(39) \qquad \rho(x) = e\,(N_d - N_a + p - n)$$

The subsequent derivations are somewhat intricate, but can be simplified in some cases :

a) <u>If the semiconductor is intrinsic</u> , equation (39) reduces to :

$$\rho(x) = e(p - n)$$

Taking into account $n_b = p_b$, the POISSON equation (38) can be written :

$$(40) \qquad \frac{d^2u}{dx^2} = \frac{1}{L^2} \sinh u$$

L being the DEBYE length :

$$(41) \qquad L = \sqrt{\frac{k\,T\,\epsilon}{2\,e^2 n_b}}$$

Integration of (40) between the boundaries $x = 0$ ($V = V_S$) and $x = \infty$ ($V = 0$) yields :

$$(42) \qquad \frac{du}{dx} = -\frac{2}{L} \sinh \frac{u}{2}$$

which can be further integrated (13) . It can be used as such for the computation of the surface charge Q_S :

$$(43) \qquad Q_S = -\int_0^{\infty} \rho(x)\,dx = -\epsilon\left(\frac{dV}{dx}\right)_{x=0}$$

which is an expression of GAUSS law. Hence ,

$$(44) \qquad Q_S = \frac{\epsilon k T}{e}\,\frac{2}{L}\,\sinh\frac{u_S}{2}$$

b) <u>When the semiconductor has to be considered as extrinsic</u>, but only with donor impurities, supposed all singly ionized to a depth ℓ , n can be supposed negligible in comparison of N_d in the depletion layer created by the negative surface charge. Then (39) takes the simpler form :

$$(45) \qquad \rho(x) = c\,N_d$$

and the twofold integration of (38) gives :

(46) $$V = - \frac{e N_d}{2 \epsilon} (\ell - x)^2$$

and

(47) $$V_S = - \frac{Q_S^2}{2 \epsilon e N_d}$$

This type of potential barrier is called parabolic or SCHOTTKY type (9).

A more elaborate model takes into account the electronic concentration n in the depletion layer, and expresses $\rho(x)$ by :

(48) $$\rho(x) = e (N_d - n)$$

The ionization equilibrium :

$$D \rightleftarrows D^+ + e^-$$

allows to write :

$$N_d n = N_{db} n_b$$

and, if the intrinsic carrier concentration is ignored,

$$N_{db} = n_b$$

Hence,

$$\frac{d^2V}{dx^2} = \frac{e}{\epsilon} (n - \frac{n_b^2}{n})$$

Substituting $n_b \exp(u)$ for n and assuming that u is small enough, the preceding equation can be replaced by :

$$\frac{d^2V}{dx^2} = \frac{1}{L^2} V$$

L having the same expression (41) as above. Hence it can be derived :

(49) $$V_S = \frac{1}{e} \sqrt{\frac{k T}{2 \epsilon n_b}} Q_S$$

where it appears that V_S is now a linear function of the charge density Q_S . This type of barrier is referred to as a MOTT barrier .

2. Outline of typical experimental techniques

a) <u>Electrical conductivity measurements</u> . An idealized sample

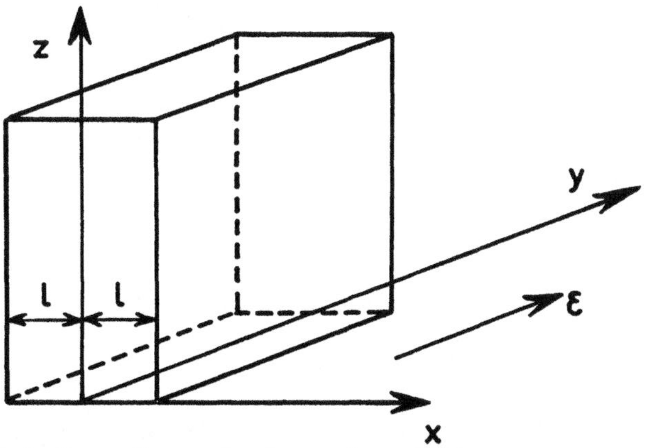

Fig. 7 : Model for electrical conductivity measurement of a single crystal

constituted by a single crystal is depicted by figure 7 . The parallelipipedic sample is located in an electrical field parallel to the Oy axis. The adsorption is supposed to take place only on the faces $x = \pm \ell$. The electrical conductivity is given by the expression :

(50) $\sigma = \mu_n e\, n$

if it can be assumed that the electrons are the majority carriers. Since n is a function of x , σ is also a function of x . If the mobility μ_n can be regarded as constant (which is a first approximation), the conductivity change between the case of no-adsorption and the case of adsorption is given by :

$$\sigma = \frac{\mu_n e}{\ell} \int_0^{\ell} n_b\,(\exp u - 1)\,dx = \mu_n e\, \Delta n_m$$

Δn_m designating the <u>average</u> variation of electron concentration. Hence ,

$$\Delta Q_S = - e\, \ell\, \Delta n_m$$

This formula, as simple as it may look, suffers from serous limitations : those coming from the above derivations (mobility invariance, adsorbate carrying one elementary charge per adsorbed particle), and those resulting from experimental deviations from the above model (when the investigated catalyst is a powder, the electrical resistance between grains may become predominant, there can be electrode polarization, etc ...) . thus, in the usual conditions of operation, this technique allows at the best an estimation of relative variations [14]. We shall see, however, in the next chapter, that, when properly used, it can yield quantitative correlations with catalytic activity.

b) Work function measurements (15). They consist in measuring the contact potential difference between the adsrobent and a reference electrode (usually made of a metal like gold, which is supposed to remain unaltered by chemisorption)

$$eV = \varphi_1 - \varphi_o$$

(φ_1, work function of the adsorbent, φ_o work function of the reference electrode).

Hence, $\Delta V = - \Delta V_S$

A widely used technique consists in building a vibrating capacitor, charge variations of which create a current, which is reduced to zero by insertion of an external bias equal and opposite to V.

c) HALL coefficient measurements (14). The HALL coefficient of semiconductors is given by :

$$(51) \qquad R = \frac{\mathcal{E}}{j\mathcal{H}} = \frac{3\pi}{8} \frac{1}{n e c}$$

where $\mathcal{E}$ and $\mathcal{H}$ are respectively the applied electrostatic and magnetic fields, j the electrical current density, c the velocity of light (fig. 8).

Polarity of the HALL effect allows the determination of the type of current carriers. If essentially simple for single crystal semiconductor samples, HALL coefficient determination for catalytic (i.e. mostly powdered) materials requires the greatest care. It has been claimed (16), (17), that it is relatively unaltered by contacts between grains and other hererogeneities in the solid. An example of its application is given by Ref. (17) : by comparing the number of adsorbed oxygen particles and the variations of the number of electrons in the solid, it has been possible to show that on Ga-doped ZnO, oxygen is chemisorbed as O_2^- between 100 and 180°C, and as O^- above 230°C.

d) Electron spin resonance spectroscopy (18). This technique is based on the fact that transitions can be made to occur by application of a magnetic field $\mathcal{H}$ between the electrons of which magnetic moment is parallel to this field and those of which magnetic moment is antiparallel to $\mathcal{H}$. If these electrons are free, their only magnetic moment comes from their spin, and the difference of energy between the two electron populations is

$$E = g\, b\, \mathcal{H}$$

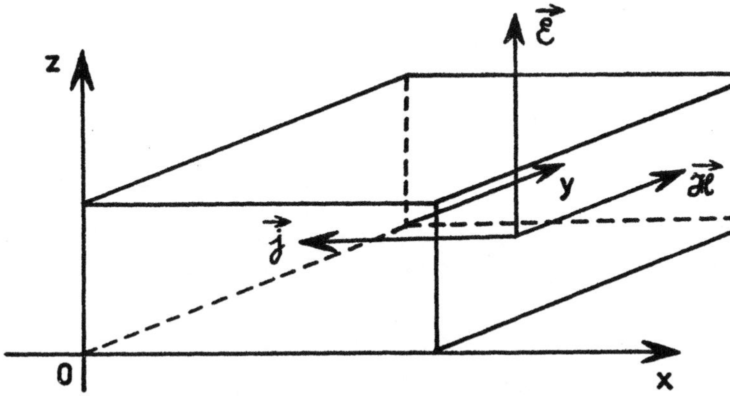

Fig. 8 : HALL coefficient measurement .

(b is here the BOHR magneton , 0.92732 x 10^{-20} erg/gauss). If the medium is exposed to electromagnetic radiation at various frequencies, absorption occurs for a frequency satisfying the equation :

$$h\nu = g\, b\, \mathcal{H}$$

with $g = 2.0023$ for free electrons. As a matter of fact, free electrons are rather exceptional . Unpaired electrons, belonging even to free atoms have orbital moments which must be coupled to their spins, which modifies the g factor. A further modification comes from the fact that in most cases atoms or molecules containing unpaired electrons belong to condensed phases, so that the spectra are markedly affected by the surroundings. Moreover, the nuclear spin can interact with the electronic one, thus making extra lines appear (hyperfine splitting) . The conduction electrons, which can be considered as "quasi-free" electrons, are expected to give a g value close to that of free electrons. For instance, the signal at $g = 1.958$ in ZnO has been attributed to conduction electrons (19) . As it is possible to determine the number of spins which give rise to the observed signal, it is possible to follow the concentration of conduction electrons in the sample as a function, for instance, of the amount of adsorbed species .

3. Application of boundary layer theories to adsorption and catalysis .

a) Energetic parameters of adsorption . In the crude model of B.L.T. , the activation energy is the energy required for an electron to be raised from the FERMI level to the bottom of conduction band at the surface , i.e. :

$$E_{ads} = E_{cS} - E_F$$

or, according to figure 6 ,

$$E_{ads} = - eV_S + E_{cb} - E_F$$

In course of adsorption , at a given temperature and for a given semiconductor $E_{cb} - E_F$ is a constant E_o . Hence ,

$$E_{ads} = E_o - eV_S$$

V_S is expressed as a function of N_S , by one of the preceding formulas. For instance, if the assumption of a MOTT barrier holds,

$$- e V_S = c N_S$$

and E_{ads} is a linear function of the adsorbed amount, as already expressed by formual (15) . This provides a physical basis for the "ELOVICH" equation.

Chemisorption energy ΔE is , according to B.L.T., the energy required for an electron to pass from the FERMI level in the solid to the surface level E_a,

$$\Delta E = E_a - E_F$$
$$= E_o - E_{des} - e V_S$$

Again, a linear variation of ΔE with N_S is found if a MOTT bar rier is admitted. This corresponds to the already mentioned formula (19) .

b) Example of a catalysis interpreted in the framework of B.L.T. . It is provided by the reaction :

$$H_{2\ g} + \frac{1}{2} O_{2\ g} \longrightarrow H_2O\ g$$

in the presence of ZnO (20) . The mechanism is supposed to be :

(I)	$O_{2\ g} + s + e^-$	$\rightleftarrows$	$O^-_{2\ ads}$
(II)	$O^-_{2\ ads} + e^-$	$\longrightarrow$	$2\ O^-_{ads}$
(III)	$O^-_{ads} + e^-$	$\longrightarrow$	$O^=_{ads}$
(IV)	$H_{2\ g} + s'$	$\rightleftarrows$	$H^+_{2ads} + e^-$
(V)	H^+_{2ads}	$\longrightarrow$	$2\ H^+_{ads} + e^-$
(VI)	$2\ H^+_{ads} + O^=_{ads}$	$\longrightarrow$	$H_2O\ _g$

Catalysis rate is the rate of H_2O production, i.e. :

(52) $$r = r_6 = k_6 \, (H^+_{ads})^2 \, (O^=_{ads})$$

As usual, it is attempted to get rid of the unknown concentrations in the rate expression by using the S.S.H. for the intermediate species. Following equations are readily obtained :

(53) $$r_5 = r_6$$

(54) $$r_3 = r_6$$

(55) $$r_4 = r_5 + r'_4$$

(56) $$2r_2 = r_3$$

(57) $$r_1 = r_2 + r'_1$$

This last equation can be further simplified if it is assumed that $r_2 \ll r'_1$. It comes to the quasi-equilibrium expression :

(58) $$(O_2^-) = K_1 \, P_{O_2} N_S \exp\left(\frac{E_F}{kT}\right)$$

by taking the energy origin so that : $E_{cS} = 0$. Here, N_S is superficial concentration of free sites s (assumed to be practically equal to the total concentration of these sites). Hence,

(59) $$r = 2\, r_2 = 2\, k_2 K_1 P_{O_2} \, N_S \exp\left(\frac{2\, E_F}{kT}\right)$$

Of course, E_F does vary as a function of the partial pressures of the reactants. This function can be made explicit by writing :

$$r = 2\, r_2 = r_5 = k_5 (H^+_{2\,ads})$$

with an expression of $(H^+_{2\,ads})$ which results from the quasi-equilibrium substituted for (55) when $r_5 \ll r'_4$:

(60) $$(H^+_{2\,ads}) = K_4 P_{H_2} N'_S \exp\left(-\frac{E_F}{kT}\right)$$

Hence,

(61) $$\exp \frac{3\, E_F}{kT} = \frac{k_5 K_4}{2\, k_2 K_1} \; \frac{N'_S}{N_S} \; \frac{P_{H_2}}{P_{O_2}}$$

Let us assume, for the sake of simplicity, that $N_S \approx N'_S$. The final rate expression is obtained as :

$$r = 2\, k_2 K_1 P_{O_2} N_S \left(\frac{k_5 K_4}{2 k_2 K_1} \; \frac{P_{H_2}}{P_{O_2}}\right)^{\frac{2}{3}} = k\,(P_{O_2})^{\frac{1}{3}} \, (P_{H_2})^{\frac{2}{3}}$$

which is in approximative agreement with experiment. What is remarkable in the preceding derivation is that the values of the partial orders do not owe anything to any dubious hypothesis on adsorption isotherms .

This mechanism can be supported by a direct determination of the free electron concentration in the catalyst by E.S.R. ([21]) .

It is observed that during catalysis $(n_o - n)^2$ is a linear function of $\log \frac{P_{O_2}}{P_{H_2}}$

Indeed, equation (61) can be written :

$$\exp-\left(\frac{E_F}{kT}\right) = A \left(\frac{P_{O_2}}{P_{H_2}}\right)^{\frac{1}{3}}$$

and can be transformed into :

$$(62) \qquad \exp\left(-\frac{e\,V_S}{k\,T}\right) = B\left(\frac{P_{O_2}}{P_{H_2}}\right)^{\frac{1}{3}}$$

If a SCHOTTKY barrier is to be dealt with, equation (47) holds, so that :

$$\frac{Q_S^2}{2\,\epsilon\,k T\,N_d} = \ln B + \frac{1}{3}\ln\left(\frac{P_{O_2}}{P_{H_2}}\right)$$

On the other hand, the electric neutrality requires that :

$$a\,Q_S + ne = N_d e$$

(a being the area per unit volume) . Finally :

$$(63) \qquad \frac{e^2(N_d - n)^2}{2\,a^2\varepsilon\,k T\;N_d} = \ln B + \frac{1}{3}\ln\left(\frac{P_{O_2}}{P_{H_2}}\right)$$

which coïncides with the experimental result if $N_d = n_o$, which means that at the origin of time, all the conduction electrons come from donor ionization .

II. WOLKENSTEIN's THEORY

As satisfying the primitive form of B.L.T. may appear in certain cases , it is obvious that it represents in general a gross oversimplification. WOLKENSTEIN ([22]) seems to have been the first to stretch out that, from the standpoint of quantum mechanics, in the complicated system formed by the adsorbate and the adsorbent, the purely ionic bond is at the best a limiting case, and that many other bonds may exist, including that formed between the solid and a neutral species. As a matter of fact,there is ample evidence for the formation of neutral oxygen species as well on ZnO (23) as on ThO_2 (6) . Adsorption of CO on the latter adsorbent has lately been followed by E.S.R. (24) . It has been shown that the paramagnetic CO species is not CO^+ , but $CO^{\delta+}$ ($\delta < 1$) . However fractional, the charge taken by CO makes the surface positive, and thus creates in the solid a negative space charge region,which is in agreement with electrical conductivity measurements. Moreover, the number of paramagnetic species created under definite adsorption conditions can be estimated by E.S.R., and shown to be about the hundredth of the overall number of adsorbed CO molecules . In other words, 99 per cent of the molecules, however chemisorbed, remain neutral. In view of these findings, WOLKENSTEIN's theory itself appears as a simplification, yet improved if compared to the preceding B.L.T. treatments .

If it assumed that a given adsorbate can give rise, per unit area, to N^o neutral species, N^- negatively charged species and N^+ positively charged species, the pertaining FERMI-DIRAC statistics can be expressed (if constant degeneracy factors are ignored), by the expressions :

$$\frac{N^-}{N^o + N^-} = \frac{1}{1 + \exp\left(\frac{E_a - E_F}{kT}\right)} \tag{64}$$

whence :

$$\frac{N^o}{N^-} = \exp\left(\frac{E_a - E_F}{kT}\right) \tag{65}$$

Similarly,

$$\frac{N^+}{N^o} = \exp\left(\frac{E_d - E_F}{kT}\right) \tag{66}$$

Many consequences may be derived from the above formulas :

1. Let us compare the FERMI level location, as derived for instance from formula (65) :

$$E_F = E_a - kT \ln \frac{N^o}{N^-} \tag{67}$$

to that given by formula (34), rewritten as :

$$E_F = E_c + kT \ln \frac{n}{N_c} \tag{68}$$

In some cases, $\frac{N^{o}}{N^{-}}$ can be held constant, so is E_a , and E_F is fixed even if n is allowed to vary. In other words, doping the bulk of the semiconductor has no influence on its adsorptive properties, although the latter do involve charge transfer. This case corresponds to what is called the "quasi-isolated"surface, and has been encountered in catalysis practice (25) . It is important to note that in such a case, finding no influence of the bulk doping on adsorption properties does not invalidate the charge transfer theory, but, on the contrary, derives from it .

2. It is possible to write otherwise the basic relationships (65) and(66) by making the relative proportions appear :

$$\eta^{o} = \frac{N^{o}}{N} \quad ; \quad \eta^{+} = \frac{N^{+}}{N} \quad ; \quad \eta^{-} = \frac{N^{-}}{N}$$

N being here the overall surface concentration of the adsorbed species. Analytical expressions of these quantities can be derived (22) from which the representation of figure 9 is obtained. In selectivity problems, it may be interesting to control the location of the FERMI level, and, to this purpose, to introduce small amounts of a non-reactive species, which is adsorbed with a constant ratio $\frac{N^{-}}{N^{o}}$ thus, according to equation (67) , anchoring the FERMI level. An excellent example has been found in the oxidation of propylene to acrolein on Cu_2O . It has been observed that the yield of acrolein is all the better as the FERMI level is higher (26) and it is possible to have it anchored at a high position by introducing traces of CH_3Br (27) .

3. A host of studies have been devoted to the influence of the FERMI level location on the catalytic properties. There is no possibility to discuss here, even briefly, their results, but it seems interesting to stress out a particular point, which concerns the possibility of making different charged species intervene in the same rate determining step. Let us take the example of CO oxidation on thoria (28) , for which it seems that the latter can be written :

$$CO_{ads} + O_{ads} \longrightarrow CO_{ads}$$

and let us take into account the most likely signs of the adsorbates. The same above step may be written under at least four different forms :

$$\text{(a)} \quad CO^{+}_{ads} + O_{ads} \longrightarrow CO^{+}_{2\ ads}$$

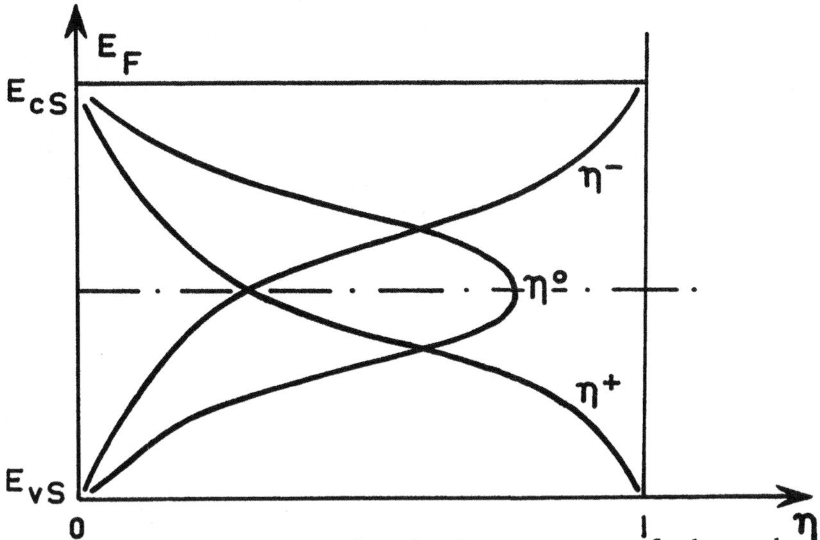

Figure 9 : Variations of relative amounts of charged and neutral adsorbed species .

(b) $CO^{+}_{ads} + O_{ads} + e^{-} \longrightarrow CO_{2\ ads}$

(c) $CO^{+}_{ads} + O^{-}_{ads} \longrightarrow CO_{2\ ads}$

(d) $CO^{+}_{ads} + O^{-}_{ads} + e^{-} \longrightarrow CO^{-}_{2\ ads}$

From the equations (65), (66) and (68), the following rate expressions can be derived :

$$r_a = k_a(CO_{ads})\,(O_{ads}) \quad \exp\left(\frac{E_d - E_F}{kT}\right)$$

$$r_b = k_b(CO_{ads})(O_{ads}) \quad \exp\left(\frac{E_d - E_{cS}}{kT}\right)$$

$$r_c = k_c(CO_{ads})(O_{ads}) \quad \exp\left(\frac{E_d - E_a}{kT}\right)$$

$$r_d = k_d(CO_{ads})(O_{ads}) \quad \exp\left(\frac{E_d - E_a + E_F - E_{cS}}{kT}\right)$$

Strikingly , FERMI level position does not influence unequivocally the reaction rate. r_a is diminished be a FERMI level rise, whereas r_d is enhanced, and the FERMI level has no influence on r_b and r_c. A recent example of such a situation has been given by Mac ARTHUR, BLISS and BUTT (29),who studied the ethanol dehydrogenation on ZnO. Such a reaction does imply charge transfer, since ethanol is chemisorbed with an increase of n-type conductivity, that is, as electron donor. This is confirmed by the influence of an electric field on that chemisorption. Catalytic rate , however, was unaffected by the application of that field. Here again, it must ne noted that the ambiguity of the role of FERMI level does not constitute an argument against the charge transfer theory, but, on the contrary, is a consequence of it.

C. COLLECTIVE PROPERTIES OF THE SOLID AND SURFACE ACTIVE CENTERS

Obviously, the usual recourse to particular models for describing catalytic reaction expresses our incapability of apprehending it in its whole. Some experimental facts are easier to be interpreted within the framework of a collective model, some others are more amenable to the consideration of localized active centers. But both standpoints look out on the same catalytic reality, and thus, must not be opposed, but reconciled, whenever it is possible. The two following examples, taken from catalytic studies on ionic oxides show it is possible .

I. ADSORPTIVE AND CATALYTIC PROPERTIES OF NON-STOICHIOMETRIC MIXED OXIDES OF URANIUM AND THORIUM

When U is used to dope thoria, the obtained solid solution is readily oxidized to give the non-stoichiometric oxide $U_xTh_{1-x}O_{2+y}$ which does not tolerate a too large excess of oxygen if it is to remain single-phased. In the following, it will be dealt only with single-phased non stoichiometric oxides $(x < 0.35)$ When their electrical conductivity σ is measured as a function of the pressure of the adsorbate CO , it is only for lower values of x $(x < 0.01)$ that a monotonous decrease of σ as a function of P_{CO} is observed, which is consistent with the sign of the charge carriers (holes), and the donor role of CO. For oxides which are more concentrated in uranium, the curve representing $\log \sigma$ as a function of $\log P_{CO}$ passes through a maximum, corresponding to a value P_m of P_{CO} which is a function of the initial composition of the mixed oxide (30) (Fig. 10) .
Following interpretation was put forward : there is first a reduction of the adsorbing surface, which occurs through two elementary steps :

$$(1) \qquad CO_g + s \rightleftarrows CO_{ads}$$

$$(2) \qquad CO_{ads} + O_\Delta \longrightarrow CO_{2\ ads}$$

Here, s designates a CO adsorption site, and O_Δ an interstitial oxygen atom at the surface. CO_2 thus formed is electron acceptor, and enhances the electrical conductivity. When surface reduction is completed, there can be an ordinary adsorption :

$$(3) \qquad CO_g + s \rightleftarrows CO_{ads}$$

This third step, which is no longer followed by a further reduction, provokes the observed decrease of conductivity .

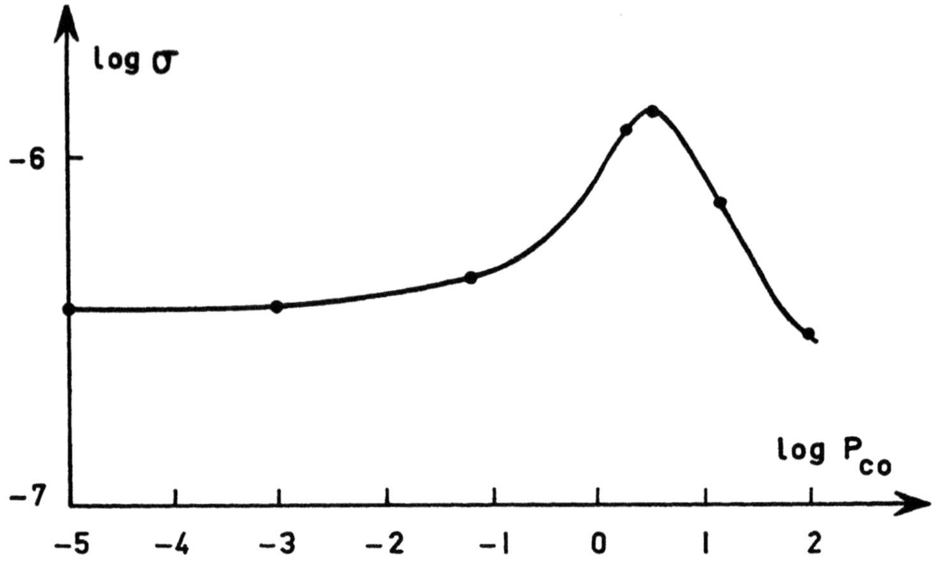

Fig. 10 : Variation of log σ as a function of log P_{CO} for a mixed oxide of initial composition $U_{0.192}Th_{0.808}O_{2.133}$. T = 300°C

Considerable evidence supports this interpretation. The most direct is provided by I.R. investigation of the adsorbent surface ([31]) (Fig. 11) .
At 1560 cm^{-1} and 1370 cm^{-1} are found the bands characterizing CO bound to normal oxygen atoms $O_{\square}$. Bands at 1520 and 1380 cm^{-1} correspond to a CO - O_{Δ} bond ; they completely disappear when the surface has been previously reduced by hydrogen. So do the shoulders at 1580 and 1310 cm^{-1} which correspond to a weak bond between CO_2 and surface .

Thus the infra-red study (local site model) wholly supports the interpretation (by a collective model) of electrical conductivity measurements. It turns out that step (2) is the rate determining step of CO oxidation on this mixed oxide. In other words, CO oxidation catalysis occurs through alternative reductions and oxidations of the surface ([32]). This can be directly checked by isotope labelling experiments. This method consists in partially substituting ^{18}O atoms for superficial ^{16}O atoms, carrying out catalysis on such a labelled oxide, and looking for $C^{16}O^{18}O$ as a product. This check requires some care, since possible exchange reactions between interstitial oxygen and gaseous O_2 or CO_2 are to be avoided. In order to minimize these exchange reactions, a temperature as low as possible (250°C) and a contact time between gas and solid not exceeding 5 min. have been chosen .
Under these conditions, exchange reactions did not take place,

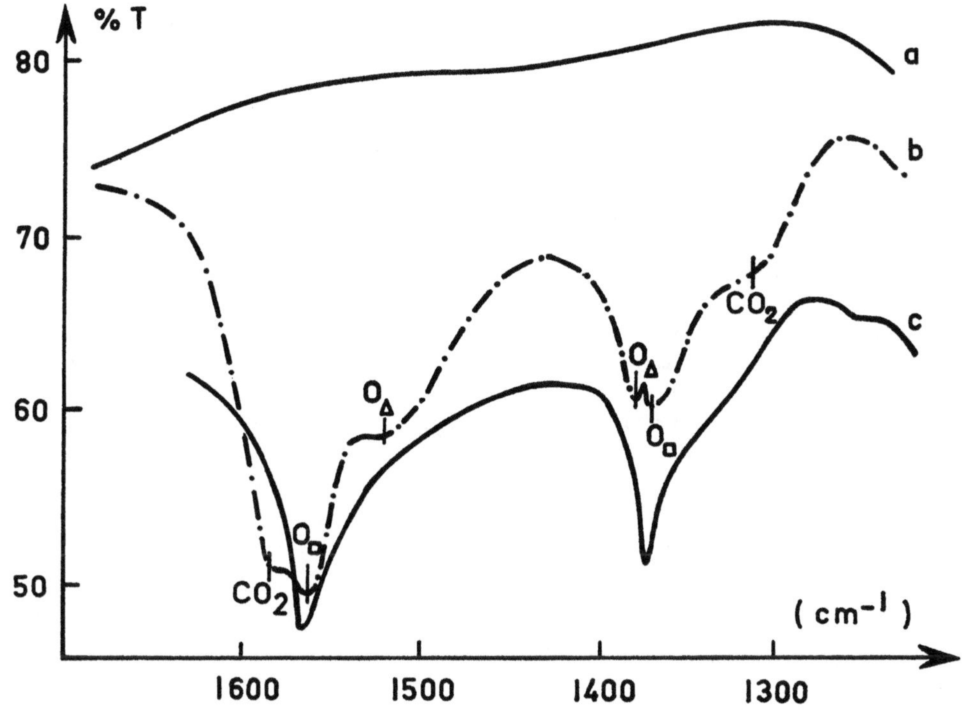

Figure 11 : Infra-red absorption spectrum of the mixed oxide: $U_{0.049}\ Th_{0.951}\ O_{2.035}$

a. Evacuated oxide

b. Oxide having adsorbed CO at a pressure greater than P_m

C. Same CO adsorption as before, but on a oxide previously treated by hydrogen .

and it has been found that practically all the ^{18}O that has left the solid was combined in CO_2 (gaseous or chemisorbed) .

II. CORRELATIONS BETWEEN CATALYTIC ACTIVITY AND ELECTRICAL CONDUCTIVITY OF NON-STOICHIOMETRIC CERIUM DIOXIDE (33)

A differential flow reactor has been used, which allowed simultaneous measurements of the electrical resistance of the catalytic powder and of its activity in CO oxidation. As previously mentioned, the former has no absolute meaning, but its relative variations, especially when directly compared to those of the latter, are valuable .

The experimental results can be summarized by the following expressions :

- for catalytic activity :

(69) $r = k\, P_{CO}^{0.84} \exp(-\frac{E}{\mathcal{R}T})$ $E = 20\ (\pm\ 1)$ kcal/mole

- for electrical conductivity :

(70) $\sigma = \sigma_o P_{CO}^{0.42}\, P_{O_2}^{-0.5} \exp(-\frac{E'}{\mathcal{R}T})$ $E'=25(\pm 1)$kcal/mole

in the presence of rectant mixtures. In the presence of an inert gas (N_2), the activation energy of conductivity is E_o=14 kcal/mole. The latter value is in close agreement with that found later for the low temperature electronic conductivity of ceria([34])

Expressions (69) and (70) can be found on the basis of following mechanism :

(1) $CO_g \rightleftarrows CO_{ads}$

(2) $CO_{ads} + O^{=}_{\square} \longrightarrow CO_{2\ ads} + \square^{=}$

(3) $CO_{2\ ads} \rightleftarrows CO_{2\ g}$

(4) $O_{2\ g} + 2\square^{=} \longrightarrow 2\ O^{=}_{\square}$

where $\square$ designates a vacancy in the anionic sublatice of CeO_2 .

If r_i designates the rate of the ith step, catalysis rate is :

(71) $r = r_2$

and S.S.H. for vacancies $\square^{=}$ enables to write :

(72) $r_2 = 2\ r_4$

or (73) $k_2(CO_{ads})(O^{=}_{\square}) = 2\ k_4 P_{O_2}\ (\square^{=})^2$

From the preceding equation (69) :

(74) $r = r_2 = k\ P_{CO}^{0.84}$

Therefore, it may be concluded that (CO_{ads}) varies as $P_{CO}^{0.84}$ and that $(O^{=}_{\square})$ does not appreciably change with the partial pressures of the reactants, and then, can be considered as constant during catalysis . Therefore ,

(75) $(\square^{=})^2 = \frac{k\ P_{CO}^{0.84}}{2\ k_4 P_{O_2}}$

Let us at present assume that the conduction electrons, which are the majority carriers, come from the ionization of anionic vacancies. The electrical conductivity can be written :

$$(76) \qquad \sigma = b\,(\square^{=})\,\exp\left(-\frac{E_o}{kT}\right)$$

E_o being the conduction activation energy at a constant concentration of $\square^{=}$. It must be pointed out, at this point, that the expression (76) does not assume any particular conduction mechanism nor any equilibrium with the gas phase (which cannot be realized, since reaction (4) is not supposed to be reversible). Thus, replacing $(\square^{=})$ by its expression (75) gives

$$(77) \qquad \sigma = c\,P_{CO}^{0.42}\,P_{O_2}^{-0.5}\,\exp\left(\frac{-\frac{E_2}{2} + \frac{E_4}{2} - E_o}{kT}\right)$$

This agrees with the experimental equation (70). Firstly, σ varies as $P_{CO}^{-0.42}$ and as $P_{O_2}^{-0.5}$. Secondly,

$$E_o = 14(\pm 1)\ \text{Kcal/mole}$$

$$E_2 = 20(\pm 1)\ \text{Kcal/mole}$$

and E_4, activation energy for the ceria reoxidation is probably very low. Hence,

$$\frac{E_2}{2} - \frac{E_4}{2} + E_o = 24\ (\pm\ 1.5)\ \text{kcal/mole;} \quad \text{whereas}$$

the value appearing in (70) is $25\ (\pm\ 1)$ kcal/mole.

This check makes at least likely the assumed mechanism involving oxygen atoms from the oxide (and not its electrons or holes). Such a mechanism is not uncommon in oxidation reactions.([35]).

There is no doubt that the detection and characterization of active centers (in the above examples superficial oxygen atoms of the solid) constitute the modern trends in catalytic research. This "localized" approach lends itself to a more "chemical" description of surface bonds. By no means it excludes the collective model of charge transfer theories. The latter suffered, till now, from an incomplete knowledge of the surface states. There is no doubt that develepments in the theory of surface physics, as well as the use of modern techniques (e.g. photoelectron spectroscopy) will contribute to fill the gap between chemical and physical standpoints, and thus, will lead to decisive progress in the field of adsorption and catalysis.

REFERENCES

([1]) M. PRETTRE and B.CLAUDEL, Elements of Chemical Kinetics, Gordon and Breach (1970)

([2]) M. BOUDART, Kinetics of Chemical Processes, Prentice Hall (1968)

([4]) J.E. GERMAIN, in " La Catalyse au Laboratoire et dans l'Industrie" , Masson (1967), p.45

([4]) M. BREYSSE , B.CLAUDEL, M.PRETTRE and J.VERON , J. Catal. 24 , 106 (1972)

([5]) D.H. BANGHAM and E.P. BURT , Proc. Roy. Soc. , A 105 , 481 (1924)
D.H. BANGHAM and W. SEVER , Phil. Mag. 49 , 935 (1925)

([6]) M. BREYSSE, Ann. Chim. 2 , 367 (1967)

([7]) A.MANY, Y.GOLDSTEIN , N.B. GROVER, Semiconductor Surfaces, North Holland (1971)

([8]) E. GORIN and H.L. RECHT, Chem. Engng Prog. 55 51 (1959)

([9]) P. AIGRAIN and C. DUGAS, Z. Elektroch. 56 , 363 (1952)

([10]) H.G. ENGELL and K. HAUFFE, Z. Elektroch. 56, 366 (1952)
57, 762 (1953)

([11]) P.B. WEISZ , J. Chem. Phys. 20 , 1483 (1952) , 21,1531 (1953)

([12]) J.E. GERMAIN, C.R. Acad. Sci. 238 , 345 (1954),J.Chim. Phys. 51 263 and 691 (1954)

([13]) G.C. DOUSMANIS and R.C. DUNCAN Jr., J. Appl. Phys. 29 , 1627 (1958)

([14]) T.J. GRAY, in "Experimental Methods in Catalytic Research" R.B. ANDERSON, Acad. Press (1968) , p. 286

([15]) P.M. GUNDRY and F.C. TOMPKINS, ibid. p. 101

([16]) H. CHON and C.D. PRATER , Disc. Faraday Soc. 41 , 380 (1966)

([17]) H. CHON and J. PAJARES, J. Catal. 14 , 257 (1969)

([18]) J.H. LUNSFORD, Adv. Catal. 22 , 265 (1972)

([19]) M. GUELTON , J.P. BONNELLE and J.P. BEAUFILS, J. Chim. Phys. 68 1122 (1971)

([20]) J.P. BEAUFILS , C.R. Acad. Sci. 263 , C , 7 (1966)

([21]) M. GUELTON, J.P. BONNELLE and J.P. BEAUFILS, C.R. Acad. Sci. 270 , C, 1766 (1970) , J. Chim. Phys. 68 , 1122 (1971)

(22) T.T. WOLKENSTEIN "The Electron Theory of Catalysis on Semiconductors" Pergamon Press (1963)

(23) P. AMIGUES and S.J. TEICHNER, Disc. Faraday Soc. 41, 362 (1962)

(24) P. MERIAUDEAU, M. BREYSSE and B. CLAUDEL, J. Catal. 35, 184 (1974)

(25) J. VERON, Ann. Chim. 4, 267 (1969)

(26) B.J. WOOD, H. WISE and R.S. YOLLES, J. Catal. 15, 355 (1969)

(27) H. WISE and L.L. HOLBROOK, J. Catal. 20, 367 (1971)

(28) M. BREYSSE, B. CLAUDEL and J. VERON, Kinetika i Kataliz, 14, 102 (1973)

(29) D.P. McARTHUR, H. BLISS and J.B. BUTT, J. Catal. 28, 183 (1973)

(30) G. BRAU, B. CLAUDEL, Y. TRAMBOUZE, H. URBAIN and J. VERON, C.R. Acad. Sci. 263, C, 844 (1966)

(31) P. PICHAT and G. BRAU, J. Chim. Phys. 66, 724 (1969)

(32) B. CLAUDEL and G. BRAU, J. Catal. 14, 322 (1969)

(33) M. BREYSSE, M. GUENIN, B. CLAUDEL, H. LATREILLE and J. VERON J. Catal. 27, 275 (1972)

(34) R.N. BLUMENTHAL and R.L. HOFMAIER, J. Electroch. Soc. 121, 126 (1974)

(35) M.J. FULLER and M.E. WARWICK, J. Catal. 29, 441 (1973)

NOTATIONS

A	constant acceptor atom
a	surfaces area proportionality constant
b	proportionality constant
c	proportionality constant velocity of light
C	concentration
D	donor atom
e	absolute value of the electron charge
$\mathcal{E}$	electrostatic field

Notations (ff.)

E	energy
f	function
g	spectral decomposition factor
$\mathcal{H}$	magnetic field
H	enthalpy
j	density of electrical current
k	rate constant
$\mathscr{k}$	Boltzmann constant
K	equilibrium constant
m	mass exponent
n	number of moles electronic density
N	particle density
p	hole density
P	pressure
q	reaction order
Q	electrical charge
r	reaction rate
$\mathcal{R}$	ideal gas constant
R	Hall coefficient
t	time
T	temperature
u	dimensionless energy ($u = \frac{eV}{\mathscr{k}T}$)
V	volume electrical potential

Greek letters

α, β, γ	proportionality constants
Δ	variation of a value
ϵ	product of the free space permittivity by the dielectric constant

(Greek letters - ff.)

η	proportion of adsorbed species
θ	surface coverage
μ	carrier mobility
ν	algebraic stoichiometric coefficient
ξ	ionisation energy
ρ	volume charge density
σ	electrical conductivity
φ	work function
χ	electronic affinity

Subscripts

a	relative to acceptor component
ads	relative to adsorption
b	bulk of the solid
c	conduction band
d	relative to donor component
des	relative to desorption
e	value at equilibrium
F	Fermi level
g	{ gas band gap
i	constituent
n	relative to electrons
o	at time $t = 0$
p	relative to holes
S	value at the surface, or per unit surface
v	valence band
V	value per unit volume

Superscripts

+	positively charged
-	negatively charged
o	non charged

DISCUSSION

DELL : In attemptions to distinguish between different adsorption models on the basis of kinetic measurements, one of the problems, surely, is the mathematical insensitivity of the various kinetic laws. A plot of log (adsorption) v. log (time) is particularly insensitive and usually leads to a linear plot.

CLAUDEL : This is quite obvious. As pointed out in the text, however, having recourse to computational techniques (e.g. least square analysis) may increase the reliability of a formal representation (within the accuracy of experimental data).

HEYNE : How was the energy of adsorption determined experimentally ? (last slide).

CLAUDEL : By means of an Eyraud-Richard microcalorimeter which allowed the simultaneous measurement of the heat evolved and of the volume adsorbed (i.e. the fraction of surface covered).
M. Breysse, B. Claudel, M. Prettre and J. Veron, J. Catal. 24, 106 (1972)

STEINBERG : I would like to ask you how you can differentiate between bulk and surface conductivity ?

CLAUDEL : Generally speaking, one has to do with both conductivities, the measured contribution of each one depending on the experimental conditions. Obviously, the most interesting one is the superficial conductivity, since it is related to the electronic concentration at the surface. According to S.R. Morrison (Adv. Catal. 7, 259 (1955) the usual experimental conditions allow us to reach mainly this superficial conductivity.

BLAKELY : In your analysis you assume the doping elements to be homogeneously distributed. Do you not think that there will be some segregation which will totally modify the analysis of the surface charge distribution and hence the excess surface conductance.

CLAUDEL : There have indeed been cases where such a segregation occurs thus making very unreliable the obtained results and their relations to any charge transfer theory. For the specific case of Ga doped ZnO, see L. Ponsolle and J.E. Germain, C.R. Acad. Sc. Paris, 263C 465 (1966).

GERISCHER : In so far as I know all the ESR measurements where ions have been detected on the surface have been performed with powders. Surface defects may there be very important for the charge transfer. Do you know analogous results, obtained with single crystals ?

CLAUDEL : Although basic solid state studies are currently performed with single crystals, I have no knowledge of any adsorption investigation carried out by ESR on single crystals.

STRUCTURE AND DEFECTS AT IONIC CRYSTAL SURFACES

J. M. Blakely

Cornell University, Ithaca, N. Y. 14853

This paper presents a discussion of our present understanding of the structure of ionic crystal surfaces. Both theoretical predictions and experimental observations will be reviewed. Emphasis will be placed on surface relaxation effects, point defects in the top monolayer and long range point defect distributions. The role of space charge regions in ionic transport and other processes will be reviewed. Attention will also be given to impurity distribution near interfaces in doped systems.

I INTRODUCTION

In this paper we will consider certain aspects of the structure of simple ionic solids which may be of importance in understanding electrical and other properties. We will primarily be concerned with deviations of the ionic structure from the ideal arrangement; the localized electronic states associated with ideal surfaces is considered in other papers. By an ideal surface with normal [hkl] we generally mean a surface formed by removing all ions/atoms whose centers lie on one side of a mathematical plane with indices [hkl] lying within the solid and keeping the relative positions of all other atoms/ions fixed. The free energy of the solid terminated in this way will have contributions proportional to the area of the surface formed. These arise primarily from changes in the potential energy due to mutual interactions of the ions and to changes in the vibrational spectrum which contribute both to the internal energy and entropy of the solid. Surfaces formed according to the above prescription may be atomically smooth (singular), have arrays of fairly widely spaced atomic steps (vicinal) or have a high density of steps and

M. Kleitz and J. Dupuy (eds.), Electrode Processes in Solid State Ionics, 83–118.

kinks (rough).

Deviations from structural ideality may be of several types. Relaxation of the atoms/ions in the topmost layers over distances which are small fractions of the interlayer spacing may produce substantial reductions in surface free energy. Calculations of such relaxations for ionic crystals has been the subject of numerous theoretical papers over several decades; we will discuss some of the results below. Surface relaxations are particularly interesting in ionic systems since the displacement of ions from their infinite crystal positions will in general produce surface dipoles in addition to those due to electronic polarization by the crystal field. Rearrangements of ions by displacements over distances of the order of interatomic spacings may also occur to produce new surface super-lattices. This is generally referred to as reconstruction; it has primarily been observed with metals (eg. Au, Pt) and semiconductors (Si) using electron diffraction techniques and does not seem to have been reported in any simple ionic system.

Real surfaces will generally also have defects analogous to those occurring in the bulk. As the temperature of an initially ideal surface is increased a certain amount of thermal disorder in the form of vacancies, interstitials and self adsorbed ions (and their complexes) will occur due to the balancing effect of the entropy contribution. The formation energies and entropies (vibrational) associated with such defects will be different from the corresponding bulk values and hence their densities will vary with depth into the crystal. Since ionic defects are of prime importance for conduction in many ionic crystals the properties of individual surface point defects are of fundamental interest in assessing surface conductance contributions.

Since the local configuration surrounding any particular surface ion is different from that deep in the crystal the binding energies associated with surface sites will in general be different from bulk sites. The ions of the crystal will therefore, in equilibrium, partition themselves between bulk and surface sites in such a way as to minimize the free energy. This distribution will usually be such as to leave the immediate surface region with a net electrostatic charge which, in an initially neutral crystal, will be balanced by a screening space charge. Because of the low density of mobile charge in most ionic crystals the screening (or Debye) lengths may be of the order of 10^3 Å or greater as compared to ~ 1 A in most metallic systems. Impurities in a different ionization state from those of the crystal will also contribute to the mobile charge and hence to the space charge region; they may also segregate to the immediate surface region where their binding energy may be different from that in the bulk crystal.

The qualitative picture which emerges from the above introductory remarks is that the surface of an ionic crystal at equilibrium may have a variety of defects associated with it and possess both localized and extended dipolar layers. We shall discuss each of the effects listed above in some detail, review some of the experimental observations on ionic surfaces and finally mention a few applications of ionic materials where the surface defects play an important role.

II Relaxation at Ionic Crystal Surfaces

The location of ion cores near singular surfaces of inert gas crystals and simple ionic crystals such as the alkali halides and simple oxides (eg. MgO) has been the subject of considerable theoretical work over a long period of time. The prime reason for the theoretical interest in these materials is that many of the bulk properties can reasonably well be described using pair potentials. For the alkali halides calculations have been made of the ion positions, electronic dipoles and surface energies for some of the most closely packed surfaces, (111) and (100). A comprehensive and critical review of earlier work in this area has been given by Benson and Yun(1).

For a defect free surface such as NaCl(100), if only normal relaxations are allowed, there are four unknown quantities for each layer of ions viz the positions and electronic dipole moments of both the cations and anions. These have been determined (at 0°K) by minimising the surface energy with respect to all of these parameters. To keep the problem tractable the number of layers allowed to relax and possess electronic dipoles must be limited; it has been found that about 5 displaced layers seem to be adequate to approximate the real situation.

The 0 K specific surface energy, e_s, may be expressed in the form

$$e_s = e_s^{\circ} + \Delta e_s + e_{vib}(0) \qquad (1)$$

where the first term is the potential energy contribution obtained by visualizing the crystal to be split into two parts with the ions in each part remaining unpolarised and with their relative spacings unchanged. The second term is the relaxation energy and is the reduction in energy due to polarization and displacement. $e_{vib}(0)$ is the change in zero point vibrational energy per unit area of surface created, and may be estimated from a modified Debye model(1). If z_λ^+ and z_λ^- are the normal displacements of cations and anions in the layer λ and μ_λ^+ and μ_λ^- the corresponding electronic dipole moments then the equilibrium positions and dipoles are obtained from the solution of the set of simultaneous equations

$$\frac{\partial(\Delta e_s)}{\partial \alpha_\lambda} = 0 \qquad \text{where} \quad \lambda = 0, 1, 2, \cdots (M-1) \tag{2}$$

and $\alpha_\lambda = z_\lambda^+, z_\lambda^-, \mu_\lambda^+$ or μ_λ^-; M is the total numbers of layers allowed to relax so that there are 4M simultaneous equations. The main difficulty in applying this scheme to real crystal surfaces arises in the choice of the pair potential required to evaluate e_s° and Δe_s; there has been much criticism of the fact that the parameters normally used in the potential are those obtained by fitting calculations to measured bulk properties. Some results of calculations for NaCl are shown in Figure (1). The type of pair potential used was of the form

$$\phi_{ij} = \left\{ \frac{e_i e_j}{r_{ij}} - \frac{c_{ij}}{r_{ij}^6} - \frac{c'_{ij}}{r_{ij}^8} + b_{ij} \exp(-r_{ij}/\rho) \right\}$$
$$+ \left\{ -e_i (r_{ij} \cdot \mu_j) \frac{1}{r_{ij}^3} + e_j (r_{ij} \cdot \mu_i) \frac{1}{r_{ij}^3} \right.$$
$$\left. - 3 (r_{ij} \cdot \mu_i)(r_{ij} \cdot \mu_j) \frac{1}{r_{ij}^3} + (\mu_i \cdot \mu_j) \frac{1}{r_{ij}^3} \right\}$$

where e_i and e_j are the charges on the ion cores, r_{ij} the distance apart of ions i and j and μ_i and μ_j their dipole moments. The terms in the first bracket represent the Coulomb, van der Waals, dipole-quadrupole and overlap repulsion contributions respectively. The second bracket consists of charge-dipole and dipole-dipole interactions. The second bracket would not generally be included in calculating bulk properties since the symmetry would disallow the occurrence of permanent electronic dipole moments. As shown in Figure (1) the maximum predicted displacements are $\sim$10% of the interlayer spacing and there should be a net dipole moment per unit area of surface even at 0°K. These calculations predict that this is of order 0.04 Debye per ion pair on the outermost plane and should produce a contribution to the electrostatic potential energy difference between the inside and outside of the crystal of $\sim$0.1 volt. The average normal field in the displaced region due to the displacement and electronic dipoles would be of order 10^6V/cm.

Diffraction methods offer the most promise for the detection of surface relaxations. Low energy electron diffraction (LEED) has been applied to a large number of metal surfaces and several semiconductors and the results of intensity versus energy measurements have been interpreted to indicate normal relaxation in several cases(2). Due to the insulating nature of most ionic crystals they have not been studied extensively by LEED although there are a number of studies now in progress(2) on MgO, Li F and

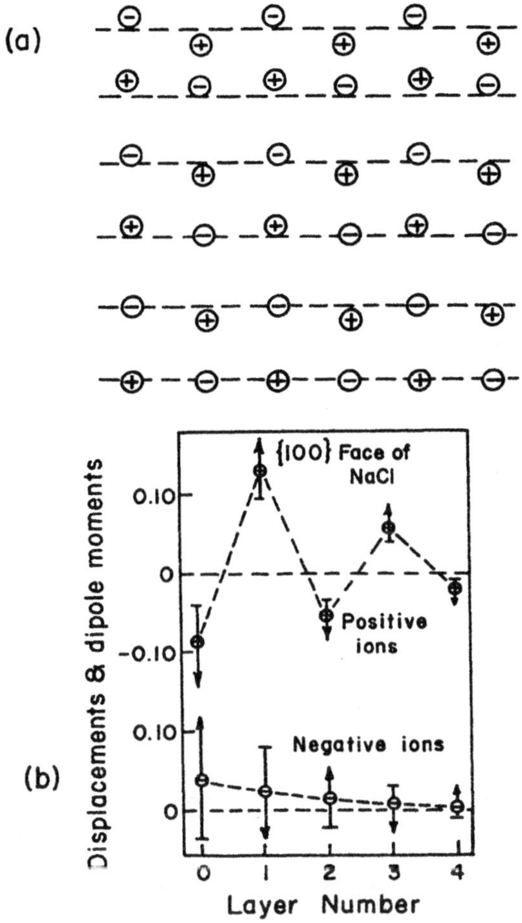

Figure (1)
(a) Schematic representation of the displacements of ions near the (100) surface of an alkali halide type crystal. The dotted lines represent the positions of successive planes in the unrelaxed crystal. (b) Calculated ionic displacements and electronic dipole moments for cations and anions in the outermost (100) planes of NaCl. The first five layers have been allowed to relax. The displacements are in units of the interplanar spacing with positive values indicating displacements in the direction of the outward normal. The direction and magnitude of the electronic dipole moments are indicated by the arrows, the unit of dipole moment being the Debye. Due to their greater polarizability the anions have much greater dipole moments than the cations; for the anions the lengths of the arrows correspond to one-tenth of the dipole moment. [From G. C. Benson and K. S. Yun, Advances in Chemistry, Series 33, 26 (1961).]

NiO. Some early LEED work on Li F[3], Figure (2), has been interpreted as being consistent with the type of normal displacement described above but the interpretation is complicated by the highly

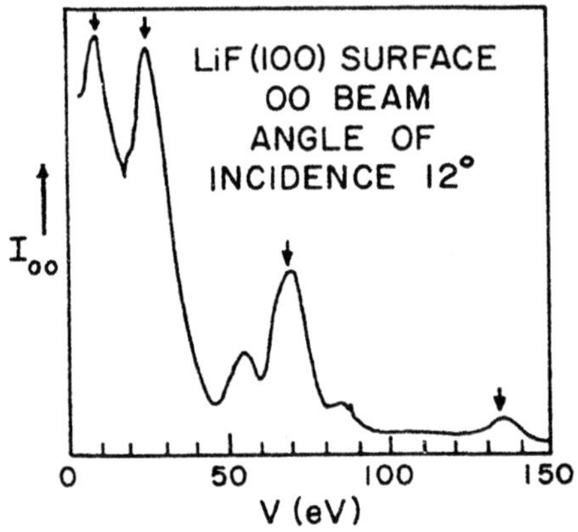

Figure (2)
Intensity versus energy plot for the (00) diffracted beam for an angle of incidence of 12° from the normal to a (100) Li F surface. In addition to the Bragg peaks, tentatively identified by arrows, there are subsidiary peaks due to multiple scattering.

dynamical nature of the scattering. Theoretical analysis including multiple scattering effects seem to indicate that the top fluorine layer is contracted about 0.1 Å toward the bulk while the top lithium layer is contracted about 0.35 Å toward the bulk so that the topmost layers of anions and cations are separated by about 0.25 Å. The agreement between theory and experiment is not however tremendously convincing in this case.

III Surface Defects

As mentioned in the introduction a variety of defects are to be expected on the outermost planes of ionic crystals and these should be of major importance in many kinetic processes occurring at these surfaces eg. surface diffusion, nucleation, evaporation, etc. Surface steps are probably the only type of surface defect which has been observed directly while the presence of others (vacancies, self-adsorbed ions, etc.) have been inferred from less direct experimental observations. Surface steps have been observed by LEED methods and by electron microscope techniques. Figure (4) shows the diffraction pattern from a surface of UO_2 containing an ordered array of steps figure (3) while figure (5) shows a replica of an NaCl surface with arrays of steps of monatomic height revealed by decoration techniques.

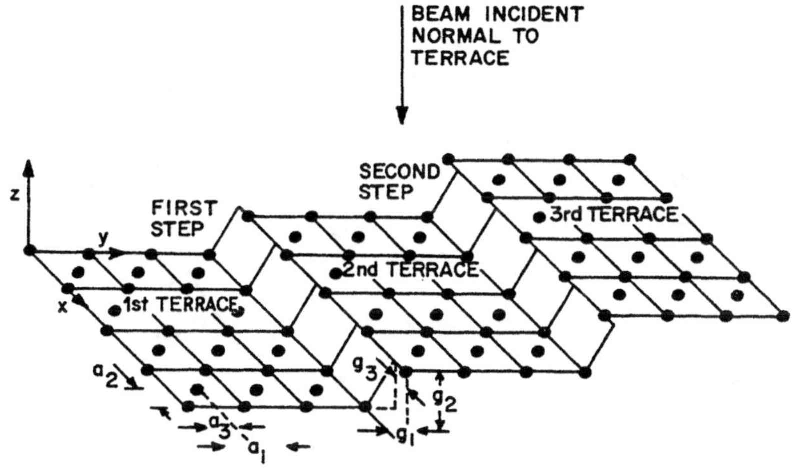

Figure (3)
Portion of a stepped surface of UO_2 showing the positions of the uranium ions.

There have been some attempts to calculate the energies associated with single ionic defects on simple alkali halide surfaces. These have been based on the use of pair potentials and (as mentioned in connection with the relaxation problem) a major difficulty arises in representing the interactions of surface ions by potentials with parameters determined by fitting bulk properties. The problem of evaluating the formation energy of a Schottky pair at a (100) surface of a NaCl type crystal may be illustrated with reference to figure (6). If we assume the initial surface to be ideal the formation energy of the Schottky pair will be equal to the work required to remove to infinity the pair of well separated ions minus the binding energy per ion pair of the crystal. The work of removal of an individual ion, W, is an extremely difficult quantity to calculate accurately even for the bulk case. This is due to the fact that it is the difference of several large energy contributions. Within the Mott-Littleton[7] scheme the work to remove a cation from the crystal, W_+, is given by

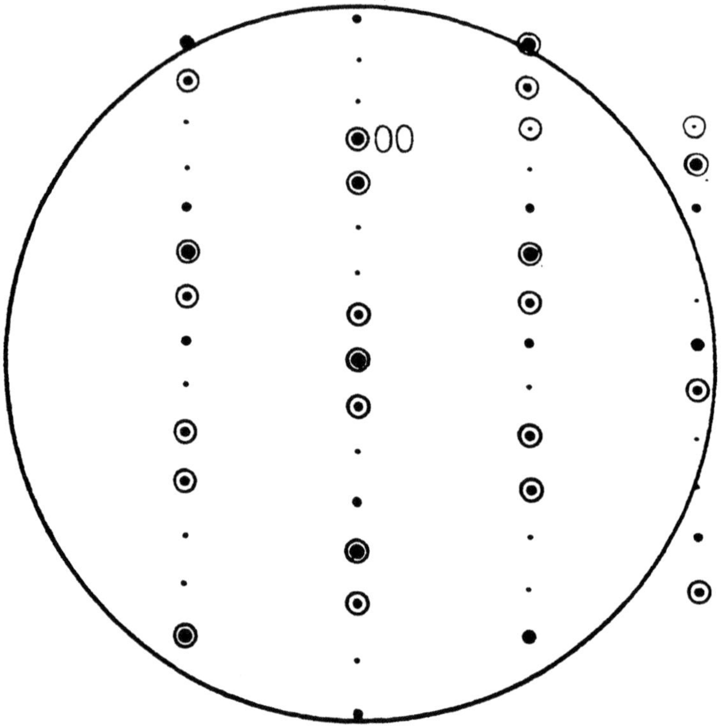

Figure (4)
Calculated electron diffraction pattern from the surface of figure (3) for normal incidence at 94 eV. The observed patterns are in good agreement with these calculations for an ordered step array. [After Ellis and Schwoebel(4)] The split diffraction spots are characteristic of the pattern from a stepped surface.

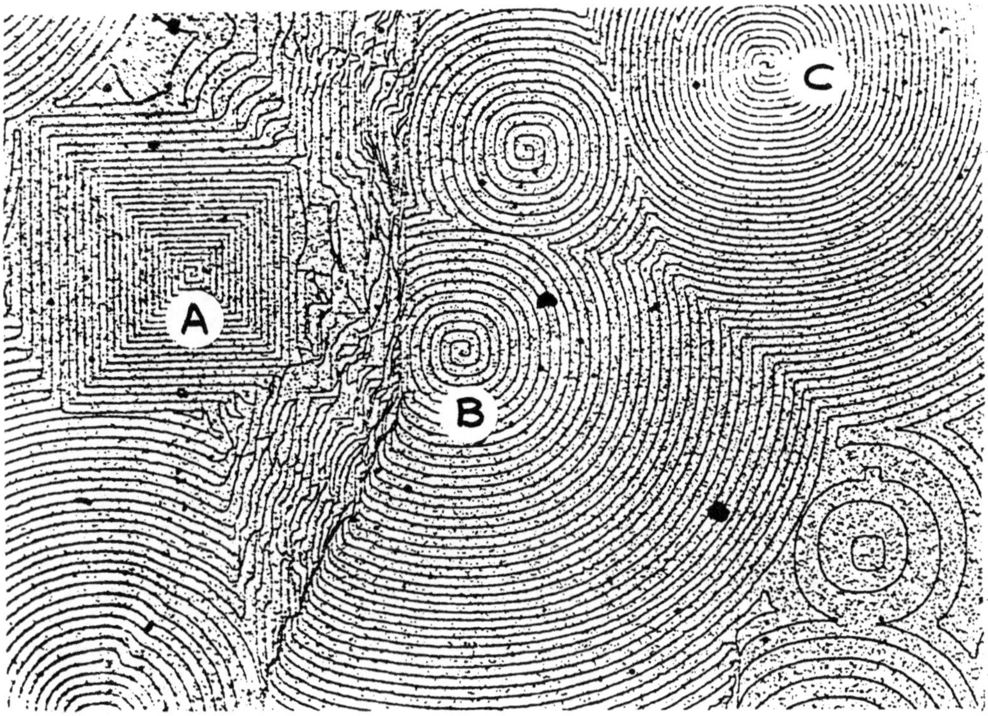

Figure (5)
Step pattern on a NaCl (100) surface. The photograph is of a replica of the surface on which the steps are decorated by gold nuclei. Some of the steps are of monatomic height. The step patterns show the presence of screw dislocations intersecting the surface at points such as A, B, C. [From H. Bethge(5)]

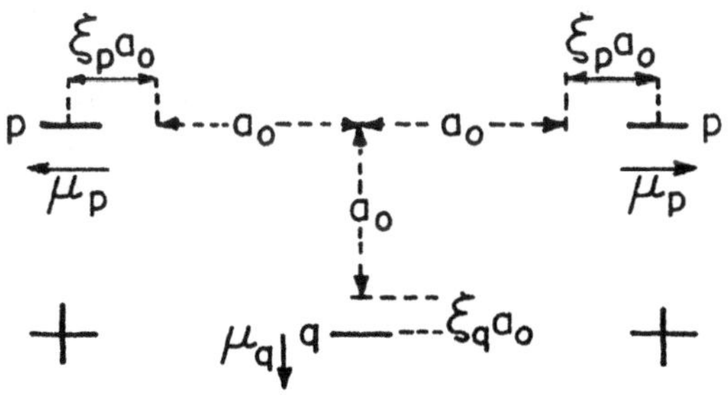

Figure (6)
Probable configuration surrounding a cation vacancy at the (100) surface of an NaCl type crystal. μ_p and μ_q are the electronic dipoles of the nearest neighbour anions and ξ_p and ξ_q their fractional radial displacements. For the calculations the polarization of the rest of the crystal was treated(6) by macroscopic electrostatic formulas.

$$W_+ = W_c + W_p + W_r \quad (3)$$

where W_c is the increase in Coulomb energy, W_p the polorization energy and W_r the increase in repulsion energy. W_c may be calculated in a straightforward way by computing the appropriate Madelung sum. The polarization and repulsion energies depend critically on the details of the relaxation around the vacant site. The relaxation depends in turn on the interionic potential parameters assumed. To give some idea of the accuracy required a simplified calculation[6] shows that for a cation vacancy at a (100)NaCl surface $W_c \sim 8.65$ eV, $W_p \sim -3.1$ eV, $W_r \sim -0.7$ eV so that according to equation (3) $W_+ \sim 4.9$ eV; similar calculations give $W_- \sim 5.2$ eV. Since the Schottky pair energy ΔE_s is given by

$$\Delta E_s = W_+ + W_- - W_L \quad (4)$$

where W_L is the binding energy per ion pair of the crystal (~ 8 eV for NaCl) we obtain $\Delta E_s \sim 2.1$ eV. The bulk value of ΔE_s computed via the same scheme gives $\Delta E_s \sim 1.9$ eV, so that to distinguish bulk and surface values would require a maximum uncertainty in the calculations of ~ 0.2 eV or about 3% in the estimate of polarization energy. It is extremely doubtful that this kind of accuracy could be acheived at present. Since experimental data on the surface value of ΔE_s is as yet unavailable we can only assume for the present that it is equal to the bulk value determined experimentally. The situation with respect to other point defects is essentially the same.

A phenomenon which has been studied from the statistical mechanical viewpoint is the spontaneous roughening[8,9,10] of singular surfaces (and steps) due to the cooperative creation of defect complexes as the temperature is increased. Quantitative predictions can however only be made on the degree of surface disorder by using simple nearest neighbor pairwise interactions figure (7); these may reasonably approximate the situation in covalent crystals and some metals but will be of little use for ionic crystals where interaction potentials are typically of very long range.

IV Surface Ionic Space Charge Regions

There is considerable evidence for the existence of space charge layers near lattice defects in ionic crystals. The experimental work has principally been concerned with the alkali and silver halides. With the alkali halides the experiments have mainly been aimed at determining the charge distribution near dislocations and grain boundaries, while the work on the silver halides arises from their use as photographic materials, although

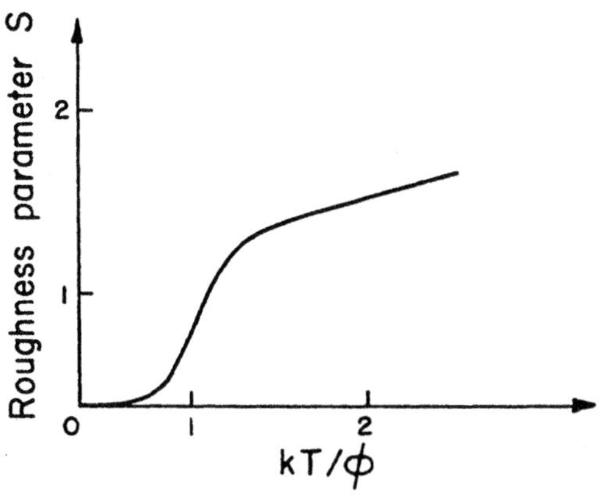

Figure (7)
Qualitative form of the surface roughness as a function of temperature for a crystal bonded by simple nearest neighbor pair interactions. The roughness is measured by the parameter S equal to the number of free bonds parallel to the surface per surface site. ϕ is the nearest neighbor bond energy. Note the precipitous increase in S as kT approaches ϕ due to cooperative effects. [After J. P. Hirth(9).]

conduction mechanisms in these crystals are also of considerable importance. A recent review of experimental data may be found in the article by Whitworth(11).

Much of the theoretical work on charge distributions in ionic crystals is based on the early ideas of Frenkel(12) who pointed out that due to differences in the work of formation of members of a point defect pair (Schottky or Frenkel), surfaces and other lattice discontinuities should have associated with them an extended dipolar configuration. A number of analyses for both pure and doped crystals have since appeared in the literature(13-18). The model which seems to contain all of the main physical ideas and lead to a tractable analysis is that in which the surface (or dislocation, etc.) is regarded as providing a limited number of sites at which the ions of the crystal may be bound with energies different from those normally associated with bulk sites. This picture is thus rather analogous to that of electronic space charge layers associated with surfaces with localized electronic states(19,20) and in fact with a change in terminology much of the analyses of the electron problem can be carried over to the ionic case.

To establish the main reason for charged configurations at ionic crystal surfaces it is perhaps worthwhile to consider first

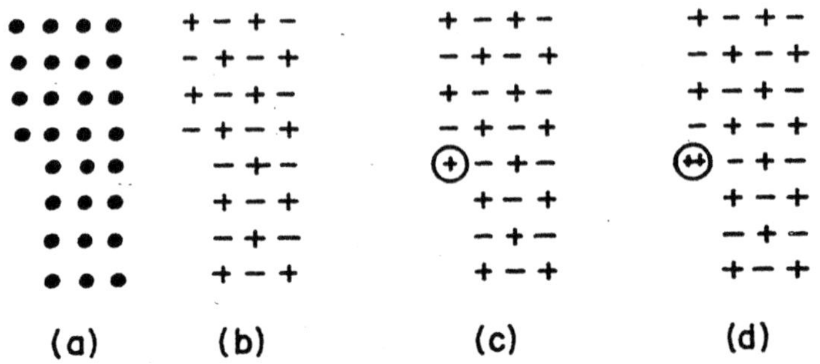

Figure (8)
Schematic representative of kink sites on a surface step. (a) depicts the one component case. Here addition or subtraction of an atom to the kink regenerates it. (b) shows a kink site on a step at a (100) surface of an NaCl or AgCl crystal. In (b) the kink site is unoccupied (by a cation), in (c) occupied and in (d) occupied by a doubly ionized impurity ion.

the creation of vacancies and interstitials at the surface of a one component crystal such as a metal or elemental semiconductor. For any general surface there will usually be, under equilibrium conditions, a certain density of kink sites associated with steps (figure 8). A vacancy or interstitial in the crystal can be created by transferring atoms either from the bulk to the kink sites or vice versa. If we imagine large numbers of both types

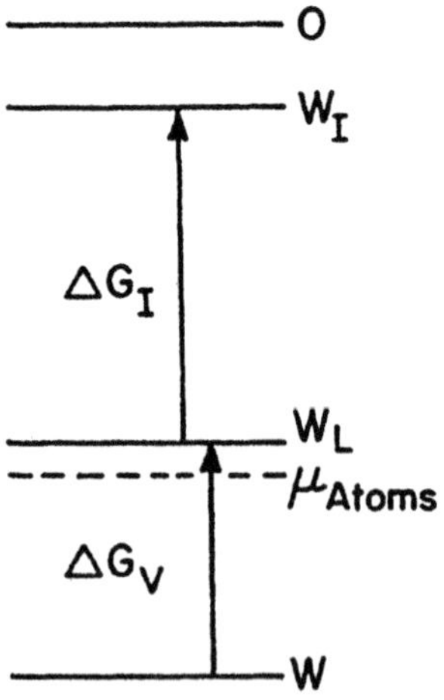

Figure (9)
Schematic energy level diagram for atoms in a one component crystal. -W is the work to remove an atom to infinity from a normal lattice site, $-W_I$ the corresponding quantity for an interstitial site and W_L the binding energy per ion of the crystal (= the work to remove an atom from a kink site).

of defects to be created in this way the surface will remain essentially unchanged whether or not equal numbers of the two types are created. Since each kink site is regenerated when each defect is created there is also no restriction on the total numbers of defects which may be formed. The densities of defects may be obtained by use of an energy diagram such as in figure (9). Here $-W$ is the work to remove an atom from the crystal to infinity leaving a vacancy, W_L the binding energy per atom of the defect free crystal, $-W_I$ the work to remove an ion from an interstitial site and μ the chemical potential of the atoms. For a one component crystal $\mu \approx W_L$[17] so that the fraction of bulk sites unoccupied is

$$\exp\left(-\frac{W-\mu}{kT}\right) \approx \exp\left(-\frac{W-W_L}{kT}\right) = \exp\left(-\frac{\Delta G_V}{kT}\right) \tag{5}$$

and the fraction of occupied interstitial sites is

$$\exp\left(-\frac{W_I-W_L}{kT}\right) = \exp\left(-\frac{\Delta G_I}{kT}\right) \tag{6}$$

With ionic crystals as may be seen by reference to figure (8) individual defects may not be created independently at crystal surfaces without producing a net change in the charged state of the surface. For this reason the details of the surface sites become important i.e., what are the densities and the binding energies associated with them. We will consider two special cases which are of interest viz. (i) a crystal with Frenkel disorder on only one sublattice and (ii) a crystal with Schottky disorder. The silver halides AgCl and AgBr closely approximate the first situation while the alkali halides are examples of the second type.

(i) Silver Halide Type Crystals

In AgCl and AgBr there is some experimental evidence that the anion sublattice remains essentially defect free over the accessible temperature range[21] and that ionic disorder is confined to vacancies and interstitials on the cation sublattice. In the bulk of the crystal charge neutrality will exist so that the numbers of silver vacancies and silver interstitials will be equal. If we represent the work to remove an ion from the two types of site by an effective "one ion" energy level diagram such as in figure (10) then the bulk interstitial and vacancy concentrations n_I and n_V are given by

$$n_I = 2N \frac{1}{\exp\left(\frac{W_I-\mu_{Ag}}{kT}\right)+1} \quad \text{and} \quad n_V = N\left(1 - \frac{1}{\exp\left(\frac{W_+-\mu_{Ag}}{kT}\right)+1}\right) \tag{7}$$

where N is the number of cation lattice sites per unit volume and μ_{Ag} the silver ion electrochemical potential. Setting $n_I = n_V$ gives after some simplification

$$\mu_{Ag} = W_+ + \frac{\Delta G_F}{2} - \frac{1}{2} kT \ln 2 \tag{8}$$

where $\Delta G_F (= (W_I - W_+))$ is the free energy of creation of a Frenkel pair. Also

$$n_I = n_V \approx \sqrt{2} N \exp\left(-\frac{\Delta G_F}{2kT}\right) \tag{9}$$

Unlike the one component crystal case the chemical potential does not correspond to the binding energy at any particular site of the crystal. The statistical problem is in fact similar to that for an intrinsic semiconductor.

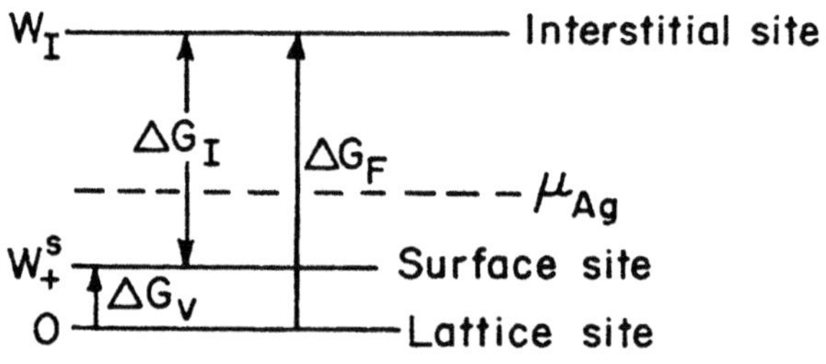

Figure (10)
Effective "one ion" energy level diagram for silver ions in silver halide crystal. For convenience we have chosen the reference zero of energy to correspond to the binding of an ion at a normal lattice site, i.e., we have set $W_+ = 0$. ΔG_F represents the Frenkel pair creation energy. Also shown schematically on this diagram is the binding energy $W_+{}^s$, of an ion at a surface site. For this particular surface site the Frenkel pair energy may be partitioned into 2 parts, ΔG_V and ΔG_I the energies to create a vacancy and interstitial respectively.

It is important to note that there is no question of uniquely partitioning the Frenkel pair creation energy between the vacancy and the interstitial; in the bulk they can only be created simultaneously. It is only when a surface is present that they may be created in unequal numbers. We now consider the necessary modifications to our scheme when the crystal is semi-infinite and bounded by a particular crystallographic plane. We suppose for the sake of simplicity that only one particular type of surface site is important in point defect creation and tentatively focus on the kink sites, (figure (8)). Let there be N_s such surface sites per unit area with energy $W_+{}^s$ relative to the normal bulk lattice site, figure (10). We may then define for this particular

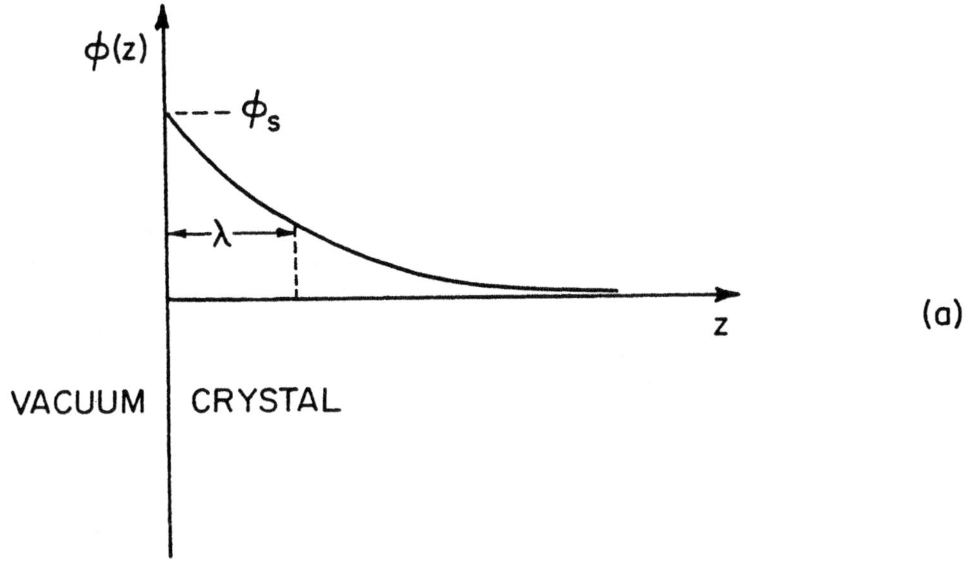

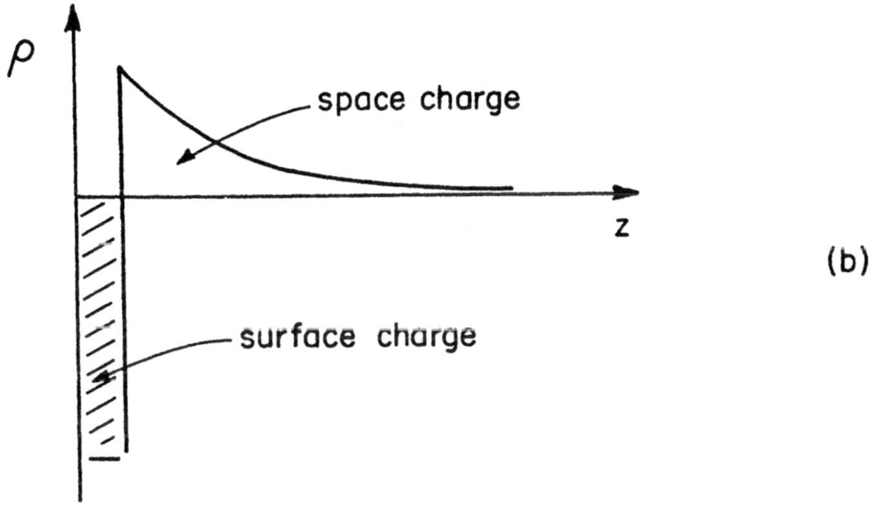

Figure (11)
(a) Schematic representation of the potential variation, $\phi(z)$, near the surface of an ionic crystal. ϕ_S is the surface potential relative to the bulk and λ the Debye length. (b) Possible charge distribution near the surface. The net charge on the surface layer compensates the space charge region.

surface the free energies of vacancy and interstitial creation ΔG_V and ΔG_I respectively. For an uncharged surface $\frac{1}{2}N_S$ of the kink sites will be occupied by Ag ions. At equilibrium the silver ions will partition themselves between bulk and surface sites in such a way that there will usually be a net charge on the surface and a compensating space charge layer, figure (11). The surface charge and potential distribution with depth, Z, may be determined in the following way. Let the zero of potential energy be taken in the bulk crystal, where the space charge density, $\rho = 0$, and let ϕ_S be the surface potential. The number of silver ions in kink sites per unit area, n_S, will be given by

$$n_s = \frac{N_s}{\exp\left(\frac{W_+^s + |e|\phi_s - \mu_{Ag}}{kT}\right) + 1} \tag{10}$$

with corresponding surface charge density $\sigma = (n_S - \frac{1}{2}N_S)|e|$. At any point in the crystal

$$n_I = 2N \frac{1}{\exp\left(\frac{\Delta G_F + |e|\phi(z) - \mu_{Ag}}{kT}\right) + 1} \tag{11}$$

and

$$n_v = N\left[1 - \frac{1}{\exp\left(\frac{|e|\phi(z) - \mu_{Ag}}{kT}\right) + 1}\right] \tag{12}$$

where the potential $\phi(Z)$ satisfies the Poisson equation

$$\nabla^2\phi(z) = -\rho/(\varepsilon\varepsilon_0) \tag{13}$$

with $\rho = (n_I - n_v)|e|$, and where ε is the dielectric constant and ε_0 the free space permittivity. The method of numerical solution is essentially as follows. We choose a value of surface potential ϕ_S. This in turn gives σ and by Gauss Law the value of the surface field

$$\left.\frac{d\phi}{dz}\right|_{z=0} = -\sigma/(\varepsilon\varepsilon_0) \tag{14}$$

These starting values of ϕ_S and $\left.\frac{d\phi}{dZ}\right|_{Z=0}$ are used to find ϕ at some incremental distance from the surface and all other values of $\phi(Z)$ are obtained by numerically solving equation (13). The correct value of ϕ_S is that which leads to satisfaction of the boundary conditions

$$\phi = 0 \quad \text{and} \quad \frac{d\phi}{dz} = 0 \quad \text{as } z \to \infty \tag{15}$$

The value of μ_{Ag} required in equations (10), (11) and (12) is given for the intrinsic case by equation (8). The parameters required for evaluating ϕ_s as a function of temperature are W_+^s, N_s and ΔG_F. ΔG_F may be determined in independent bulk conductivity measurements; for AgCl, $\Delta G_F \approx (1.44-9.4\ kT)$ eV.[22] The form of the surface potential variation with temperature is shown in figure (12) for particular values of ΔG_F and the surface site energy W_+^s and various values of N_s the density of surface sites. In the low temperature region the surface potential is independent of surface site density and corresponds to the situation depicted in Figure (13a) i.e., the chemical potential coincides with the energy of the surface site. As the temperature is increased the surface kinks become saturated and the space charge potential diminishes since insufficient charge can be accommodated to maintain the surface field. It should be emphasised that figures (12 a,b) apply to somewhat arbitrairly chosen values of W_+^s or ΔG_V; surfaces of different crystallographic orientation will in general have different values of W_+^s. ϕ_s will then vary from one plane to another and in any finite macroscopic crystal electrostatic

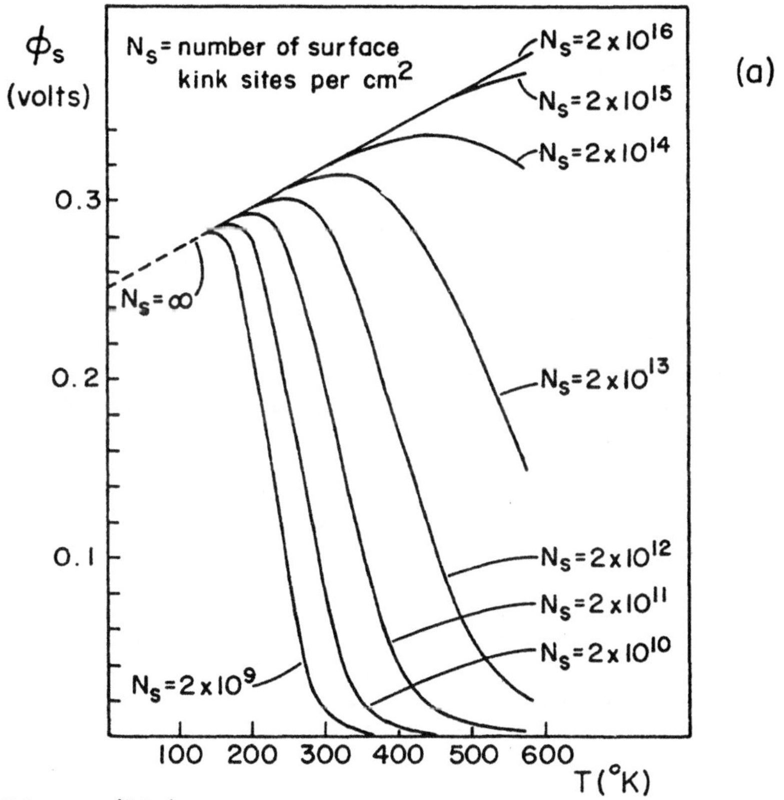

Figure (12a)

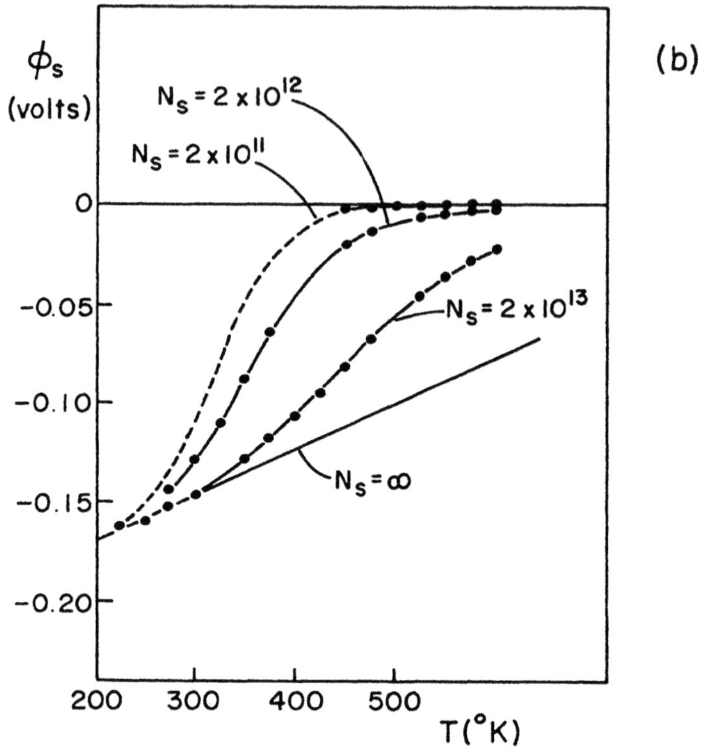

Figure (12b)

Figure (12 a,b)

Calculated potential difference across the space charge region of a silver halide crystal as a function of temperature. In (a) ΔG_F has been taken equal to (1.44 - 9.4 kT), the value appropriate to AgCl(22) and ΔG_V = 0.8 - 7.7 kT. N_s is the density of kink sites for silver ions. The curve of $N_s = \infty$ corresponds to previous calculations(15,16) neglecting surface effects. In (b) ΔG_F has been taken to be (1.16 - 9.4 kT) and ΔG_V equal to 0.8 - 7.7 kT, the values probably appropriate to (111) AgBr surfaces.(25)

fields will exist outside the crystal in much the same way as occurs with a metallic crystal with an orientation dependent work function.

We might also note that for different choices of the parameter W_+^s it is possible that the potential of the surface goes through zero at some finite temperature. This would be an intrinsic isoelectric point i.e., a temperature at which no space charge region exists. We shall discuss isoelectric points more fully in connection with doped crystals.

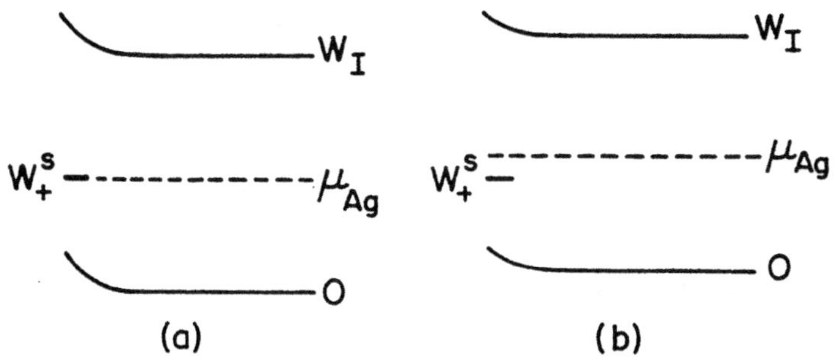

Figure (13)
Schematic potential distribution at (a) low temperatures and (b) high temperatures for a silver halide surface.

(ii) NaCl Type Crystals

This case has certain features which are similar to the one component crystal case and also to the silver halide Frenkel disorder case. If only Schottky disorder exists the creation of vacancies of either type necessarily involves a free surface (or other source and sink). In the electrically neutral bulk of the crystal anion and cation vacancies must be present in equal numbers and hence the properties of the source will be unimportant so long as the addition/subtraction of anion-cation pairs at the source does not change its overall average configuration. An energy level diagram analogous to that in figures (9) and (10) is shown for the NaCl case in figure (14). Here $-W_+^s$ and $-W_-^s$ are the energies to remove a cation and an anion respectively from the surface site. [We again assume for simplicity that only one type of surface site is important.] In this case the surface sites will be regenerated on removal or addition of a pair of surface ions. In general $W_+^s \neq W_-^s$ and $W_- \neq W_+$ but these quantities are related through the equations

$$W_+^s + W_-^s = W_L \qquad (16)$$

and

$$-(W_+ + W_-) + W_L = \Delta G_s \qquad (17)$$

where W_L is the binding energy per ion pair of the crystal and ΔG_s is the change in free energy of the crystal on formation of a Schottky pair. If we consider only bulk properties for the moment and neglect surface effects the free energy per unit volume of the crystal may be expressed as

$$G = G^\circ + n\Delta G_s + 2kT \ln\left[\frac{(N'+n)!}{N'!\,n!}\right] \qquad (18)$$

where G° is the value for the defect free crystal, n is number of cation or anion vacancies per unit volume and N' the number of cation or anions per unit volume. Setting $(\partial G/\partial n)_{P,T,N'} = 0$ for equilibrium gives

$$\frac{n}{N'} = \exp\left(-\frac{\Delta G_s}{2kT}\right) \qquad (19)$$

and from the definitions of the chemical potentials μ_{Na^+} and μ_{Cl^-} (i.e., for example $\mu_{Na^+} = (\frac{\partial G}{\partial N'_{Na^+}})_{T,P,N'_{Cl^-},n_-}$) we obtain

$$\mu_{Na^+} = -W_+ + \frac{1}{2}\Delta G_s \qquad (20)$$

$$\mu_{Cl^-} = -W_- + \frac{1}{2}\Delta G_s \qquad (21)$$

as indicated in figure (14). From equations (16), (17), (20) and (21) it is readily shown that the differences in energy between the chemical potentials of the two ions and the surface site energies are equal in magnitude and opposite in sign; it is therefore clear that the same potential (relative to a bulk zero) with align the surface site energies with their respective chemical potentials. The method of obtaining the equilibrium surface potential and space charge distribution is essentially the same as in the silver halide, or Frenkel disorder, case. The surface charge density is again given by $(n_s - \frac{1}{2}N_s)|e|$ where n_s,

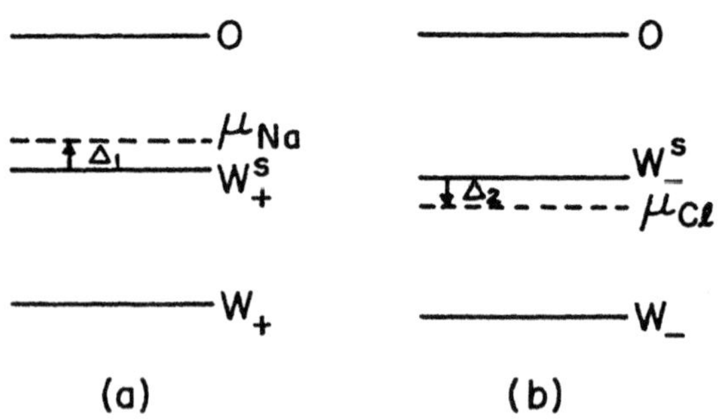

Figure (14)
Binding energy diagram for (a) cations and (b) anions in an NaCl crystal. $-W_+$, $-W_-$ are the energies to remove a cation and anion respectively from normal lattice sites and $-W_+{}^s$ and $-W_-{}^s$ the corresponding values for surface sites. μ_{Na} and μ_{Cl} are the ion chemical potentials. Note that $\Delta_1 = -\Delta_2$.

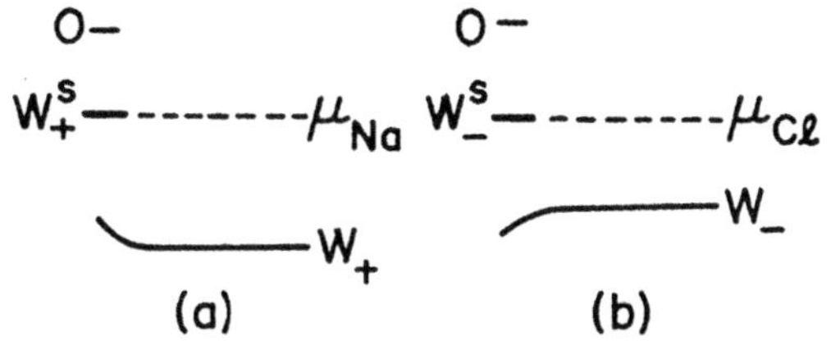

Figure (15)
Schematic potential distribution at low temperatures at a NaCl type surface. Note that at the equilibrium potential distribution both the cation and anion chemical potentials coincide with the energies of the surface sites.

is in turn, the number of kinks occupied by Na ions and is given by an equation similar to (10). The density of charge in the space region ρ is given by

$$\rho = -|e|(n_+ - n_-) \tag{22}$$

where n_+ and n_- are the densities of cation and anion vacancies, given by equations similar to (12). The method of obtaining ϕ_s and the variation of ϕ with depth is essentially the same as in the Frenkel disorder case with similar conclusions. At low temperatures the surface sites will be unsaturated and their energies will coincide with their respective chemical potentials, figure (15).

(iii) Crystals Containing Divalent Impurities

Impurities ions with valence different from those of the host lattice ion play an extremely important role in determining the conduction[23] and surface properties of ionic crystals. From our present point of view the main effect of divalent impurities is to change the density of mobile ionic charge in the bulk and the surface space charge configuration. We shall consider explicitly the silver halide case. One way of viewing the effect of the divalent impurities is to consider how the chemical potential of the silver ions is affected by the impurity additions. In the low temperature region the surface potential is simply determined by the difference between this chemical potential and the energy of the surface sites. When the impurity ions have very low mobility in the crystal it may be a good approximation to regard them as uniformly distributed in calculating the space charge distribution. This considerably simplifies the analysis.[18] However the more general case of equilibrium impurity distribution can also be treated; we shall outline the analysis of this more general case. As before we shall assume that the binding energies of ions differ only in the topmost layer from those characteristic of the bulk crystal and shall again consider the kink sites as the important surface sites. A kink site may now be unoccupied (i.e., a Cl^- kink) occupied by a silver ion or by a divalent impurity, (figure (8)).

The effect of the divalent impurity ions (eg. Cd^{++} in AgCl) may be represented in a formal way within our energy level scheme by the introduction of a new level at energy B above that associated with a normal lattice site as shown in figure (16). B is then the energy required for dissociation of a divalent ion-vacancy complex; it may alternatively be viewed as the energy of a silver ion next to the impurity. In the bulk of the crystal where charge neutrality exists we have

$$\rho = 0 = (n_I - n_v + n_i^+)|e| \tag{23}$$

where n_i is the number of impurity ions per unit volume and n_i^+ the number dissociated from vacancies (we refer to these as ionized impurities). n_I and n_V are given by equations similar to (7), while for n_i^+ we have

$$n_i^+ = \frac{n_i}{\exp\left(\frac{B-\mu_{Ag}}{kT}\right) + 1} \tag{24}$$

The silver ion chemical potential is then obtained by numerical solution of equation (23). The result for different choices of the parameters B, n_i are shown in figure (17). At low temperatures the silver ion chemical potential has the limiting value ½B (figure (18)) as expected from the analogous electron problem and approaches the intrinsic value, equation (8), at sufficiently high temperatures. It is possible using only this set of curves

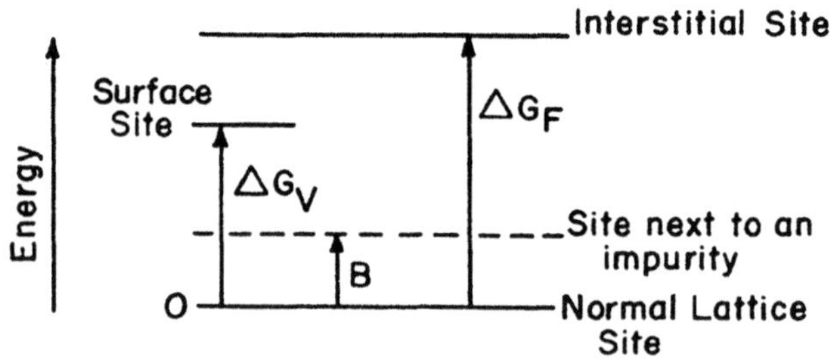

Figure (16)
Energy level diagram for a silver halide type of crystal containing divalent cation impurities. -B is the binding energy of a vacancy-divalent cation complex. The separation of a divalent cation from its associated vacancy is referred to as ionization in the text.

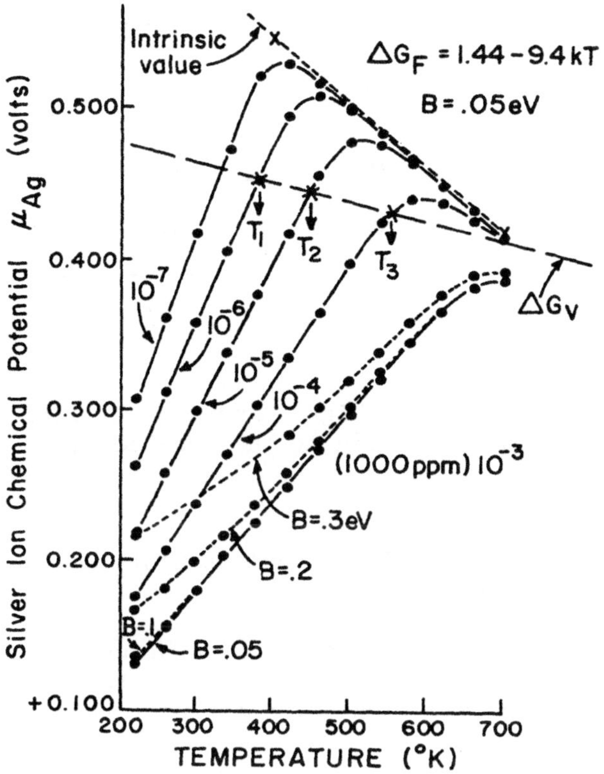

Figure (17)
Chemical potential μ_{Ag} of Ag ions in a AgCl crystal doped with divalent impurities to the level indicated. For this calculation we have taken B = 0.05 eV [the effect of varying B is shown by the dotted curves for a doping level of 10^3 ppm]. A method of interpreting isoelectric points is indicated. If T_1, T_2, T_3 are the isoelectric points for impurity levels of 1, 10 and 10^2 ppm the corresponding ΔG_V is represented by the line through the intercepts with the μ_{Ag} curves.

to obtain an idea of the magnitude and sign of the surface potential particularly in the low temperature region. The value of ϕ_s is simply the difference between μ_{Ag} and the surface site, silver ion binding energy (ΔG_V in Figure (16)). In addition the coincidence of μ_{Ag} with the surface site energy will correspond to the extrinsic isoelectric point. Alternatively the experimental determination of isoelectric temperatures for various doping levels may be used to determine the binding at the surface site. Since $\mu_{Ag} = \Delta G_V$, when $\phi_s = 0$ the experimental isoelectric temperatures T_1, T_2, T_3.... are then used to obtain the straight line representing ΔG_V as a function of T. Such data is not yet available for free surfaces; however Kim[24] has determined isoelectric

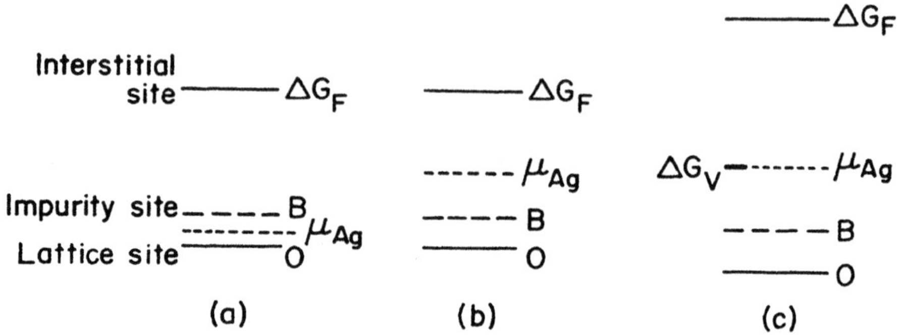

Figure (18)
(a,b) Diagrams illustrating the low and high temperature limits of figure (17). (c) depicts the situation at the isoelectric temperature.

points for dislocations in AgCl crystals at various $CdCl_2$ doping levels and his data has been interpreted in this way.

The formal solution of the space charge problem proceeds in a manner similar to previous cases. At any point where the potential is ϕ the number of ionized impurities per unit volume is

$$n_i^+ = \frac{n_i(\phi)}{\exp\left(\frac{B + |e|\phi - \mu_{Ag}}{kT}\right) + 1} \tag{25}$$

where $n_i(\phi)$ is the total impurity concentration at a point where the potential is ϕ. Since the kink sites may be occupied by either silver ions or impurity ions the number of silver ions in kink sites n_s becomes[18]

$$n_s = \frac{N_s \exp\left(-\frac{W_+^s + |e|\phi_s + \mu_{Ag}}{kT}\right)}{1 + \exp\left(\frac{\mu_{Ag} - W_+^s - |e|\phi_s}{kT}\right) + \exp\left(\frac{\mu_i - W_i^s - 2|e|\phi_s}{kT}\right)} \tag{26}$$

where W_i^s is the binding energy of an impurity ion at a kink relative to the value for a normal lattice site, $W_+^s = \Delta G_V$ and μ_i is the impurity chemical potential. Correspondingly the number of impurity ions at kinks would be

$$n_i^s = \frac{N_s \exp\left(\frac{\mu_i - W_i^s - 2|e|\phi_s}{kT}\right)}{1 + \exp\left(\frac{\mu_{Ag} - W_+^s - |e|\phi_s}{kT}\right) + \exp\left(\frac{\mu_i - W_i^s - 2|e|\phi_s}{kT}\right)} \tag{27}$$

The surface charge density σ is then given by

$$\sigma = |e|\left(n_s + 2n_i^s - \tfrac{1}{2} N_s\right) \tag{28}$$

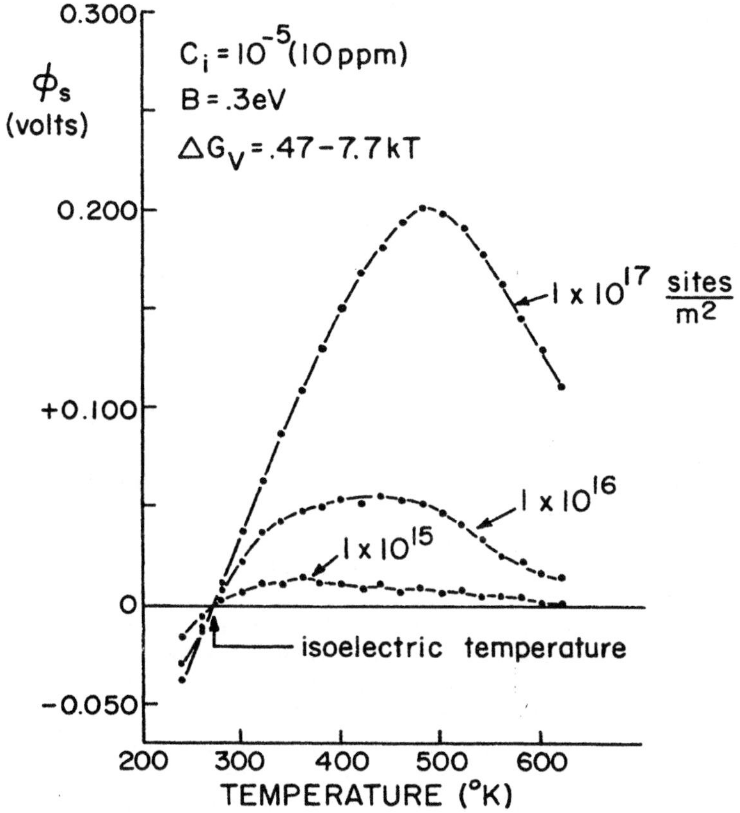

Figure (19)
Surface potential ϕ_s as a function of temperature and kink site density when impurities are allowed to redistribute to their equilibrium distribution. The parameters chosen were: impurity concentration C_I = 10 ppm, B = 0.3 eV, ΔG_V = 0.47 - 7.7 kT (= $W_+{}^s$), ΔG_F = 1.4 - 9.4 kT and have set $W_i{}^s$ equal to $\Delta G_V(=W_+{}^s)$. Note that the isoelectric temperature does not vary with the density of available surface sites.

The impurity chemical potential is obtained from the conservation condition

$$n_i = (\text{\# associated with vacancies}) + (\text{\# unassociated}) \qquad (29)$$

In order to obtain numerical results we must of course assume values for the parameters, B and $W_i{}^s$ in addition to those already discussed. The actual results of the computations must therefore be regarded as being purely illustrative of what may happen rather than as applying to any particular case. With this in mind the form of surface potential versus temperature variation is shown in figure (19). The low temperature behavior of ϕ_s is primarily due to changes in μ_{Ag} with T while the high temperature part is due to kink saturation effects.

As seen from equation (27) the surface concentration of impurity ions depends critically on the difference in binding at the surface from the bulk. The space charge concentration of divalent ions on the other hand depends mainly on the sign of the surface electrostatic potential. Thus, various impurity concentration profiles are possible i.e., enriched/depleted surface monolayers combined with enriched/depleted space charge regions. As discussed later experimental surface analytical techniques have difficulty in distinguishing these possibilities.

(iv) Experimental Observations on Surface Ionic Space Charge Layers

To this point we have mainly been discussing free surface effects. However it should be clear that the approach outlined here is sufficiently general that it may be applied with minor modifications to other defects such as grain boundaries or dislocations. Much of the experimental work has concentrated on the charge configurations at dislocations since this is of considerable importance in determining the plastic flow characteristics. A detailed review of this work, which has mainly been concerned with the alkali halides, has been given recently by Whitworth(11). It appears that the dislocation cores are normally negatively charged with the space charge containing an excess of cation vacancies. However detailed studies over wide temperature and doping level ranges do not yet seem to have been made.

For the study of surface space charge regions a number of experimental approaches would seem to be feasible. For this problem we may again draw from experience with electronic space charge layers at free semiconductor surfaces. In that case the experiments which have proven most useful(19,20) have involved measurements of surface conductance both with and without applied normal fields, surface potential studies by Kelvin probe techniques, photovoltage measurements and photoemission studies. Some of these techniques are applicable to ionic crystal surfaces although only relatively few studies of this nature have been carried out. Due to their extremely high electrical impedance there has been little systematic work on the alkali halides and most of the recent experiments have been with the silver halides whose electrical conductivity is greater by factors of about 10^7. The surface properties of the silver halides are of course of interest in understanding photographic processes in small crystals as discussed later.

Baetzold and Hamilton(25,26) have studied the excess surface conductance of silver halide evaporated films. Figure (20) shows

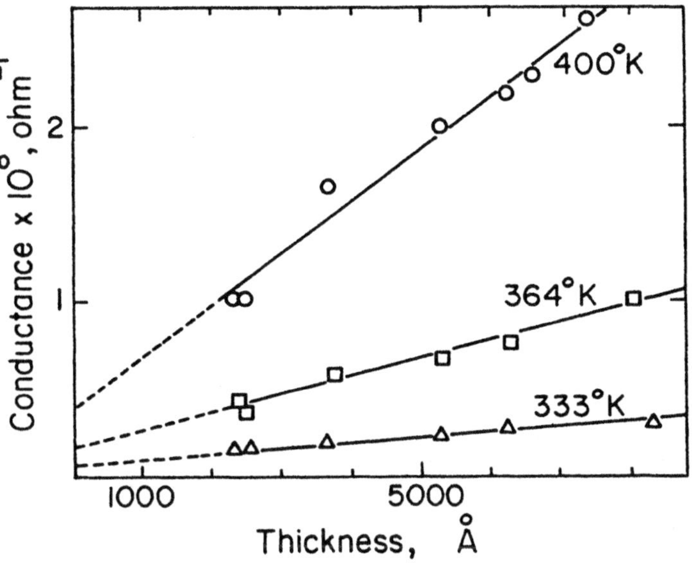

Figure (20)
Conductance versus thickness for (111) AgBr films evaporated on mica. The effect of temperature on conductance is also indicated. The intercept on the vertical axis at zero thickness corresponds to the excess surface conductance. [From Baetzold and Hamilton(25).]

the film conductance as a function of thickness for (111) AgBr showing clearly an excess surface contribution. Through an interpretation of their data based on the type of analysis presented above they conclude that for the (111) surface of both AgBr and AgCl the surface layer is negatively charged while the space charge layer contains an excess of Ag interstitials and is positively charged. In our present notation, their data give for AgBr (111) $\Delta G_V = 0.85$ eV and for AgCl(111), $\Delta G_V = 0.69$ eV.

Measurements of the variation with temperature of the surface potential, relative to a metallic, reference surface have also been made on a number of surfaces of the silver halides, using the vibrating capitor method(27). A schematic diagram of the experimental arrangements is shown in figure (21a) and the circuit details in figure (21b). Since the values of the surface space charge potential, ϕ_S, can only be derived from the measurements if all other junction potentials are known the absolute values of ϕ_S determined in this way have large uncertainties. However the main result of these studies so far has been the demonstration that for both AgCl and AgBr the value of ϕ_S and its temperature variation are sensitive to surface orientation,(27) as predicted from our model.

Some preliminary photovoltage measurements on AgCl have been

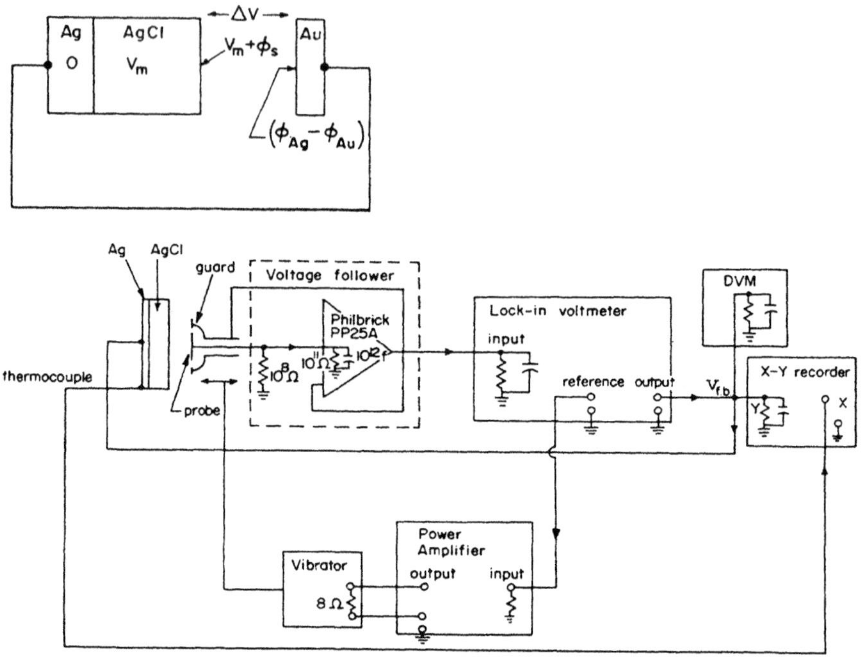

Figure (21)
(a) Schematic diagram of the experimental arrangement for the vibrating capacitor technique. Choosing the reference potential zero in the bulk metal the bulk of the silver halide is at potential V_m and its surface at $V_m + \phi_s$. The surface potential of the Au reference is equal to the work function difference $\Phi_{Ag} - \Phi_{Au}$.
(b) Circuit diagram for the surface potential measurements.

carried out.(28) However the interpretation of these measurements is quite complicated due to the occurrence of a significant Dember effect and the formation of photolytic silver during light exposure.

Most of the spectroscopic techniques currently being used for surface composition analysis(29) should be appropriate for studying the composition of the immediate surface region in ionic crystals and the variation of doping concentration with depth. For the former Auger spectroscopy(29,30) is currently the most used while Rutherford backscattering promises to be generally useful for profile determination. Few studies of equilibrium impurity segregation to ionic surfaces have so far been carried out although there are a number of observations on systems of considerable practical importance. For example, fracture surfaces of β - alumina have been examined by Auger spectroscopy(31,32) and recently the segregation of Ca to grain boundaries in this material has been detected in our laboratory. Figure (22) is an Auger electron

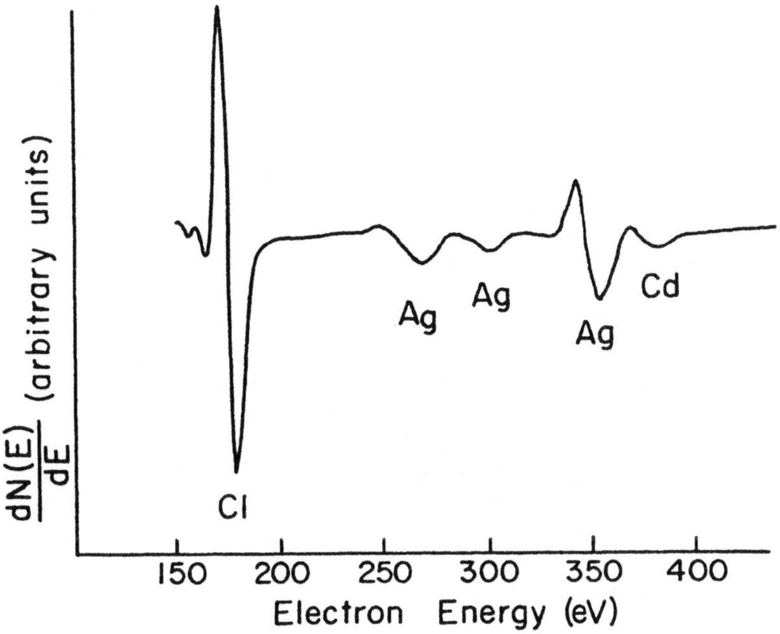

Figure (22)
Portion of an Auger spectrum from a (111) surface of a Cd doped AgCl crystal taken with a cylindrical mirror analyser. The primary beam has a diameter of ∿1 mm and is incident at ∿27° to the surface. Primary current 10μA, modulation voltage 2.8 V peak to peak, frequency 15 kHz, sweep rate 1 V/sec.

spectrum from a (111) surface of a Cd doped AgCl crystal. The magnitude of the Cd concentration in this case is ∿30 times the average bulk value. However attempts to correlate the temperature variation of this concentration with that predicted by the theory outlined have not been successful due to the irreversible photolytic effects of the exciting electron beam on the silver halide surface. Since the Auger electron escape depth in this material is probably ≳ 10 Å it is not possible from one measurement to conclude whether the impurity is concentrated on the topmost layer or in the space charge region.

It is clear that the techniques of surface physics are just beginning to be applied to ionic crystatls. We should certainly expect signficant progress in the near future on obtaining experimental data on the structure and properties of these surfaces.

(v) Effect of Space Charge Regions on Properties

Surface and other interface space charge regions have been

invoked most frequently to explain observations on ionic conduction, mass transport and latent image formation. The experiments of Baetzold and Hamilton[25] mentioned above have demonstrated the existence of excess ionic conductance parallel to a free surface. In polycrystalline materials there may be large effects on conduction and diffusion processes from the space charge regions surrounding grain boundaries. Some of the relevant experiments on grain boundary diffusion contributions have recently been reviewed by Yan et al[33]. The experimental situation is that there are still rather conflicting results probably due to significant effects of small impurity concentrations. The extent of the region of enhanced or diminished diffusivity associated with grain boundaries and dislocations in ionic materials will generally be much greater than in metals, so that significant overlapping of short circuit zones will occur at relatively large grain sizes. For example with a Debye length of order 10^3 Å the equilibrium space charge regions will overlap appreciably at a grain size of order 1μ and in that case the material would have no regions with bulk ionic transport properties. For ionic materials we should consider the boundary as consisting of two fairly distinct zones, figure (23). The core region of atomic dimensions is probably a region of enhanced diffusivity for both anions and cations; on either

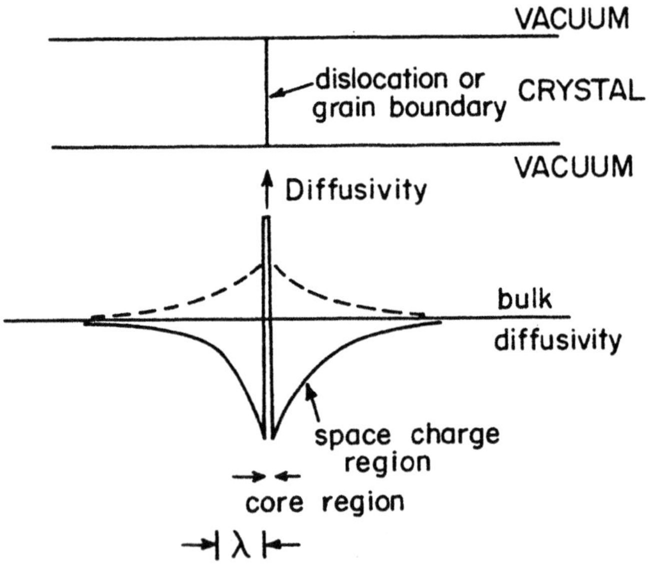

Figure (23)
Possible form of the diffusivity as a function of position near a grain boundary (or dislocation) in an ionic solid. Two regions are distinguished: the core and the space charge region. In the latter the diffusivity may be enhanced (broken curve) or suppressed (solid curve).

side of the core region is a diffuse layer in which the diffusivity is either enhanced or diminished depending on the mode of diffusion and the sign of the space charge potential. We may obtain an estimate of the grain boundary space charge contribution for a special case as follows. For cation self diffusion by an interstitial mechanism in AgBr the excess flux carried by the space charge region parallel to a planar grain boundary would be

$$\frac{D}{kT} 2 \int \left[n_I(z) - n_I(\text{bulk}) \right] dz \; . \; \text{grad}\, \mu_+ \tag{30}$$

where D is the interstitial diffusivity, $n_I(z)$ the interstitial concentration at distance z from the boundary and μ_+ the cation electrochemical potential. For fixed gradient in electrochemical potential the excess flux is thus proportional to the net interstitial excess in the space charge layer. For AgBr at 400°K the bulk concentration of interstitials is $\sim 10^{18}$ cm^{-3} and the excess charge of the space charge region at a free (111) surface has been found[25] to be $\sim 10^{13}/cm^2$; if we take this to apply also to an internal boundary we see that the space charge regions in material with 1000 Å grains (eg. a thin film with columnar structure) would enhance the conductivity by a factor of two. Further enhancement may of course arise from core contributions.

The temperature dependence of the excess grain boundary flux is likely to be rather complicated particularly for impure crystals as may be seen by reviewing the space charge temperature variation discussed in the previous section. On passing through the isoelectric point the space charge contribution may even change sign while the core contribution may also have a complex temperature dependence. Since most of the relevant parameters are not yet known for those systems on which boundary diffusion information exists a quantitative comparison with space charge theory is not yet possible.

Mass transport processes in ceramics are of interest due to their importance in materials preparation techniques. The phenomenon of sintering by diffusion is reasonably well understood for metallic systems.[34] For 2-component ionic systems two additional features must be considered. (i) In order to produce an overall transfer of material a net flux of each component will occur, the components having in general unequal mobilities. These fluxes are interdependent since, locally, certain concentration ratios must be maintained. (ii) The concentration of defects will vary with distance from the free surface. As an example of the incorporation of these effects we may consider the changes in morphology of a nearly planar surface to which Mullins'[35] theory of mass transport may be applied. For the two component case, eg. NaCl, the equation describing the surface evolution by volume diffusion processes may be written as[36]

$$\frac{\partial z_s}{\partial t} \approx \frac{D_2}{kT} \exp\left(-\frac{|e|\phi_s}{kT}\right) \frac{\partial}{\partial z}\left(\mu_1 + \mu_2\right) \quad (31)$$

where z_s is the surface coordinate, D_2 is the diffusivity of the more slowly moving element and μ_1 and μ_2 are the chemical potentials of the two species. If $|\phi_s|$ is of the order of a few tenths of a volt then the exponential term accounting for the space charge is $\sim 10^{-2}$ at 300°C and therefore represents a significant correction. In calculating diffusion coefficients and activation energies from observations on surface mass transport processes (sintering) account must therefore be taken of the space charge layer.

As noted previously the interest in surface space charge regions in the silver halides arises mainly from the use of these materials in the form of small crystallites in photographic emulsions. Current theories of latent image formation are still largely based on the general schemes proposed by Mott and Gurney(37) and by Mitchell(38). On illumination photoelectrons and photoholes are created in the crystal; some electrons are trapped at special sites and subsequently combine with mobile silver interstitials. A recurrence of this process at the same site leads to the formation of silver particles (the latent images) which subsequently act as the nuclei for chemical reduction of the silver halide crystal. The latent image centers do not occur homogeneously throughout the volume of large crystals and are in general found to be concentrated near lattice imperfections such as surfaces, grain boundaries and dislocations. There are a number of ways in which the surface properties may enter in the photolysis process. Electron or hole traps, essential ingredients of the Mott-Gurney and Mitchell theories may be associated with intrinsic surface electron states, defects in the surface region or extrinsic surface states due to adsorbed or segregated dopants. There has also been considerable speculation on the role of surface electrostatic fields on the process of latent image formation(39). It has been suggested that the space charge region serves to separate the photogenerated electrons and holes thus preventing their recombination and also provides an enhanced interstitial silver ion concentration. Simple calculations based on the experimental results on the silver halides shows that in a small grain the total number of interstitials may be increased by a factor of $\sim 10^3$ over that extrapolated from the bulk interstitial densities. Surface fields may also be important in sensitization by adsorbed molecules since the position of the electron energy band edges relative to the levels of the adsorbed molecules will depend on the potential distribution near the surface.

References

1. Benson, G. C. and Yun, K. S., "Surface Energy and Surface Tension in Crystalline Solids" in The Solid-Gas Interface, Vol. I (ed. E. A. Flood) Dekker, NY, 1967.

2. Strozier, J. S., Jepson, D. W. and Jona, F., "Surface Crystallography", in Surface Physics of Materials, Vol. I (ed. J. M. Blakely) Academic Press, NY, 1975.

3. McRae, E. G. and Caldwell, C. W., Surf. Sci. 2. 509 (1964).

4. Ellis, W. P. and Schwoebel, R. L., Surf. Sci. 11, 82 (1968).

5. Bethge, H., in Molecular Processes on Solid Surfaces, (ed. E. Drauglis et al) McGraw Hill, 1969.

6. Blakely, J. M. and Li, C. Y., J. Phys. Chem. Solids, 26, 1863 (1965).

7. Mott, N. F. and Littleton, M. J., Trans. Faraday Soc., 34, 485 (1938).
Fumi, F. G. and Tosi, M. P., Disc. Faraday Soc., 23, 93 (1957).

8. Burton, W. K., Cabrera, N., and Frank, F. C., Phil. Trans. Roy. Soc., A243, 299 (1951).

9. Hirth, J. P., in Energetics in Metallurgical Phenomena, Vol. II (ed. W. M. Mueller), Gorden and Breach (1965).

10. Leamy, H. J., Gilmer, G. H. and Jackson, K. A., in Surface Physics of Materials, (ed. J. M. Blakely) Academic Press, NY (1975).

11. Whitworth, R. W., Adv. in Phys., 24, 203 (1975).

12. Frenkel, J., Kinetic Theory of Liquids, Dover Publications, (1955).

13. Lehovec, K., J. Chem. Phys., 21, 1123 (1953).

14. Eshelby, J., Newey, C., Pratt, P. and Lidiard, A., Phil. Mag., 3, 75 (1958).

15. Kliewer, K. L. and Koehler, J. S., Phys. Rev. 140, A1226 (1965).

16. Kliewer, K. L., J. Phys. Chem. Solids, 27, 705 (1965).

17. Poeppel, R. B. and Blakely, J. M., Surf. Sci. 15, 507 (1969).

18. Blakely, J. M. and Danyluk, S., Surf. Sci. 40, 37 (1973).

19. Many, A., Goldstein, Y. and Grover, N. B., Semiconductor Surfaces, North Holland, (1965).

20. Frankl, D. R., Electrical Properties of Semiconductor Surfaces, Pergamon, (1967).

21. Fouchaux, R. D. and Simmons, R. O., Phys. Rev., 136, A1664 (1964).

22. Abbink, H. C. and Martin, D. S., J. Phys. Chem. Solids, 27, 205 (1966).

23. Crawford, J. H. and Slifkin, L. M., in Point Defects in Solids, Vol. I, Plenum Press, (1972).

24. Kim, J. S., Ph.D. Thesis, Univ. N. Carolina, (1967).

25. Baetzold, R. C. and Hamilton, J. F., Surf. Sci., 33, 461 (1972).

26. Baetzold, R. C., J. Phys. Chem. Solids, 35, 89 (1974).

27. Blakely, J. M. and Danyluk, S., Photog. Sci. and Eng., 18, 489 (1974).

28. Kim, J. S. and Blakely, J. M., Materials Science Center Report #1196, Cornell Univ., Ithaca, NY (1969).

29. Park, R. L., "Chemical Analysis of Surfaces" in Surface Physics of Materials, (ed. J. M. Blakely) Academic Press, (1975).

30. Harris, L. A., J. Appl. Phys., 39, 1419 (1968).

31. Gjostein, N. A., and Chavka, N. G., J. Testing and Evaluation, 1, 183 (1973).

32. Stoddart, C. T. H. and Hondros, E. D., Trans. and J. Brit. Cer. Soc., 73, 61 (1974).

33. Yan, M. F., Cannan, R. M., Bowen, H. K. and Coble, R. L., "Space Charge Contribution to Grain Boundary Diffusion", (to be published).

34. Herring, C., in Physics of Powder Metallurgy, (ed. W. E. Kingsto McGraw Hill, NY (1951).

35. Mullins, W. W., in Metal Surfaces, Amer. Soc. Metals, Cleveland, (1963).

36. Blakely, J. M. and Li, C. Y., Acta. Met., 14, 279 (1966).

37. Mott, N. F. and Gurney, R. W., Electronic Processes in Ionic Crystals, Clarendon Press, (2nd ed.) (1948).

38. Mitchell, J. W., Reps. Prog. Phys., 20, 433 (1957).

39. Matejec, R., Meissner, H. D. and Moisar, E., in Progress in Surface and Membrane Science, (ed. E. Danielli) 6, 1 (1973).

DISCUSSION

HEYNE : Did you experience interference from photodecomposition ? (during photovoltage measurement).

BLAKELY : Yes. The effect is smaller at liquid nitrogen temperature but nevertheless may still be important.

DUPUY : Why don't you mention specially dislocations which are important sources of electrostatic defects in ionic crystals ? (i.e. surrounding space charge...)

BLAKELY : The analysis for dislocations is in fact the same in principle. The geometry of the space charge distribution is of course cylindrical in an isotropic crystal. A recent paper by Whitworth reviews the experimental situation for dislocations.

KLEITZ : Is there any limitation in the Kelvin probe measurement due to a low crystal conductivity ? What could be a reasonable value for the limiting conductivity ?

BLAKELY : For low conductivity crystals,it is necessary to use a negative feedback arrangement to maintain the field outside the crystal at zero. Highly insulating crystals will also show long term drift effects due to the relaxation time for transfer of real charge to and from the surface. Crystals such as NaCl are extremely difficult to work with.

KROGER : 1. What is the relaxation time of the establishment of the Cd-enrichment at a freshly cleaned surface ?

2. Do you know of a way to distinguish between the contributions of the core and the space charge region to the excess surface conductivity ?

BLAKELY : 1. We have been unable to measure this experimentally since the silver halides do not fracture by cleavage even at liquid nitrogen temperature. Sputtering with inert gas ions leads to irreversible effects and hence cannot be used to produce a fresh surface.

2. These are rather difficult quantities to separate experimentally. I believe that in work of Baetzold and Hamilton they found that the excess surface conductance was not modified by depositing thin layers of other materials on the surface ; they argued that this indicated that surface monolayer transport was unimportant.

WAGNER : What is the purpose of adding gaseous chlorine to your "Kelvin Probe System" containing Ag/AgCl ? How thick was the AgCl ?

BLAKELY : Pretreatment with Cl_2 was necessary to achieve reproducible and reversible results. It probably removes any excess Ag produced during specimen preparation. The sample thickness was of the order of a few mm.

RALEIGH : Could you discuss the nature of the (1,1,1) surface on AgBr as it pertains to surface conduction ? it would seem that the (1,1,1) surface must be all cations, all anions or patches of the two and that analysis of results on such surfaces would be difficult.

BLAKELY : The (III) surface consisting of large-regions of cations/anions in the surface monolayer is probably unstable with respect to the formation of microfacets of (100) type.

THE ELECTROCHEMICAL DOUBLE LAYER IN SOLID ELECTROLYTES

Douglas O. Raleigh

Science Center, Rockwell International
Thousand Oaks, California 91360 USA

1. INTRODUCTION

When a d.c. voltage is applied across an electrochemical cell, the response of its electrodes can be considered in terms of two possible modes of idealized behavior. These are illustrated in Fig. 1. The left-side electrode is composed of the parent metal M(s) of an electrolyte for the conduction of a cation species M^+. Because of the presence of M(s) and M^+ at the electrode interface, an equilibrium $M(s) \rightleftharpoons M^+ + e$ is assumed to obtain. If establishment and maintenance of the equilibrium is ideally rapid, then the response of the interface to a nonequilibrium field is to drive the equilibrium predominantly in one of its component directions and result in a basically nonresistive transfer of charge across the interface. This is the well-known behavior of the ideally reversible electrode.

The right-side electrode metal, M', is assumed to be chemically unreactive towards the electrolyte and to be incapable of relating to it via an electrochemical equilibrium. At sufficiently cathodic potentials, the metal M(s) can be electrodeposited on its surface. At sufficiently anodic potentials, some anionic species can be discharged. In an intermediate range of applied voltage, generally determined in a solid electrolyte by its free energy of formation, ion discharge is not possible, and ions accumulate at the electrode surface. This is the behavior of the ideally polarizable electrode.

In practice, most electrodes display some intermediate behavior with a given electrolyte. Establishment and maintenance of equilibrium at essentially reversible electrodes involves finite

M. Kleitz and J. Dupuy (eds.), Electrode Processes in Solid State Ionics, 119–147.

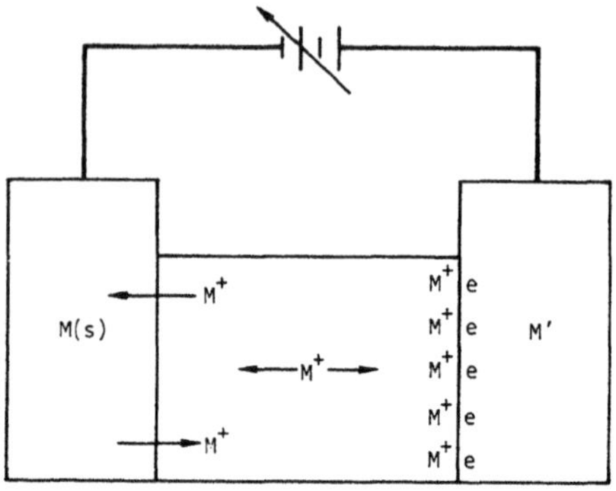

Fig. 1 Schematic of cell with reversible electrode and polarized electrode.

rates. At essentially polarizing electrodes, there is indeed a first-order current blockage over an appreciable voltage range, but effects such as M - M' surface alloy formation and M diffusion in M' can result in transient or low-level discharge currents. Experimentally, however, "current-carrying" electrodes are studied by essentially voltammetric methods and "current-blocking" electrodes by essentially capacitative methods. The study of current-blocking electrodes, with which this paper is concerned, can provide valuable insights into the manner in which charge is stored at the interface between an ionic- and an electronic-conducting medium and hence into the fundamental character of such interfaces.

The most convenient electrochemical arrangement for studying current-blocking electrodes is in fact the one schematically illustrated in Fig. 1. The range of polarizing potentials lies ideally from electrode M' being just anodic (positively charged) of M(s), to sufficiently anodic to permit the discharge of anions. Over this range, there might be a broad variation in the character of the ion layer at the M' surface. At the weakly anodic end, we might expect an accumulation of M^+ ions (the case illustrated), and an equal and opposite electronic charge within the electrode. At strongly anodic potentials, accumulated (or, basically, "abandoned") anions might predominate. The resultant change in interface charge storage characteristics could be considerable. It is important, then, to explore the interface over this full range of potential. Symmetric-electrode cell arrangements, such as M'|M^+(electrolyte)|M', do not permit this possibility, and generally result in a poorly defined character of the electrode interfaces.

In the arrangement of Fig. 1, various polarizing potentials are applied to the cell and the M^+(electrolyte)|M' interface charge storage characteristics are studied in one of several ways. A small "step" voltage can be applied to the cell, and the current i(t) and its time integral Q(t), corresponding to the recharging of the interface capacitance, can be determined. If the M(s) electrode is indeed ideally reversible, then the cell system will behave electrically as the resistance of the electrolyte in series with the impedance of the M^+(electrolyte)|M′ interface. If, more likely, the M(s) electrode is not ideally reversible, there are a number of well-known electrochemical techniques, such as the use of potentiostatted three-electrode systems (1), for making i(t) and Q(t) behave as though it were, and for minimizing the effective electrolyte resistance, when necessary. Such experimental details are outside the scope of this paper. In the work to be discussed, they have been adequately taken care of, unless otherwise indicated. Additional methods for studying the charge storage capability of the M^+(electrolyte)|M′ interface in the schematic cell arrangement of Fig. 1 are by the measurement of the a.c. cell impedance as a function of the applied d.c. biasing potential, and by various voltammetric methods such as cyclic voltammetry (1).

2. ELECTROSTATIC CHARGE STORAGE AT SOLID-ELECTROLYTE INTERFACES

The discussion above is in essence applicable to any type of electrolytic medium. Solid ionic conductors, however, have some obviously distinctive characteristics as electrolytes. It is of interest to consider how these might affect their interface characteristics with electrodes.

Figure 2 shows one type of solid electrolyte lattice, assumed to be in contact with the ideally inert surface of a current-blocking metallic electrode. For the present, assume that only electrostatic forces are operative. The cations are taken to be the mobile species in the electrolyte and to migrate via Frenkel defects (equal bulk concentrations of cation interstitials and vacancies). A polarizing voltage, applied to the cell arrangement of Fig. 1, basically ends up in steady-state across the blocking interface, minus a constant voltage drop across the reversible M(s)|M^+(electrolyte) interface.

Despite the vastly different nature of the media, the basic situation differs little in principle from that in aqueous solutions. There is a diffuse space charge layer, defined by Poisson-Boltzmann distributions of the positive (interstitial) and negative (vacancy) carriers in the electric field, which are truncated at $x \leqslant d_c$, the distance of closest approach of the mobile species to the electrode surface. This truncation has the effect of placing a parallel-plate "inner layer" capacitor of plate separation d_c in series with the diffuse layer capacitance, giving the usual

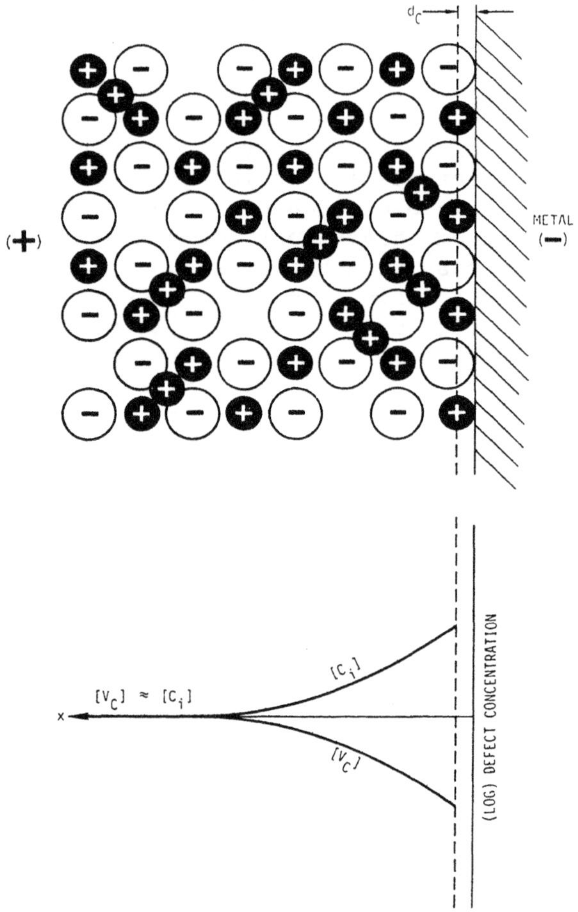

Fig. 2 Schematic of Frenkel-conducting solid electrolyte on negative polarized electrode and resulting defect distributions. Defect concentrations exaggerated for illustration.

"two-capacitor" picture of the double layer (2). The inner layer capacitance per unit area is independent of voltage and bulk defect concentration and should be given by

$$C_i = \kappa\varepsilon_o/d_c \quad , \tag{1}$$

where κ is the effective dielectric constant. For small interface voltage drops, the diffuse layer capacitance should be given by:

$$C_d = \left(\frac{2\kappa\varepsilon_o e^2 N_D}{kT}\right)^{\frac{1}{2}} \cosh(e\Delta\phi_d/2kT) \tag{2}$$

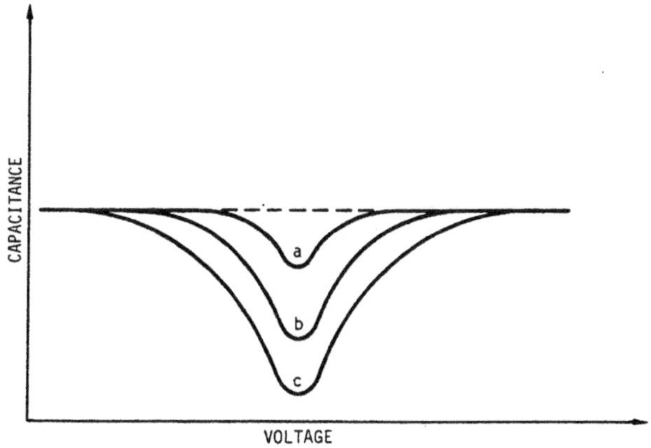

Fig. 3 Qualitative potential dependence of interface capacitance resulting from series contributions of inner-layer and diffuse layer capacitance. a, b and c show the effects of successively decreasing bulk ionic defect concentrations.

where $\Delta\phi_d$ is the portion of the interface potential drop across the diffuse layer ($x > d_c$) and N_D is the bulk defect concentration. For $e\Delta\phi_d >> kT$, other expressions for C_d should apply (3).

Because C_i is constant and C_d increases exponentially with $\Delta\phi_d$, C_d should only dominate the series interface capacitance in a restricted range about $\Delta\phi_d = 0$, a range which is increasingly narrowed with increasing values of N_D. The expected potential dependence of the total interface capacitance is shown qualitatively in Fig. 3. The basic differences from aqueous solutions are that generally (though not always) only one ionic species is appreciably mobile, moving through a rigid lattice of the other ions, and that polar solvent molecules are absent. The effect of the latter is that solvation of individual ions and the electrode surface is not present. For solid electrolytes where conduction is by thermally generated bulk lattice defects (Frenkel, Schottky), the concentration of current carriers is generally low enough to anticipate diffuse double layer effects. For electrolytes such as $RbAg_4I_5$ and $\alpha-AgI$, where all the ions of one type are thought to be simultaneously mobile, the large value of N_D should assure that there is no detectable diffuse layer capacitance and that in fact all the ionic double layer charge resides in one or two ion layers at or near the electrode surface. For both diffuse- and inner-layer capacitance, charging should be intrinsically rapid, involving the RC time constant of the cell (4).

We consider now the experimental evidence for diffuse-layer or inner-layer electrode capacitance in ionic solids. Dating from the earliest studies of solid-state ionic conduction, time-dependent currents and frequency-dependent impedances were observed which were attributed to ion discharge blocking and diffuse space charge accumulation at the electrodes. A number of theoretical treatments of such effects have been proposed (5-10).

More recently, several experimental studies have been carried out with pure and doped alkali halide single crystals, using extremely careful preparative conditions, to try to ascertain whether a quantitative fit to diffuse space charge theory could be obtained (11-14). Symmetric electrodes, generally graphite or noble metals, were employed. However, only partial agreement with theory was obtained, and later studies indicated possible interferences from interfacial air gaps (15), chemical interactions with the electrodes (16), and insulating surface layers (17).

Perhaps most importantly, the use of symmetric-electrode cells in these studies did not allow the d.c. voltage dependence of the interface capacitance to be determined in an unambiguous manner. The applicability of Eq. 2 could not be investigated. In general, the artificial (and generally incorrect) assumption was made that $\Delta\phi = 0$ for a solid ionic conductor in contact with a graphite or noble metal electrode in the absence of an externally applied potential. The temperature and dopant level dependence of the experimental capacitance was then compared with the factor $(2\kappa\varepsilon_o e^2 N_D/kT)^{\frac{1}{2}}$ or its equivalent. Other aspects of this work are discussed elsewhere (3,18). To the author's knowledge, the validity of Eq. 2 for a solid ionic conductor with a low enough bulk defect concentration to expect predominant diffuse space charge capacitance has not yet been shown.

During the last decade, several measurements of interface capacitance have been carried out with essentially the experimental arrangement of Fig. 1 (19-27). As indicated above, the voltage dependence shown schematically in Fig. 3 has not yet been observed. In several studies, however, low, voltage-independent, fast-charging capacitances have been observed that might be attributed to electrostatic inner layer capacitance. Hull and Pilla (21) found a value of 7.8 $\mu f/cm^2$ for the graphite|AgI pressed-powder interface at room temperature, constant over the voltage range between the start of Ag discharge and I_2 discharge. Raleigh (26a) demonstrated that this type of interface was microscopically smooth, and reported similar fast, voltage-independent capacitances for similarly prepared graphite|AgBr and graphite|$2Me_4NI \cdot 13AgI$ interfaces at room temperature. Armstrong and Mason (23) reported low, almost voltage-independent capacitances for α and β-AgI|vitreous carbon interfaces at elevated temperatures. Bert, Picot, and Dupuy (25) reported low, voltage-independent capacitances for carbon electrodes on AgCl at 25-200°C.

The use of Eq. 1 assumes the existence of a parallel-plate-type capacitance with one "plate" a layer of electronic charge at the electrode surface and the other a parallel plane at d_c, the distance of closest approach of the charge centers of the current-carrying ions. The ionic double layer charge appears as a slight excess of cations over anions, or vice versa, in the interfacial ion layer. If d_c is simply the radius of the charge-carrying ions, then $\kappa = 1$ seems appropriate and d_c values of 1 to 3Å result (inversely) in capacitances of about 3 to 9 $\mu f/cm^2$. (A useful form of ε_o here is 8.85 μf-Å/cm^2.) This is the range of capacitances seen in the above studies.

In applying such a simple model, however, several caveats seem necessary. In the work on pressed-powder interfaces (26a), Raleigh noted that the capacitance for graphite on AgBr was smaller by almost a factor of two than on AgI and $2Me_4NI \cdot 13AgI$. Clearly, d_c cannot be the radius of the silver ion in both cases. A very real possibility for any type of electrode is the presence of some insulating interfacial layer. We shall discuss in the next section cases where an electrolyte component is chemisorbed. Adsorbed atmospheric impurities and the presence of an oxidized surface layer are also possibilities, for both graphite and noble-metal electrodes. Besides affecting d_c, such layers may introduce a nonunity κ. If we now introduce microscopic surface roughness as yet another variable, we see that a number of combinations of such factors may combine to give the appearance of an ideal inner-layer capacitance. The finding, then, of a small, voltage-independent electrode capacitance in a solid-electrolyte cell seems only to indicate some sort of parallel-plate-type interface charging.

An important factor may make it difficult to see diffuse-layer electrode capacitance in ionic solids. A distinctive aspect of solid electrolytes is the general presence of grain boundaries and planar dislocations. In electrolytes with low enough bulk defect concentrations to permit diffuse space charge effects, these features can act as highly localized high-conductivity paths. As such, they should tend to shunt current through the diffuse layer to the electrode surface, having the effect of a resistance in parallel with C_d. In this respect, capacitance-voltage data on low-conductivity solids such as the alkali halides would be extremely interesting.

3. CHEMISORPTIVE ELECTRODE CAPACITANCE

The series sum of Eqs. 1 and 2 predicts rather small capacitance values ($\lesssim 10$ $\mu f/cm^2$) for solid-electrolyte interfaces with purely electrostatic charge storage. For voltage drops of electrochemical interest, the corresponding surface charge density represents a quite small fraction of a monolayer. For inner-layer capacitance, $q_i = \varepsilon_o \Delta\phi_i / d_c$, where $\Delta\phi_i$ is the voltage drop across the inner layer. While such capacitances have been observed in some systems, others with apparently well-prepared interfaces have shown up to hundreds of $\mu f/cm^2$, representing an appreciable fraction of a charge monolayer. In these cases, where the reproducibility of the capacitance and its voltage dependence seem to eliminate impurity and faradaic effects, one expects chemisorptive charge storage. Since the surface charge density is considerably larger than can be explained by electrostatic forces, one invokes an additional, short-range chemical force between ions and the electrode surface.

If we consider Fig. 2 again, with a chemisorptive attraction between the cations and the electrode added, the distance d_c may be shortened slightly, but this is not necessary. The most significant change is that the parallel-plate-capacitor model of the inner layer is no longer valid, because $q_i > \varepsilon_o \Delta\phi_i / d_c$. In aqueous systems, one envisions specific adsorption in terms of ions breaching a hydration sheath to move closer to the electrode surface, but this does not seem relevant to fused and solid-electrolyte systems. In these, it seems necessary to define specific adsorption by an excess of q_i over $\varepsilon_o \Delta\phi_i / d_c$.

If electrostatic and chemisorptive forces are comparable, i.e., if the excess of q_i is modest, separation of the associated capacitances would seem difficult. If chemisorption, however, is relatively strong, it should result in a substantial change in the interfacial-layer electrolyte composition over the voltage range of interest. The interface capacitance should then be dominated by it and should be simply related to the voltage derivative of the adsorption isotherm of the ion involved. In this case, the adsorptive charging resembles a large capacitance in parallel with other sources of interface capacitance, since the ions involved are free to distribute themselves between the electrolyte bulk and interface sites, irrespective of other capacitance sources.

With moderate chemisorption, one may picture the ionic inner layer charge as simply an accumulation or depletion of the current carrying defects in the first ion layer of an otherwise undisturbed electrolyte lattice. With strong chemisorption, however, a distinct "interface structure" may result. If only current-carrying ions are chemisorbed, there is nothing to prevent a flow of them to the interface that fully covers the electrode surface. Likewise, if the noncurrent-carrying ion in a binary electrolyte is strongly chemisorbed, the current-carrying ions may flow out of the interface region, and local rearrangement could result in electrode coverage by the noncurrent-carrying ion.

In the voltage range of full coverage, the layer of surface-adsorbed ions would function as a "dielectric" for the current-carrying electrolyte ions on one side and the electronic electrode charge on the other. In this range, a fast, parallel-plate type capacitance should result, given by $C = \kappa\varepsilon_o/(d_c+\delta)$, where δ is the effective thickness of the chemisorbed layer. This picture, with anions as the chemisorbed layer, has been suggested to describe the capacitances of silver halide-graphite interfaces (26). At voltages where the coverage becomes less than full, an enlarged, adsorptive capacitance would be expected, with possible slow charging associated with the need for structural rearrangement of the interface layer. The latter effects were not seen in the studies with graphite electrodes, probably because graphite-halogen ion chemisorption is expected to be strong and graphite-silver ion adsorption much weaker.

An additional type of chemisorption is where both the anions and cations of a binary electrolyte are strongly adsorbed on an array of sites covering the electrode, and the change in their ratio with voltage results in a large adsorptive capacitance. Lorentz and coworkers (28) have presented a number of cases in aqueous solutions where electrodeposition of a cation is preceded by adsorption of an approximate monolayer of the ions at somewhat less cathodic voltages. If anions can be similarly adsorbed at more anodic potentials and the range of complete polarization is not too large (as in silver halides), it might be reasonable to assume that the electrode is always covered by some combination of adsorbed anions and cations.

In what follows, we will offer experimental evidence for such a situation and present a crude, first-order theory that fits the experimental voltage dependence of capacitance and gives about 2/3 the observed magnitudes. We will then explore refinements of the theory that give better agreement with the observed magnitudes and a fuller picture of the physical situation. The usefulness of Poisson's Equation in deriving interface voltage profiles for various charge distributions will be discussed, using a simple model. Finally, the effects of temperature on the interface roughness factor and interfering impurity currents will be discussed.

4. COMPETITIVE CHEMISORPTION

A. Experimental Evidence

For carefully prepared AgBr|Pt interfaces, previous work (22) in our laboratory yielded a large, voltage-dependent capacitance that was the same above and below the AgBr melting point and the same for single-crystal and polycrystalline Pt. Figure 4 shows the a.c. capacitance-voltage curves at 1 kHz on the two sides of the AgBr melting point (423°C). Supplemental experiments indicated that the capacitance upturns at the ends of the voltage range

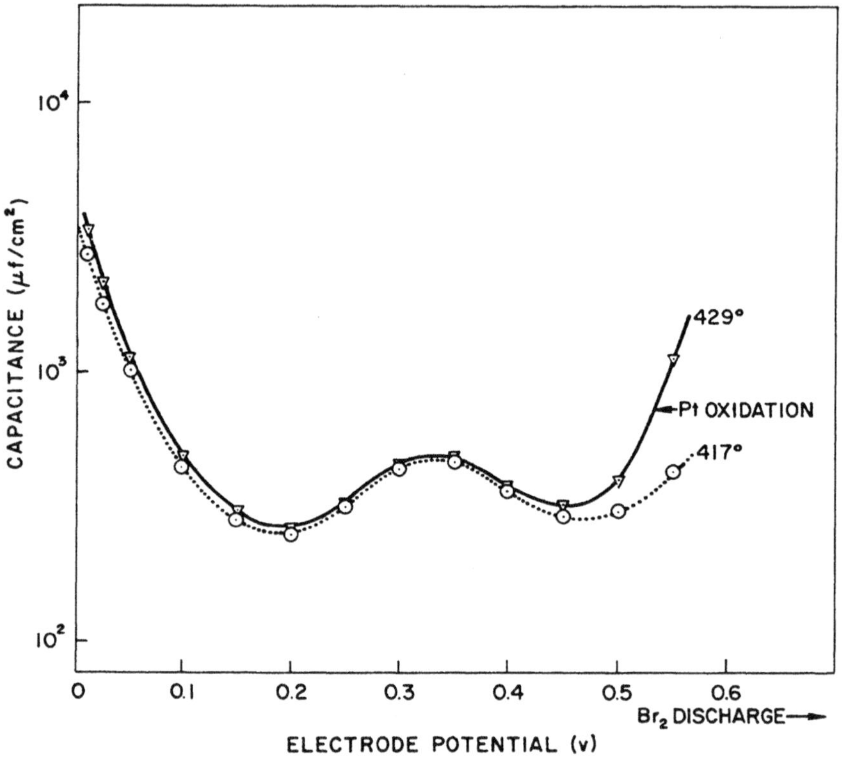

Fig. 4 A.C. capacitance at 1 kHz of Pt electrode in liquid (429°) and solid (417°) AgBr electrolyte as a function of voltage with respect to the Ag potential.

were due to faradaic processes (Ag submonolayer predeposition at the cathodic side), and that the central peak region represented true capacitative charging, i.e., double layer capacitance. At lower temperatures to 300°, the charging rate decreased, but the magnitude and voltage dependence of the total capacitance (as determined by chronocoulometry) remained nearly the same, despite a 40-fold decrease in electrolyte conductivity.

We suggested that the large double layer capacitance was due to charge storage by a layer of strongly adsorbed electrolyte ions on the Pt whose nature was independent of the state and conductivity of the electrolyte. We assumed that the electrode was always fully covered by some combination of anions and cations and that the capacitance resulted from the change of their ratio on traversing the voltage range. A rough calculation showed that such a model could provide the required amount of charge transfer, but we did not derive a theoretical capacitance-voltage expression.

Since this work, more recent studies suggest that this type of capacitance mechanism may obtain for other halide electrolytes with platinum electrodes. A follow-up study in our laboratory (26a) established the same type of capacitance-voltage behavior, with vastly reduced charging rates, for AgBr|Pt and AgI|Pt at room temperature, while graphite electrodes gave a much smaller, fast-charging. voltage-independent capacitance. The slow room temperature kinetics with platinum electrodes can be readily explained with the above model, since it involves both anion and cation motion in the charging process, and anions in silver halides are in general only appreciably mobile near the melting point, becoming progressively immobilized on cooling.

In addition, results by Armstrong and Mason (23) show similar shapes of capacitance-voltage curves for $RbAg_4I_5$|Pt at 22° and 95°C and for AgI|Pt at 200°. It seemed worthwhile, then, to attempt to develop a simple theoretical "adsorption capacitance" model and compare it with available data.

B. First-Order Theory

We consider competitive chemisorption of anions and cations at a fixed array of adsorption sites on an electrode surface in terms of a voltage-dependent ideal thermodynamic equilibrium. For simplicity, we consider systems with sufficiently high electrolyte conductivity that all the interface voltage drop is across a single layer of chemisorbed ions at the polarized electrode surface. This seems a reasonable initial assumption for most conductive electrode materials and for the fused and solid silver halides cited above.

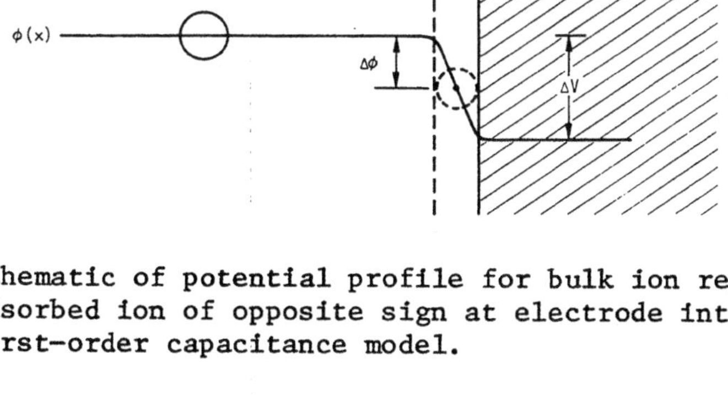

Fig. 5 Schematic of potential profile for bulk ion replacing adsorbed ion of opposite sign at electrode interface in first-order capacitance model.

Figure 5 shows the potential profile at the interface for an ion in the bulk electrolyte replacing an oppositely charged ion at an adsorption site. The capacitance-determining equilibrium is

$$C^+(\text{bulk}) + A^-(\text{ads}) \rightleftharpoons C^+(\text{ads}) + A^-(\text{bulk}) \tag{3}$$

and involves the voltage and the chemisorption strengths of anions and cations. Equating electrochemical potentials for this equilibrium, we have

$$\eta_C(\text{bulk}) + \eta_A(\text{ads}) = \eta_C(\text{ads}) + \eta_A(\text{bulk}) \tag{4}$$

where $\eta_i = \mu_i + z_i F\phi(x)$, where $\phi(x)$ is the local electrostatic potential. Then, for either ion,

$$\eta_i(\text{ads}) - \eta_i(\text{bulk}) = \mu_i(\text{ads}) - \mu_i(\text{bulk}) + z_i F(\phi_{\text{ads}} - \phi_{\text{bulk}})$$

$$\equiv \Delta\mu_i(\text{ads}) + z_i F\Delta\phi \tag{5}$$

where $\Delta\mu_i(\text{ads})$ may be called the free energy of adsorption and $\Delta\phi$ is the potential difference between the charge center of an ion on the adsorbed layer and in the bulk. For simplicity, we assume $2\Delta\phi = \Delta V$. Making these substitutions, we obtain for a uni-univalent electrolyte

$$\Delta\mu_{A(\text{ads})} - \Delta\mu_{C(\text{ads})} = 2F\Delta\phi = F\Delta V \quad . \tag{6}$$

For the adsorption process, we make the ideal solution assumption

$$\Delta\mu_i = \Delta\mu_i^\circ + RT\,\ell n\,N_i = \Delta H_i^\circ - T\Delta S_i^\circ + RT\,\ell n\,N_i \tag{7}$$

that the process is characterized by a well-defined standard enthalpy and entropy of adsorption plus a configurational entropy term for the mole fraction N_i of the species i in the adsorbed layer (the ion concentrations in the electrolyte being large and invariant). This gives us

$$\Delta\mu^\circ_{A(\text{ads})} - \Delta\mu^\circ_{C(\text{ads})} + RT\,\ell n\,\frac{N_A}{N_C} = F\Delta V \quad . \tag{8}$$

For convenience, define ΔV_o as the interface potential drop ΔV when $N_A = N_C = 0.5$. Then $\Delta\mu^\circ_{A(\text{ads})} - \Delta\mu^\circ_{C(\text{ads})} = F\Delta V_o$ and hence

$$RT\,\ell n\,\frac{N_A}{N_C} = F(\Delta V - \Delta V_o) \tag{9}$$

In practice, one would measure the electrode capacitance by the use of a polarization cell such as

$$\overset{\longleftarrow E \longrightarrow}{(-)\text{Ag}\,|\,\text{AgBr}\,|\,\text{Pt}(+)}$$

where the other electrode is reversible to the electrolyte, where various polarizing potentials E of the indicated polarity are applied across the cell, and the capacitance is measured by applying a small potential step or a.c. signal. Then, since the potential drop across the reversible electrode is an invariant, ΔV above is related to E simply by $E = (\text{const.}) + \Delta V$. If E_o is the value of E when $\Delta V = \Delta V_o$, i.e., at $N_A = N_C$, then it is convenient to define the cell voltages with respect to E_o, i.e., $\varepsilon \equiv E - E_o$ and Eq. 9 reduces to

$$RT \ln \frac{N_A}{N_C} = F\varepsilon = RT \ln \frac{1-N_C}{N_C} \tag{10}$$

or

$$N_C = [1 + \exp(\varepsilon F/RT)]^{-1} = 1-N_A \tag{11}$$

which gives the composition of the adsorbed layer at voltages ε with respect to the voltage at which $N_A = N_C = 0.5$.

If Γ_o is the fixed concentration of adsorption sites on the electrode, each of which is always occupied by an anion or cation, then the "adsorption capacitance" is given by

$$C = -e(z'_{mA} \frac{d\Gamma_A}{d\varepsilon} + z'_{mC} \frac{d\Gamma_C}{d\varepsilon}) = -e\Gamma_o(z'_{mA} \frac{dN_A}{d\varepsilon} + z'_{mC} \frac{dN_C}{d\varepsilon}) \tag{12}$$

where Γ_A and Γ_C are the concentrations of specifically adsorbed anions and cations. $-z'_{mA}$ and $-z'_{mC}$ are the electronic charges on the electrode associated with an adsorbed anion and cation, respectively. Lorenz et al. (28) call z'_{mA} and z'_{mC} "charge coverage coefficients" and Schultze and Vetter (29) refer to "electrosorption valences". In principle, we might expect $z'_{mC} = -z'_{mA} = 1$ for a uni-univalent electrolyte, but wide deviations from the ion charge number have been observed in aqueous solutions and various explanations have been advanced (29).

Since $dN_A/d\varepsilon = -(dN_C/d\varepsilon)$, we have

$$C = -e\Gamma_o(z'_{mC} - z'_{mA}) \frac{dN_C}{d\varepsilon} \equiv -e\Gamma_o \Delta z'_m \frac{dN_C}{d\varepsilon} \tag{13}$$

and using Eq. 11 to obtain $dN_C/d\varepsilon$, we have finally

$$C = \frac{e\Gamma_o \Delta z'_m F}{RT} \; \frac{\exp(\varepsilon F/RT)}{[1 + \exp(\varepsilon F/RT)]^2} \tag{14}$$

where $-\Delta z'_m$ is the change in the electrode metal charge when an absorbed anion is replaced by a cation. Figure 6 shows plots of $(C/e\Gamma_o \Delta z'_m)$ vs. ε from Eq. 14 at 100° and 400°C. The capacitance-voltage relationship is symmetric about a maximum at $\varepsilon = 0$, corresponding to $N_A = N_C$. The peak sharpens with decreasing temperature, with the overall charge transfer $\int_{-\infty}^{\infty} C\, d\varepsilon$ remaining constant and equal to $e\Gamma_o \Delta z'_m$.

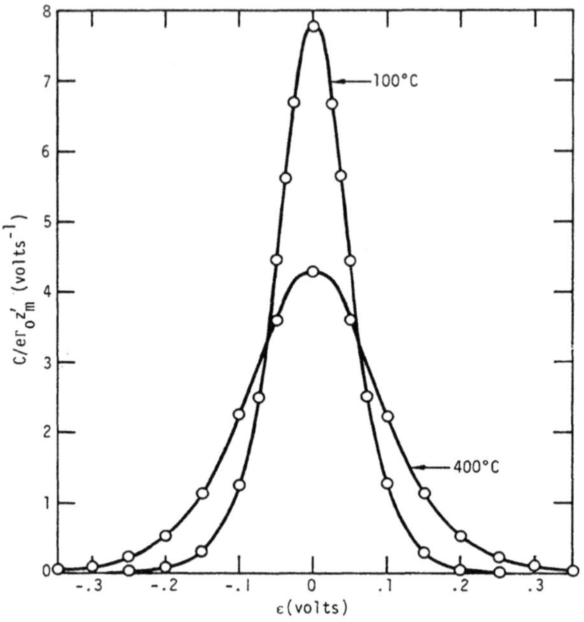

Fig. 6 Adsorption capacitance (in reduced units) as function of voltage with respect to PZC at two temperatures, from first-order theory (Eq. 14).

Some characteristics of the model are of interest. Note that the shape of the $C(\varepsilon)$ curve is independent of the anion and cation adsorption free energies, $\Delta\mu^{\circ}_{A(ads)}$ and $\Delta\mu^{\circ}_{C(ads)}$. These enter only in that their difference is equal to $F\Delta V_o$, and hence determines the cell voltage at which the capacitance maximum occurs.

A questionable assumption of the treatment is the constancy of $\Delta\mu^{\circ}_{A(ads)}$ and $\Delta\mu^{\circ}_{C(ads)}$ over the full coverage range, admittedly made for simplicity. We might expect the binding energy of an adsorbed ion to be greater when the neighboring adsorption sites are occupied by oppositely charged ions. This would have the effect of requiring larger values of $|\varepsilon|$ to effect full conversion to ions of one sign, and would result in a broadening and flattening of the $C(\varepsilon)$ curve.

Since adsorption is taken to be strong and of monolayer-level density, we assume that the fixed array of adsorption sites involves the surface crystallographic structure of the electrode. The adsorbed-layer "superlattice" structures of Lorenz and coworkers (28) are envisioned, and will be discussed for the case of AgBr|Pt in the following section.

C. Application to Experiment Data (AgBr|Pt)

When a polycrystalline Pt wire electrode is electropolished and annealed, faceting may be expected to produce essentially a statistical mixture of the <100>, <110>, and <111> low-index crystal faces as its surface. To estimate a Γ_o for such a surface, we must identify a structure-related pattern of adsorption sites for each type of face and weight the resultant site densities appropriately.

Lorentz et al. (28) have proposed that ions such as $T\ell^{+1}$ chemisorb on Ag° (where $r_{T\ell^{+1}} > r_{Ag^{\circ}}$) in the form of a (2 × 2) superlattice (30), and have calculated the charge density $e\Gamma_o$ associated with adsorption on the <100>, <110>, and <111> surfaces. Since Pt is isostructural with Ag and has almost the same lattice parameter, we have adopted this model for the present system. A (2 × 2) superlattice adsorption pattern on Pt allows the sites to be occupied interchangeably by Ag^+ or Br^-, as we require. Figures 7, 8, and 9, following Lorentz et al., show the <100>, <110>, and <111> surfaces, respectively, of Pt, the appropriate (2 × 2) site structures, and the sizes of an Ag^+ and a Br^- ion on each site.

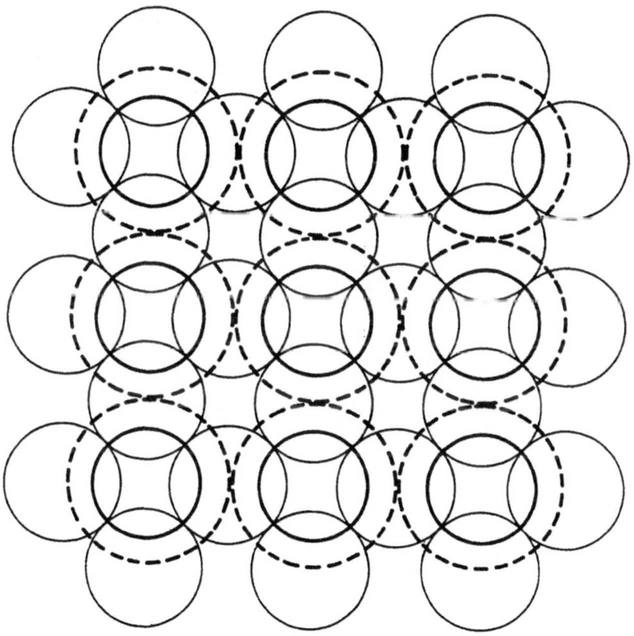

Fig. 7 Ideal (2 × 2) superlattice structure of Ag^+/Br^- adsorption sites on <100> Pt substrate. Solid bold circles are Ag^+ ions; dashed bold circles Br^- ions. Assumed radii: Ag^+ = 1.26Å, Br^- = 1.95Å, Pt = 1.38Å.

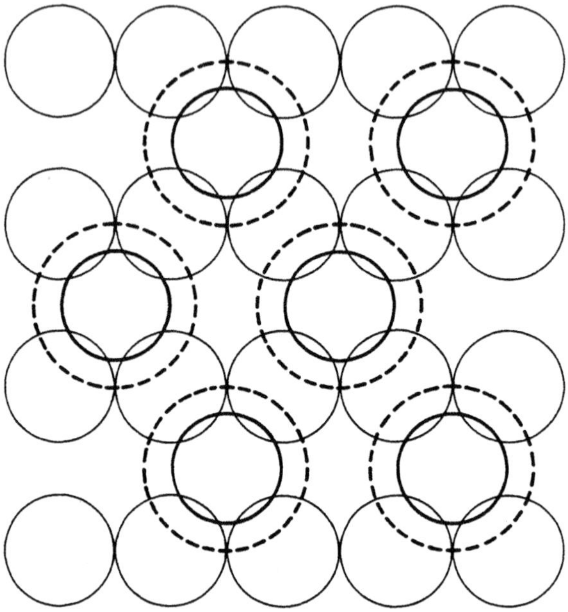

Fig. 8 Ideal (2 × 2) superlattice structure of Ag^+/Br^- adsorption sites on <110> Pt substrate. Legend and radii as in Fig. 7.

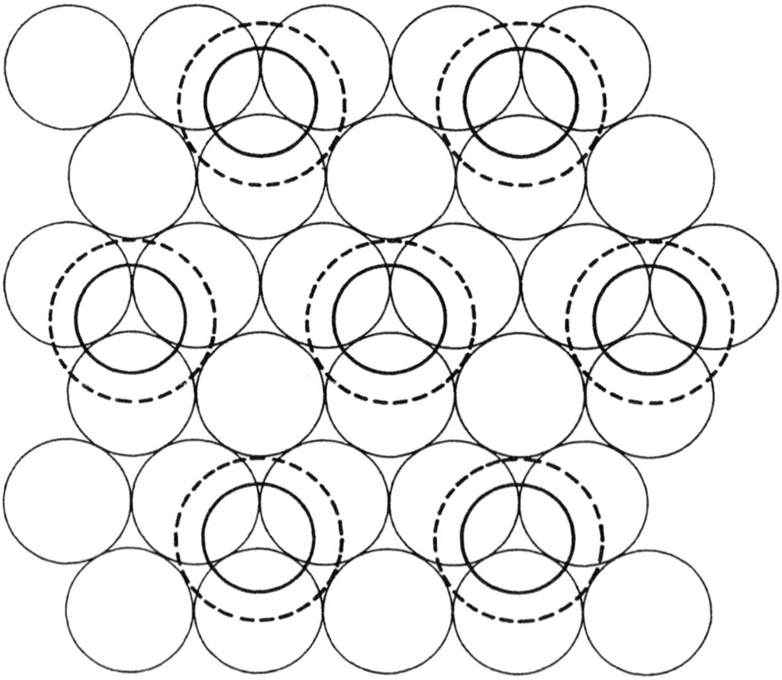

Fig. 9 Ideal (2 × 2) superlattice structure of Ag^+/Br^- adsorption sites on <111> Pt substrate. Legend and radii as in Fig. 7.

Note that dense packing of anions occurs on the <100> face and a more "dilute" adsorbate density on the other faces. The corresponding values of $e\Gamma_o$ for Pt at 423°C are:

<100>: 115 μcoul/cm^2
<110>: 81.5 μcoul/cm^2
<111>: 66.5 μcoul/cm^2 .

The unit cell of a cubic material such as Pt has 6 equivalent <100> planes, 12 <110> planes, and 4 <111> planes. If we average the above densities in this weighting ratio, we arrive at an estimate for $\Gamma_o e$ of 88 μcoul/cm^2.

To compare the theory with experimental results, we see if the capacitance expression of Eq. 14 can be fitted to the central region of Fig. 4 with a reasonable value of $\Delta z_m'$. In Fig. 10, the 417° experimental AgBr|Pt capacitance-voltage curve of Fig. 4 is replotted on a linear scale. On the same graph, the theoretical $C(\varepsilon)$ curve is plotted, with $\Delta z_m'$ chosen such that the capacitance at $\varepsilon = 0$ equals the experimental capacitance at the midpoint

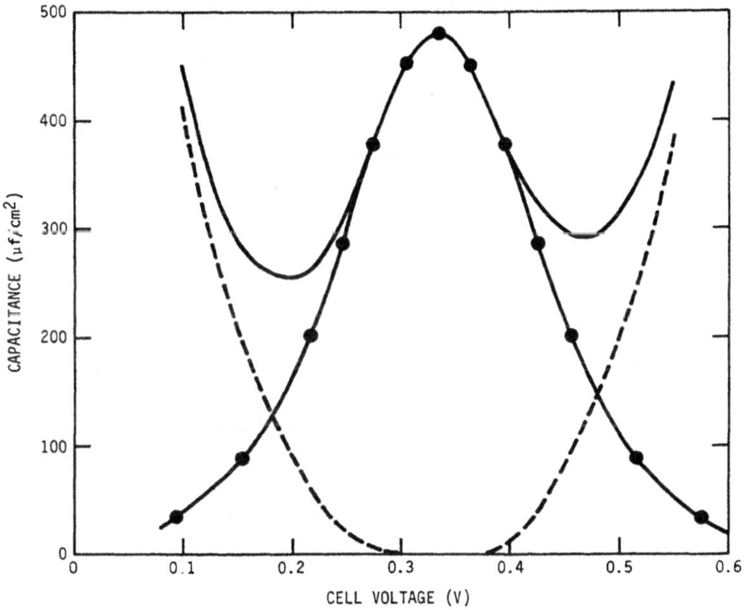

Fig. 10 Comparison of experimental (Fig. 4) and theoretical (Eq. 14) capacitance-voltage curves for AgBr|Pt interface at 417°C. Solid curve is experimental; solid with dots is theoretical. Dashed curves are theory-experiment difference curves. $\Delta z_m'$ in Eq. 14 was chosen to match experimental capacitance at maximum.

maximum. The dashed lines are the difference curves between theoretical and experimental values, which we attribute to faradaic pseudocapacitance at the anodic and cathodic ends of the voltage range. Equation 14 is seen to follow the mid-region shape of the experimental curve very well. This can be seen not only in the region of obvious congruence, but from the fact that the difference curves rise smoothly and quasi-exponentially towards the anodic and cathodic limits with no points of inflection, as one expects for faradaic pseudocapacitance.

The value employed for $\Delta z_m'$ was 1.31, or about 2/3 the "ideal" value of 2.0. One possible explanation of this result might be that Γ_0 is lower than assumed, because of a less dense superlattice structure than (2 × 2). The (2 × 2) structure, however, seems optimal, since (on a <100> Pt surface) it has about the same ion density as a <111> lattice layer in AgBr, which is the densest permissible packing for a layer of Br^- ions. A less dense adsorption site spacing would correspond to a considerable drop in electrolyte density at the electrode surface, while a denser spacing would not permit free interchange between occupation by Ag^+ and Br^-.

An alternate explanation is that some sort of "image force" may affect the electrolyte directly behind the adsorbed ion layer, so that an adsorbed anion, for instance, would have some preference for a cation as a 2nd-layer near neighbor. This, in fact, now brings us to the need for modifications in the first-order theory. We will see that such an "image force" layer of the appropriate size appears naturally as a consequence of the need to minimize the interfacial free energy.

D. Further Theoretical Considerations

An important consideration neglected in the above treatment is that of "superequivalent adsorption" (31). We had assumed implicitly that the net charge of the adsorbed ion layer is always matched by an equal and opposite electronic charge at the surface of the electrode. In fact, if there is an electronic charge per unit area, q_M, on the electrode, the tendency of chemisorbed ions is to give an opposite charge q_{CS} larger than $-q_M$, because of the chemisorptive force. Interface electroneutrality is then met by a difference charge, $q_E = -(q_M + q_{CS})$ which appears in a compact or diffuse ion layer in the electrolyte behind the adsorbed layer.

To examine this problem for the present case, consider that an isolated piece of platinum is immersed in an AgBr melt just above its melting point. Assume both cations and anions are strongly adsorbed, one more strongly than the other. Then because of the ionic strength of the medium, all of the difference charge q_E will be in a single ion layer directly behind the adsorbed ion layer. Assume also that no spontaneous discharge of anions or cations can occur, so that faradaic effects may be ignored. Then we may take $q_M = 0$, either because we have grounded the metal before immersing it, or because we have adjusted its potential with respect to another electrode to give $q_M = 0$ and then disconnected the lead-wire. If we then ignore anion-cation size differences to assume for simplicity the same number of ions in the two ion layers, we have a situation in which preferential adsorption results in an excess of one type of ions in the chemisorbed (CS) layer, balanced by the same excess of the other ion in the adjacent electrolyte (E) layer; $q_{CS} = -q_E$.

If N_i is the mole fraction of cations in either layer, we take the total free energy of ions in the two layers as:

$$\begin{aligned} G = \Gamma_o \{ & N_{CS}\mu_C^\circ + (1-N_{CS})\mu_A^\circ + N_E\mu_C^\circ + (1-N_E)\mu_A^\circ + N_{CS}\Delta\mu_{C(ads)}^\circ \\ & + (1-N_{CS})\Delta\mu_{A(ads)}^\circ + N_{CS}RT\,\ell n\,N_{CS} + (1-N_{CS})RT\,\ell n\,(1-N_{CS}) \\ & + N_E RT\,\ell n\,N_E + (1-N_E)RT\,\ell n\,(1-N_E) + N_{CS}F\phi_{CS} - (1-N_{CS})F\phi_{CS} \\ & + N_E F\phi_E - (1-N_E)F\phi_E \} \end{aligned} \tag{15}$$

where ϕ_{CS} and ϕ_E are the local electrostatic potentials, taken at the respective midpoint planes of the CS and E layers. Then, since the mole fraction of cations in either layer must equal the mole fraction of anions in the other, we may define $N \equiv N_{CS} = 1 - N_E$. Making this substitution, differentiating with respect to N, and equating to zero to minimize G, we obtain

$$\Delta\mu_{C(ads)}^\circ - \Delta\mu_{A(ads)}^\circ + 2RT\,\ell n\,[N/(1-N)] + 2F(\phi_{CS}-\phi_E) = 0 \tag{16}$$

We may relate $(\phi_{CS}-\phi_E)$ to N either by using $q_{CS} = e\Gamma_o(2N-1) = C(\phi_{CS}-\phi_E)$, where C is the parallel-plate capacitance between the CS and E midpoint planes, or, more accurately, by Poisson's Equation, whose use will be developed momentarily. The result of the latter is

$$(\phi_{CS}-\phi_E) = \frac{3}{4}\,\frac{de\Gamma_o}{\varepsilon_o}(2N-1) \tag{17}$$

where d is the average ionic diameter (r_C+r_A). Inserting this result in Eq. 16 yields

$$\Delta\mu_{A(ads)}^\circ - \Delta\mu_{C(ads)}^\circ = 2RT\,\ell n\,\frac{N}{1-N} + \frac{3}{2}\,\frac{Fe\Gamma_o(2N-1)d}{\varepsilon_o}\ . \tag{18}$$

This equation may be solved numerically for N, but more easily by inserting various values of N and calculating the required $\Delta\mu_{A(ads)}^\circ - \Delta\mu_{C(ads)}^\circ$ values. If the latter is done, we find, interestingly, that for any value of N not very close to 0.50, the second term on the right dominates and requires unrealistically high values of $\Delta\mu^{\circ}_{A(ads)} - \Delta\mu^{\circ}_{C(ads)}$. For instance, using values of Γ_o and d appropriate for the AgBr|Pt interface, we calculate that $N = 0.6$ requires $\Delta\mu^{\circ}_{A(ads)} - \Delta\mu^{\circ}_{C(ads)} \approx 220$ kcal/mole.

The situation, in fact, is that $\Delta\mu^{\circ}_{A(ads)} - \Delta\mu^{\circ}_{C(ads)}$ is the driving force for the charge separation in the two layers that finite $q_{CS} = -q_E$ values represent, and charge separations involving any appreciable fraction of a monolayer require prohibitive amounts of electrical free energy. A complementary viewpoint is that Eq. 17 gives $\phi_{CS} - \phi_E = 4.8$ V for $N = 0.6$ for the AgBr|Pt system, a clearly excessive interface potential for electrode adsorption effects. The result is that at $q_M = 0$, N(cations $\approx$ N(anions) for any reasonably different anion and cation adsorption energies.

It might seem from this that similar considerations for $q_M \neq 0$, i.e., over a range of applied potential, would prohibit <u>any</u> type of chemisorptive capacitance on energetic grounds. With $q_M \neq 0$, however, there is another variable with which to minimize the interfacial free energy. We shall see in fact that low-energy ion-chemisorption configurations <u>are</u> permitted, and actually lead to a much-improved agreement with experimental AgBr|Pt capacitances. To show how this occurs, we introduce an interface model for employing Poisson's Equation to calculate the interfacial voltage drop and the conditions of its minimization.

Figure 11 is a schematic of the interface region for a particular charge distribution. An externally applied voltage with respect to a second electrode results in a negative value of q_M.

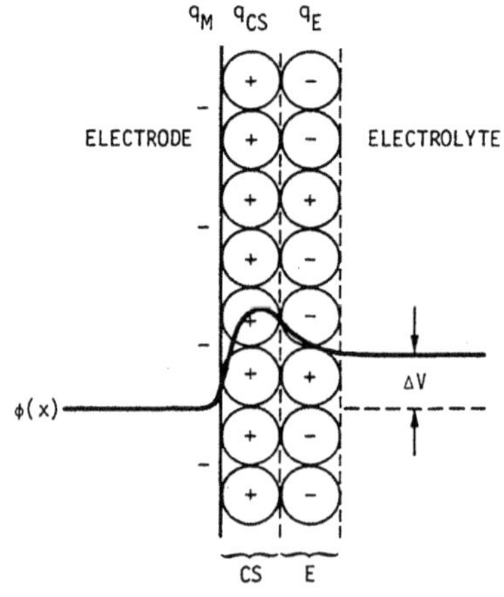

Fig. 11 Schematic of interface region and potential profile for negative electrode charge layer q_M, larger positive chemisorbed-ion charge layer q_{CS}, and difference charge layer q_E in strong electrolyte.

Superequivalent cation adsorption gives a larger positive value of q_{CS}, with electroneutrality being met by a negative difference charge in the E layer. We apply Poisson's Equation to calculate ΔV:

$$\frac{d^2\phi}{dx^2} = -\frac{\rho}{\varepsilon_o} \quad . \tag{19}$$

To calculate ρ, we make the simplifying assumption that within each ion layer there is a uniform time-average charge density given by q/d, where d is the layer thickness, as before. Then double integration within a layer gives $\phi(x)$ as the segment of a parabola, where the boundary conditions determine its course. Since Eq. 19 relates only to the curvature of $\phi(x)$, it allows both $\phi(x)$ and $d\phi/dx$ to be continuous through an abrupt ρ change from q_{CS}/d to q_E/d at the CS|E boundary. When q_{CS} and q_E have opposite signs, a point of inflection results.

Consider the left-to-right course of $\phi(x)$ across the interface region. We assume that the interface is fully polarized, so $\phi(x)$ is horizontal in the electrode and electrolyte bulk. The charge q_M is confined to a very narrow electrode surface region, so it imparts intense curvature to $\phi(x)$, but a generally negligible $\Delta\phi$. Its main effect is to set the value of $d\phi/dx$ that must be matched on the CS side. Single integration of Eq. 19 gives

$$\left.\frac{d\phi}{dx}\right|_x - \left.\frac{d\phi}{dx}\right|_o = -\frac{1}{\varepsilon_o}\int_o^x \rho(x)\,dx \quad . \tag{20}$$

But this integral through the electrode up to its surface is simply q_M. Thus, $d\phi/dx$ at the interface depends only on q_M, not its density distribution. Another result of this equation is that since $q_M = -(q_{CS} + q_E)$, integration over the entire interface region gives $(d\phi/dx)_\infty = (d\phi/dx)_{-\infty}$, since the full integral sums to zero. Thus, $\phi(x)$ emerges horizontal on the right side as required, with the difference between initial and final levels giving ΔV.

Defining $\phi(x) = 0$ at the electrode surface, using $\rho_{CS} = q_{CS}/d$, and integrating Eq. 19 twice, subject to $(d\phi/dx)_o = -q_M/\varepsilon_o$, we obtain

$$\phi(x)_{CS} = -\frac{1}{\varepsilon_o}\left[\frac{q_{CS}}{2d}x^2 + q_M x\right] \quad . \tag{21}$$

Then at $x = d$, the CS|E interface,

$$\phi(d) = -\frac{d}{\varepsilon_o}\left[\frac{q_{CS}}{2} + q_M\right] = \Delta\phi_{CS} \tag{22}$$

where $\Delta\phi_{CS} = \phi(d) - \phi(0)$. Since $d\phi/dx = 0$ at the electrolyte side of the E layer, we may evaluate $\Delta\phi_E$, the potential drop across the E layer, by right-to-left double integration

$$\Delta\phi_E = -\frac{q_E}{d\varepsilon_o}\int_o^d dx'\int_o^{x'} dx' = -\frac{q_E}{d\varepsilon_o}\left[\frac{1}{2}x'^2\right]_o^d = -\frac{q_E d}{2\varepsilon_o} \tag{23}$$

where x' is a variable of integration that increases positively to the left from $x'=0$ at the right side of the E layer. Then the total interface voltage drop, ΔV, is given by

$$\Delta V = \Delta\phi_{CS} - \Delta\phi_E = -\frac{d}{\varepsilon_o}\left[\frac{q_{CS}}{2} + q_M - \frac{q_E}{2}\right] \qquad (24)$$

whence, using the interface neutrality condition $q_E = -(q_M + q_{CS})$, one obtains

$$\Delta V = -\frac{d}{\varepsilon_o}\left[q_{CS} + \frac{3}{2}q_M\right] \quad . \qquad (25)$$

If we now ask what ratio of q_{CS} to q_M minimizes ΔV and hence the interfacial electrical free energy, it is apparent by inspection that it is $q_{CS} = -\frac{3}{2}q_M$.

When this ratio obtains exactly, we see that $\Delta V = 0$. Consider, however, the general situation. An externally applied voltage results in an electrode charge q_M. Current conservation delivers an equal and opposite ionic charge $(q_{CS} + q_E)$ into the adjacent electrolyte via the external circuit and a second electrode. We ask what distribution of this charge between the CS and E layers minimizes the total (chemical plus electrical) free energy, given the chemisorptive free energies $\Delta\mu^\circ_{C(ads)}$ and $\Delta\mu^\circ_{A(ads)}$ in the CS layer. We see that arbitrarily small ΔV values are possible when $q_M \neq 0$, despite q_{CS} values up to the equivalent of a full monolayer of charge. Equation 25, however, and the calculations from Eq. 18 show that q_{CS}/q_M ratios substantially different from -3/2 should give unrealistically large interface voltage drops and electrical free energy values for available chemisorption energies. The situation $q_{CS} \approx -\frac{3}{2}q_M$ would then seem to be a general one for systems such as in Eq. 8, where the electrical free energy is in balance with modest chemisorption and configurational free energies.

This result is of interest, since it now formalizes the "image charge" considerations mentioned earlier. We see that Eq. 14 now refers to $dq_{CS}/d\varepsilon$. But $dq_M/d\varepsilon$ is the experimentally measured capacitance, and its value is indicated to be 2/3 that predicted for $dq_{CS}/d\varepsilon$, as we have indeed observed. The "image charge" q_E is approximately equal to $-1/3\ q_{CS}$ and results from free energy minimization of the system.

Recapitulating, our calculations indicate that when a polarizable electrode such as Pt is placed in a fused or solid electrolyte such as AgBr, where both anions and cations tend to chemisorb, and the external voltage is set to give $q_M = 0$, the condition $N_C \approx N_A$ obtains in the CS layer for any reasonable value of $\Delta\mu^\circ_{A(ads)} - \Delta\mu^\circ_{C(ads)}$. For other voltages, resulting in $q_M > 0$ or $q_M < 0$, the voltage dependence of the measured interface capacitance agrees well with Eq. 14 for the case of AgBr|Pt, but better agreement with observed capacitance values is obtained by introducing that amount of "image charge" in the E layer that minimizes the interfacial free energy. In this, the use of Poisson's Equation with a simple interface model appears to be useful.

5. TEMPERATURE DEPENDENT EFFECTS

In measuring the electrode capacitance of solid electrolytes, it has been found in at least several instances that an interface showing a well defined, fast-charging capacitance at one temperature shows more poorly defined charging behavior at other temperatures (21,22,26), or indeed shows a poorly defined capacitance over the whole temperature range of investigation. In this section, we will discuss some of the responsible factors and methods for taking them into account.

We have noted already that purely electrostatic interface charging, whether dominated by diffuse or inner layer capacitance, should in principle be well defined and intrinsically rapid, involving the RC time constant of the cell. In the case of high frequency measurements or very fast RC constants, capacitance "dispersion" effects due to ionic atmosphere relaxation, finite defect pair equilibration rates, and shunting by geometric cell capacitance need also be considered. Such effects, however, are outside the scope of this article and have been adequately reviewed elsewhere (6,7,32,33). Intrinsically rapid charging (on the time scale of milliseconds or slower) is expected not only for diffuse and inner layer capacitance, but for any type of "parallel plate" capacitance where an effectively insulating surface layer (such as of permanently chemisorbed ions) acts as a "dielectric" between mobile electrolyte ions and the electrode. In all these cases, charging is presumed to be accomplished solely by mobile ions or ion defects in an undisturbed crystal structure.

If charging, however, involves a change in the interface composition that is too great to be accomplished solely by defect motions, the result may be a voltage-dependent "interface structure", such as we have described for competitive chemisorption. In this case, charging should involve the motion of more than one type of ion, and if a low mobility ion is involved, the charging process should be intrinsically slow. We have noted that this becomes increasingly the case for AgBr on Pt electrodes when cooled from the AgBr melting point.

To allow for such possibilities, it is highly useful to study a polarized interface charging process over at least several decades of frequency or time. It is felt that the best technique for doing this is potential-step chronocoulometry. In this, a small voltage step is applied to a polarization cell and the integrated charging current Q(t) may be directly displayed over many decades

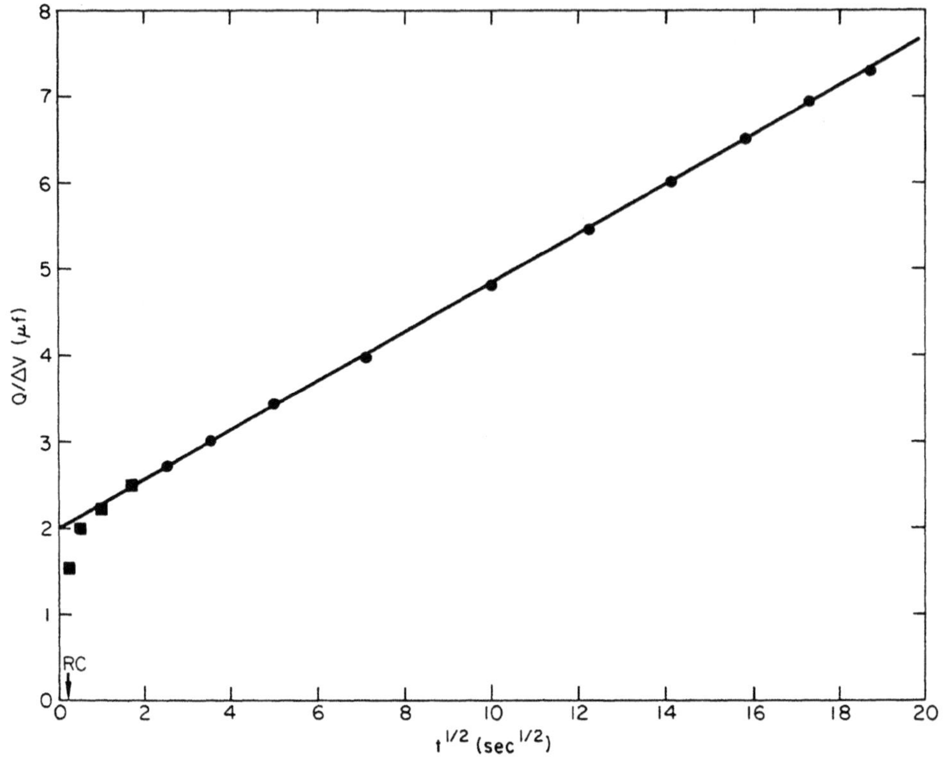

Fig. 12 Example of potential-step chronocoulometric curve (Q vs. $t^{1/2}$) with impurity diffusion current adding to interface capacitive charge.

of time by oscillographic and recorder techniques. This method was used to show the ideally rapid room temperature charging of the AgI|graphite pressed-powder interface and the ill-defined character of AgI|Pt charging under similar conditions (26a). Other types of capacitance dispersion measurements, such as a.c. impedance as a function of frequency, generally tend to give a less complete picture because of the relative inconvenience of measurements over a comparable time scale and the possible missing of "fine structure" (34) with point-by-point measurements.

A second temperature-dependent effect is the presence of an impurity diffusion current in the interface charging process. This has now been observed in at least several studies (22,26b). A potential-step chronocoulometric curve that rises to a horizontal Q(t) "plateau" at one temperature displays at a higher temperature a continuing rise with time on a Q vs. t plot or a quasi-exponential rise at long times on a Q vs. log t plot. The effect is poorly reproducible from cell to cell, increases strongly with temperature, and is often reduced by overnight "annealing" of the cell at temperature under a polarizing bias. In such cases, plots of Q vs. $t^{1/2}$ may be quite useful.

A "successful" (i.e., linear) plot of this type is shown in Fig. 12 for the AgI|C pressed-powder interface at 75°C (26b). The linearity of the plot for times larger than several RC allows a "fast-charging" capacitance component to be extracted from the data by extrapolation to $t = 0$. Again, the usefulness of the chronocoulometric method is seen. In the illustrated system, the temperature dependence of the Q vs. $t^{1/2}$ slope yielded a linear Arrhenius plot, giving evidence that a single diffusing impurity species was involved.

Finally, two significant effects of temperature that have been observed are permanent interface roughening by electrolyte grain growth and electrode-electrolyte interpenetration. AgBr|C pressed-powder interfaces that were observed to be microscopically smooth at room temperature developed irreversible progressive roughening on heating to elevated temperatures (26b). Grain growth in the polycrystalline electrolyte appeared to be responsible. Chronocoulometric capacitance measurements were used to determine interfacial areas that had increased up to 17-fold. At still higher temperatures, the solid AgBr "soaked" into fracture damage in the pressed graphite by capillary action. The resulting interface showed the typical capacitative charging behavior of a porous electrode. Because of the common use of graphite or carbon as a solid-electrolyte electrode material, it would be well to be aware of such possibilities, especially with nonrefractory electrolytes such as the silver, copper, and alkali halides.

6. REFERENCES

(1) "Techniques of Electrochemistry", E. Yeager and A. J. Salkind, eds., Wiley-Interscience, New York, 1972.

(2) See, for instance, R. Parsons, "Electrode Double Layer", in "Encyclopedia of Electrochemistry", C. A. Hampel, ed., Reinhold, New York, 1964.

(3) D. O. Raleigh, Phys. Stat. Sol. (a) 4, 215 (1971).

(4) D. C. Grahame, J. Am. Chem. Soc. 68, 301 (1946).

(5) G. Jaffe, Ann. Phys. 16, 217, 249 (1933); H. Chang and G. Jaffe, J. Chem. Phys. 20, 1071 (1952); G. Jaffe and C. Z. Lemay, ibid. 21, 920 (1953).

(6) J. R. Macdonald, Phys. Rev. 92, 4 (1953); J. Chem. Phys. 29, 1346 (1958); Trans. Farad. Soc. 66, 943 (1970); J. Chem. Phys. 54, 2026 (1971); ibid. 58, 4982 (1973); J. Electroanal. Chem. 53, 1 (1974).

(7) R. J. Friauf, J. Chem. Phys. 22, 1329 (1954).

(8) T. M. Proctor and P. M. Sutton, J. Chem. Phys. 30, 212 (1959).

(9) R. Meaudre and G. Mesnard, J. Phys. C7, 1271 (1974).
(10) S. Lanyi, J. Phys. Chem. Solids 36, 775 (1975).
(11) A. R. Allnatt and P.W.M. Jacobs, J. Phys. Chem. Solids 19, 281 (1961).
(12) P.W.M. Jacobs and J. N. Maycock, J. Chem. Phys. 39, 757 (1963).
(13) A. R. Allnatt, P.W.M. Jacobs, and J. N. Maycock, J. Chem. Phys. 43, 2526 (1965).
(14) J. H. Beaumont and P.W.M. Jacobs, J. Phys. Chem. Solids 28, 657 (1967).
(15) D. Miliotis and D. N. Yoon, J. Phys. Chem. Solids 30, 1241 (1969).
(16) D. F. Gibbs and B. W. Jones, J. Phys. C2, 1392 (1969).
(17) J. M. Wimmer and N. M. Tallan, J. Appl. Phys. 37, 3728 (1966).
(18) D. O. Raleigh, J. Phys. Chem. Solids 29, 261 (1968).
(19) D. O. Raleigh, J. Phys. Chem. 70, 689 (1966); 71, 1785 (1967).
(20) R. D. Armstrong et al., J. Electroanal. Chem. 27, 158 (1970); ibid. 48, 47 (1973).
(21) M. N. Hull and A. A. Pilla, J. Electrochem. Soc. 118, 72 (1971).
(22) D. O. Raleigh and H. R. Crowe, J. Electrochem. Soc. 118, 79 (1971).
(23) R. D. Armstrong and R. Mason, J. Electroanal. Chem. 41, 231 (1973).
(24) A. V. Joshi, p. 173, "Fast Ion Transport in Solids", W. van Gool, ed., North Holland, Amsterdam, 1973.
(25) J. Bert, J. L. Picot, and J. Dupuy, Phys. Stat. Sol. (a) 19, 119 (1973).
(26) D. O. Raleigh, J. Electrochem. Soc. 121, 632, 639 (1974).
(27) J. Goldman and J. B. Wagner, J. Electrochem. Soc. 121, 1318 (1974).
(28) W. J. Lorenz, H. D. Hermann, N. Wuthrich, and F. Hilbert, J. Electrochem. Soc. 121, 1167 (1974).
(29) W. Schultze and K. J. Vetter, J. Electroanal. Chem. 44, 63 (1973).
(30) An adsorption site mesh equal to two unit cell spacings of the substrate crystal structure in both the X and Y directions. For discussion of this nomenclature, see G. A. Somorjai, Catal. Revs. 7, 96 (1972).
(31) See, for instance, J. O'M. Bockris and A.K.N. Reddy, "Modern Electrochemistry", Vol. II, pp. 638-9; Plenum Press, New York, 1970.
(32) P. H. Sutter and A. S. Nowick, J. Appl. Phys. 34, 734 (1963).
(33) R. D. Armstrong and W. P. Race, J. Electroanal. Chem. 33, (1971).
(34) See, for instance, Reference 22, Fig. 11.

DISCUSSION

DICKINSON : I do not understand why you use the term chemisorption. In a solid electrolyte, chemisorption must surely involve either some degree of charge transfer between the ion and the electrode or some movement of the ion from its normal lattice site. Your model does not involve either of these phenomena, but you still describe the ions as being specifically adsorbed.

RALEIGH : I understand chemisorption to be adsorption with the formation of a chemical bond. The bond may be covalent, and may or may not have polar character, so could involve any degree of charge transfer. The first order theory described does not directly involve an assumed degree of charge transfer. The second-order calculations assume no charge transfer, but could be repeated for various degrees of charge transfer. Operationally I would use the term specific adsorption to describe any net ion-charge concentration on the electrode surface that is too large to be accounted for by the well-known expressions for electrostatic attraction. Because solid electrolytes involve no hydration sheath on the electrode, there should be no appreciable reduction in the distance of closest approach of an ion when a chemisorption force adds to the normal electrostatic force.

KROGER : I understand that with a chemical type interaction, one type of ion can assemble at the interface. But is this enrichment dependent on the applied field ?

RALEIGH : Yes, the chemisorbed ion experiences, with respect to the bulk, both a chemisorption energy and an electrostatic energy (so long as it maintains even a partial effective ionic charge). The electrostatic energy, $z_i F \phi_i$, favors the chemisorption of anions at anodic potentials, cations at cathodic potentials, and eventually overrides the chemisorptive energy.

GERISCHER : In case of chemisorption of ions, the pure electrostatic picture breaks down in explaining the differential capacities. While in electrolytes with solvation of the ions, there is a possibility that the ion approaches closer to the electrode surface under the influence of chemisorptive forces by leaving its solvation shell, it has been shown even there that in most s systems the chemisorbed ions are partially discharged (Vetter and Schultze). This means in my opinion that "chemisorbed ions" can better be described as chemisorbed atoms with a polar bond to the surface. The polarity of this chemisorptive bond varies with the voltage applied and this together with the concentration dependence of the chemisorption isotherm gives the very high values of the apparent capacity over a wide voltage range.

RALEIGH : I agree that the electrostatic model breaks down when ion chemisorption is involved and said this in my lecture. The combined effects of ion concentration and degree of charge transfer in the chemisorbed species suggests one model for accounting for high capacitances in these systems. In my lecture, I offer a somewhat different model which I believe explains other aspects of the interface capacitance more satisfactorily.

ARMSTRONG : For specific adsorption in the general case, we must have <u>both</u> partial charge transfer to some species and a degree of surface rearrangement. I think that it would be rather difficult to make a quantitative fit to experimental data without making some drastic simplifying assumptions (e.g. that the partial charge transfer is zero).

MACDONALD : Why should there not still be a series inner layer present in the case you discussed of chemisorption <u>without</u> appreciable partial charge transfer ?

RALEIGH : The series two-capacitor (diffuse, inner) model of electrochemical double layer charging is only a valid model for purely electrostatic interface charging. Chemisorptive capacitance involves a thermodynamically determined partitioning of charge carriers between the electrolyte bulk and electrode surface sites. Since bulk ions have free access to the electrode surface, irrespective of the character of any diffuse charge layer, this partitioning gives rise to an adsorption isotherm whose voltage derivative may be considered a capacitance in <u>parallel</u> with diffuse space charge capacitance.

ETSELL : In your second model ($q_M \neq 0$), do you feel it is reasonable to minimize ΔV since the system will, of course, minimize the total free energy ? It is quite improbable that these minima will coincide.

RALEIGH : I carried out the minimization of ΔV to show that the electrical free energy, which was quite large for any appreciable charge separations in the $q_M = 0$ case, can be brought to arbitrarily low values in the $q_M \neq 0$ by an appropriate partitioning of the ionic charge between q_{CS} and q_E. It is of course the total free energy that would be minimized. The ΔV minimization permits the electrical free energy to be brought into quality with modest chemical free energy values. In this situation, q_{CS}/q_M should have approximately the same value as when ΔV alone is minimized, because of the great sensitivity of the electrical free energy to this ratio.

SPACE CHARGE POLARISATION*

J. Ross Macdonald

Department of Physics and Astronomy
University of North Carolina
Chapel Hill, North Carolina 27514 U.S.A.

ABSTRACT

Space charge polarisation is first defined and those aspects of it which will and will not be considered at length are distinguished. Basic equations of charge motion in solids and liquids are discussed for the situation where a single mobility is an adequate approximation for each mobile species. Rather general boundary conditions at the interface between the material of interest and its electrodes are introduced and intrinsic-carrier recombination is considered. Theoretical results are first discussed for quasi-static differential capacitance as a function of applied dc potential, both without and with some account of specific adsorption. Small-signal ac results and analysis methods are then described. After consideration of Maxwell-Wagner interfacial polarisation, the remainder of the paper deals with detailed ac response results for a homogeneous material and shows how bulk, surface reaction, recombination, and diffusion effects may be differentiated and analyzed.

I. INTRODUCTION

Space-charge polarisation (SCP) is a widely encountered component of the electrical behavior of solids and liquids. In order to define it unambiguously, it is desirable to start with the more general phenomenon of electric polarisation.

---

*Preparation of this manuscript supported by the U.S. National Science Foundation (Grant No. DMR75-10739).

M. Kleitz and J. Dupuy (eds.), Electrode Processes in Solid State Ionics, 149–183.

Electric polarisation arises from charge separation. Two kinds of electric polarisation are readily distinguished. Crudely speaking, dielectric polarisation involves charge separation on an atomic or molecular scale, while SCP is associated with greater separations. Dielectric polarisation $\vec{P}$ is defined as the dipole moment per unit volume of a material. It may involve both permanent and induced dipoles or either separately. $\vec{P}$ is usually defined for a homogeneous material or at least for a homogeneous region of macroscopic dimensions of the material. For an isotropic or cubic medium exposed to an applied macroscopic electric field $\vec{E}$, $\vec{P}$ and $\vec{E}$ are parallel and one can introduce the scalar $\chi \equiv \vec{P}/\vec{E}$, a quantity known as the polarisability or dielectric susceptibility. In CGS-Gaussian units, which will be used herein, the dielectric constant associated with χ is $\varepsilon \equiv 1 + 4\pi\chi$. It is, properly, an intensive property of the material considered, the same for any region of greater than molecular dimensions and thus independent of sample volume.

One of Maxwell's equations may be expressed as $\vec{\nabla}\cdot\vec{D} = 4\pi\rho$, where $\vec{D}$ is the electric displacement, given by $(\vec{E} + 4\pi\vec{P})$, and ρ is the net unbalanced charge per unit volume, the local space charge. For a homogeneous material, $\vec{\nabla}\varepsilon = 0$, and one obtains the following form of Poisson's equation

$$\vec{\nabla}\cdot\vec{E} = (4\pi/\varepsilon)\rho \ . \tag{1}$$

Here all the dielectric polarisation has been subsumed in ε and the SCP manifested in ρ. Thus the bulk dielectric properties of the material are accounted for by the intensive quantity ε which will here be taken, as usual, independent of ρ and $\vec{E}$. The present work will, for simplicity, deal with a homogeneous, isotropic material, usually in a one-dimensional approximation. It will be assumed that all frequency dispersions associated with ε occur appreciably above all SCP frequencies of interest. This criterion requires that the lowest-frequency dielectric dispersion region occurs appreciably above the radial frequency, $\omega_D = \tau_D^{-1}$, associated with the dielectric relaxation time of the bulk material, τ_D.

One might be tempted to conclude from Eq. (1) that SCP only occurs where $E \neq 0$. But this is clearly an insufficient condition. Ordinary ohmic current flow, as in a metal, involves $E \neq 0$ but no significant charge separation. Only when $\vec{\nabla}\cdot\vec{E} \neq 0$ is there net charge separation and thus space charge. But $\vec{\nabla}\cdot\vec{E} \neq 0$ does not occur when $E \neq 0$ in the bulk region (no electrode effects) of a homogeneous isotropic material held at constant temperature throughout. It is generally the presence of inhomogeneities and/or phase changes within the material or at its surfaces which leads to $\rho \neq 0$. Here, I shall primarily be concerned only with homogeneous, isotropic bulk material at constant temperature, and thus with SCP associated with surface boundaries and electrode regions. Nevertheless, it will be shown that SCP which arises from surface region discontinuities can under certain circumstances extend throughout the entire region between electrodes of a material of interest. Further, the existence of SCP requires inhomogeneity or discontinuity of a special kind, since in principle ohmic electrodes attached to a homogeneous, isotropic material need generate no SCP even though there is a phase transition at the interface from experimental material to (different) electrode material. From another point of view, if an electrode is reversible (infinitely rapid electrode kinetics) for all current carriers, it will fix the thermodynamic potentials of these carriers[1] and no SCP will occur. Since most SCP of interest arises from boundary phase discontinuities, it is clear that detailed boundary conditions are of crucial importance in determining SCP and its concomitant effects.

The usual experimental arrangement for the investigation of the electrical behavior of solids involves a parallelepiped of material, often in the form of a thin layer, between two plane, parallel, frequently identical electrodes. The electrodes are usually metallic but may sometimes be liquid electrolytes involving one or more of any ionic species which are mobile in the solid. Just as in liquid electrolyte experiments, it is sometimes desirable that all processes of interest be concentrated in the material near one electrode, the working electrode. The boundary parameters (and/or shape) of the other electrode, the indifferent electrode, are then taken so that $\rho \simeq 0$ in its neighborhood and its influence on the electrical behavior of the system is usually negligible compared to that of the working electrode. Plane, parallel, equal size electrode arrangements are convenient for comparing experimental results with the predictions of theoretical models which approximate the situation by considering only one-dimensional current flow. All quantities herein which are proportional to electrode area, such as capacitance, will be given in specific form, per unit area, but will, nevertheless, usually be referred to as capacitance, conductance, etc.

Current-carrying entities in the electrodes and in the material under investigation may be of a variety of types: electrons and/or holes, positive and/or negative ions, and positive and/or negative defects such as cation vacancies. In addition, immobile charges may be present with which the mobile charge carriers may or may not recombine. In the present work, it will usually be assumed that the electrodes are electronic conductors and that the material being studied is essentially non-electronic, involving a single species of positive and a single species of negative ionic and/or defect charge carrier. Many of the theoretical results to be described are not limited to these conditions, however.

The question of detailed boundary conditions at an electrode is a delicate one still containing much terra incognita. Omitting the possibility of rectification, which will be done throughout this paper, one can distinguish a continuum of possible boundary effects for a given species of current carrier, ranging from complete blocking, through partial blocking, finally to zero blocking, ohmic, or open electrodes. Completely open or reversible electrodes generally involve an extremely large (infinite in the limit) heterogeneous rate constant for the charge species involved. Detailed discussion of electrode-contact characterization for various charge carrier conditions has been given by Kröger[2]. In the present work, the above continuum will be described by means of dimensionless boundary condition parameters, r_p and r_n, for positive and negative mobile charge species, respectively. These parameters may be related to thermally activated heterogeneous rate constants or to surface recombination rates. Although they may sometimes be complex and frequency-dependent in ac situations, such possibilities will not be further considered herein. When $r_p = 0$ for a given positive species at a given electrode, the contact is completely blocking for this species. At the other extreme when $r_p = \infty$, the contact is completely open. It is worth mentioning that complete blocking is an idealized, limiting condition, but one nevertheless that is well approximated in many experimental situations. It requires an infinite potential barrier between electrode and experimental material, which precludes mass transfer either over the barrier or by tunnelling through it. In this limit, the electrode and the material are not in thermodynamic equilibrium, and the electrochemical potential thus need not be the same in both.

Since the subject of SCP is an extremely broad one, it is necessary to limit its scope for the present discussion. First, no further consideration will be given to ordinary dielectric response, although electrical behavior arising from the presence of electric dipoles may often be confused with SCP behavior and vice versa, especially in blocking electrode situations and especially when a distribution of relaxation times is invoked to explain putative dielectric response behavior. Incidentally, it is worth mentioning that the expressions frequently used to describe a distribution of thermally activated relaxation times are inappropriate when applied over a range of temperatures and when the enthalpy involved in the thermally activated dipole behavior is itself distributed[3].

Second, the present paper will not deal with current injection into insulators leading to space-charge limited currents. One and two carrier injection situations have been discussed quite thoroughly in the past[4-6]. Although the present work does deal with positive and negative carrier injection through the boundary conditions discussed above, it will always involve situations (which may be semi-insulating) where there is a non-zero bulk-region concentration of positive and negative charge (with $\rho = 0$) in thermal equilibrium under unperturbed conditions. Such a condition is much more applicable to solid state ionics than is the one considered in the conventional space-charge limited current situation, which invariably considers only electron and/or hole injection, usually deals with bulk rather than electrode effects of these injected carriers, and generally ignores intrinsic carrier generation.

Further, no consideration of ordinary transient response will be included. Much useful information can be obtained from charging and discharging measurements at different temperatures[3]. When the system responds in a linear range, however, the same information can be obtained from steady-state frequency response measurements[7,8]. On the other hand, in the nonlinear response region this is no longer the case, and it is inappropriate to derive frequency response results from a Fourier or Laplace transform of transient response results.

Finally, no discussion will be given of the powerful technique of ionic thermoconductivity, also known as thermally stimulated depolarisation[9-11]. This procedure is particularly useful in observing the relaxation of the polarisation of dipoles and/or the release of mobile charge from traps in the material. Neither of these processes is included in the present models, but thermally stimulated depolarisation measurements should certainly be made in parallel with ordinary transient and frequency response measurements whenever possible in order to characterize the system investigated more fully.

There are still many SCP areas of interest not excluded by the above limitations, particularly static and steady state situations. The static, zero-current condition with time-independent electric field applied requires complete blocking of mobile carriers at at least one electrode. SCP can only be inferred under such static conditions by potential (or field) probe techniques. Alternating and direct steady-state non-zero current situations are of more interest because more can generally be learned from them. Note that a non-zero purely displacement current can flow across completely blocking contacts (if the capacitance of the blocking region or layer is non-zero) in response to an impressed alternating potential, since no mass flow between electrode and material is then involved. In steady-state ac and dc situations, any current-induced temperature variation in the material itself and between the bulk material and mobile carriers will be neglected, a particularly good approximation when the carriers are ionic rather than electronic.

The present work deals only with quasi-static and steady-state conditions. It applies in situations where charge motion may be adequately described by means of mobilities and diffusion coefficients. Thus, certain hopping/percolation processes are excluded[12,13]. Further, it is assumed that the Einstein relation, $D_i = (kT/ez_i)\mu_i$, holds between diffusion coefficient, D_i, and mobility, μ_i, of a charge carrier of valence number z_i. Here k is Boltzmann's constant, T is the absolute temperature, and e is the proton charge. The models to be described also assume that D_i and μ_i are independent of field, concentration, and position.

Two other idealizations, which like the above have usually been made in most past calculations, are inherent in the present approach. One is the neglect of all inertial effects of the charge carriers, effects which may tend to make them contribute a negative term to the effective dielectric constant of the material below their plasma frequencies[14]. The present models apply primarily, however, at frequencies so much lower than the plasma frequency that one would expect that other effects would dominate and the present approximation would be quite adequate. For reference, the plasma frequency of Ag^+ ions at a concentration of 10^{18} cm^{-3} is of the order of 45 MHz.

A more stringent approximation under certain conditions is the neglect of discreteness-of-charge effects[15,16], including the finite size of charge carriers such as ions[17]. This approximation may be expected to be weakest when the charges under consideration are completely blocked and when they may be specifically adsorbed at an electrode as well. This matter will be further considered in Section III.

Section II is concerned with the basic equations which describe the idealized situation defined by the above discussion. Then, in Section III the differential SCP capacitance of a system which may involve specific adsorption is considered as a function of dc bias. Section IV is concerned with ac response as a function of frequency in a variety of situations. First, certain aspects of the Maxwell-Wagner theory of interfacial polarisation are discussed; then, some data analysis methods are considered. Finally, Section IV-4 deals with the varieties of frequency response possible when a single species of mobile positive charge and a single species of mobile negative charge are present in a material, the mobilities and valences of the charges are arbitrary, and many different boundary conditions are possible. Because of the restriction to a single species of positive and a single species of negative charge, the situation considered corresponds to the unsupported electrolyte condition in liquid electrolytes, that which leads to maximum space charge.

II. BASIC EQUATIONS

The basic equations consistent with the discussion of the last section will be given here in one-dimensional form. It will usually be assumed that symmetrical plane, parallel electrodes are placed at $x = 0$ and $x = \ell$. The situation to be considered involves an arbitrary amount of dissociation of intrinsic neutral centers of concentration $n_c(x)$, which arises from an originally unperturbed, homogeneous, neutral bulk concentration N. It will also be assumed that there may be present a homogeneous concentration, $N_e \equiv N_D^+ - N_A^-$, of immobile charged impurity centers associated with extrinsic conduction. Here N_D^+ and N_A^- are the individual concentrations of charged donor and acceptor centers. It will be assumed for simplicity that any impurity centers present are fully ionized at all temperatures of interest. Thus, although the treatment may include generation-recombination of intrinsic centers, it takes all extrinsic centers always fully charged. This is equivalent to assuming shallow donor and/or acceptor energy levels and not too low temperatures. When $N_e = 0$, one deals with intrinsic rather than extrinsic conduction.

Let z_p and z_n be the valence numbers of positive and negative charge species of concentrations p and n and mobilities μ_p and μ_n. The current equation is

$$I = I_p + I_n + (\varepsilon/4\pi)\frac{\partial E}{\partial t} , \qquad (2)$$

where

$$I_p \equiv ez_pj_p , \qquad I_n \equiv -ez_nj_n , \qquad (3)$$

and

$$j_p \equiv (\mu_p pE - D_p \frac{\partial p}{\partial x}) , \qquad j_n \equiv -(\mu_n nE + D_n \frac{\partial n}{\partial x}) . \qquad (4)$$

Poisson's equation becomes

$$\frac{\partial E}{\partial x} = (4\pi/\varepsilon)\rho , \qquad (5)$$

where

$$\rho \equiv e(z_p p - z_n n + N_e) . \qquad (6)$$

The material will be taken electrically neutral ($\rho = 0$) in its unperturbed bulk region under thermal equilibrium conditions.

Since the field E is related to the electrostatic potential ψ by $E = -(d\psi/dx)$, the applied voltage across the electrodes is given by

$$V_a = -\int_0^{\ell} E(x)dx \ . \tag{7}$$

The continuity equations[18] may be written in the following generalized form[19]

$$\frac{\partial p}{\partial t} = -z_{pL}\frac{\partial n_c}{\partial t} - \frac{\partial j_p}{\partial x} \ , \tag{8}$$

$$\frac{\partial n}{\partial t} = -z_{nL}\frac{\partial n_c}{\partial t} - \frac{\partial j_n}{\partial x} \ , \tag{9}$$

and

$$\frac{\partial n_c}{\partial t} = -k_1 n_c + k_2 (n^{z_{nL}})(p^{z_{pL}}) \ . \tag{10}$$

In the above, k_1 is a temperature-dependent generation coefficient and k_2 is a recombination coefficient. For generality, it has been assumed that an intrinsic neutral center dissociates into z_{pL} positive charges each of valence number z_p and z_{nL} negative charges of valence number z_n. The original neutrality of the center then requires that $z_{pL}z_p = z_{nL}z_n$. In all the work to be discussed here, whenever intrinsic generation-recombination is considered, z_n and z_p will be taken equal and $z_{nL} = z_{pL}$ taken unity in order to avoid the need to consider m-body mass-action recombination with $m > 2$.

Since one of Maxwell's equations leads directly to the condition $\vec{\nabla}\cdot\vec{I} = 0$, the total current (including both conduction and displacement terms) is space invariant at all times even in dynamic situations. This is often a useful condition in the one-dimensional case since it allows one to write

$$I = \langle I \rangle \equiv \ell^{-1}\int_0^{\ell} I(x)dx \tag{11}$$

for the current path $x = 0$ to $x = \ell$. The averaged form of I is often simpler to use in obtaining final results than the original form. Equations (2)-(11), the Einstein relation between D_i and μ_i, and appropriate boundary conditions define the idealized situation considered. Note that no mobile charged aggregates have been included. Any immobile charged aggregates may be included in N_e. Some conditions which need to be met for the diffusion terms in Eqs. (4) to be appropriate have been given elsewhere for electronic charge carriers[20]. It should also be noted that models of the present genre are usually limited to conditions which ensure that maximum mobile charge concentrations are small compared to the concentration of intrinsic atomic or ionic sites in the material.

The boundary conditions considered herein are generalized forms of those suggested by Chang and Jaffé[21] which involve the dimensionless r_p and r_n parameters already mentioned. Using subscripts "L" and "R" to denote left and right electrodes, one can write these conditions for the conduction current components I_p and I_n as

$$I_{pL} = -r_{pL}(ez_pD_p/\ell)[p_L - p_{eL}] , \tag{12}$$

$$I_{nL} = r_{nL}(ez_nD_n/\ell)[n_L - n_{eL}] , \tag{13}$$

$$I_{pR} = r_{pR}(ez_pD_p/\ell)[p_R - p_{eR}] , \tag{14}$$

and

$$I_{nR} = -r_{nR}(ez_nD_n/\ell)[n_R - n_{eR}] . \tag{15}$$

In the above, the subscript "c" denotes thermal equilibrium. Thus, the boundary conditions are linear in the boundary concentration perturbations from the zero-current equilibrium situation. Such linearity is necessary for small perturbations around equilibrium but cannot necessarily be expected to hold accurately for large perturbations. Note, however, that a boundary concentration perturbation must be zero when the corresponding r_i ($i = n$ or p) is infinite. Also when $r_i = 0$, the current is zero even for large perturbations. Thus, the above expressions may be expected to apply more broadly near these limits.

Let us now consider that the applied potential difference V_a is made up of a possibly large static part V_{ao} and a small perturbation part δV_a, so that $V_a = V_{ao} + \delta V_a$. Then the current, field, and concentrations will also each divide into a static part and a perturbed part. But since the basic equations governing the system are nonlinear, it will not generally be possible to write I, and the other quantities, in the separated form $I = I_o(V_{ao}) + \delta I(\delta V_a)$ with a linear dependence of δI on δV_a. If δV_a is sufficiently small [generally $\ll (kT/e)$], however, the system may be linearized and such separation justified.

The static and differential capacitances of the system are of particular interest and relevance for completely blocking electrodes. If the total amount of stored SCP charge at $V_a = V_{ao}$ is designated q_o, then the static or integral capacitance is $C_s \equiv q_o/V_{ao}$. Similarly, if at a given $V_a = V_{ao}$ an additional small perturbation δV_a is applied, the differential capacitance is given by $C_d \equiv (dq/dV_a) = [d(V_a C_s)/dV_a \cong \delta q/\delta V_a$, where δq is the change in q arising from δV_a. The differential capacitance is usually of more interest than is C_s.

The small-signal ac response of the system may be calculated from the above equations around the bias point V_{ao} by taking, for example, $\delta V_a = V_1 \sin(\omega t)$, where V_1 is the amplitude of the sinusoidal perturbation of radial frequency ω. When $V_1 << (kT/e)$, all higher-order harmonics in the Fourier expansion of quantities such as I, E, n, and p may be neglected compared to the fundamental, and the admittance $Y(V_{ao}) \equiv \delta I/\delta V_a$ is a well-defined quantity which may be calculated from the above equations. The explicit separation of the equations into a nonlinear set depending only on V_{ao} and a linearized ac perturbation set dependent on $\delta V_a = V_1 \exp(i\omega t)$ is necessary to solve the equations in the most expeditious way[22]. First the nonlinear dc equations must be solved, then the resulting steady-state inhomogeneous fields, concentrations, etc. must be used in the perturbation set of equations to obtain the final ac current or admittance. In general, this sort of sequential calculation must be done numerically by computer simulation. The equations can, however, be solved exactly for small-signal ac response when $V_{ao} = 0$ and the equilibrium state of the system is one with $E_o(x) = 0$ for all x[22]. Results of such a solution for completely symmetrical electrodes will be discussed in Section IV-4.

There is one further complication which needs consideration. It causes $E_o(x)$ not to be zero near surfaces even in thermal equilibrium. Although an isolated ionic single crystal such as AgCl may be taken macroscopically neutral, Frenkel[23] pointed out long ago that in thermal equilibrium the surface of the crystal should be charged and this charge balanced by a diffuse space charge of opposite sign adjacent to the surface. In this instance, the above non-ideal surface dipole arises from any difference in the energy required to form positive and negative defects in the crystal[23-26]. Even the presence of an essentially completely blocking electrode adjacent to the surface may be expected, however, to modify the surface-region dipole through imaging effects. The presence of a partially blocking electrode, which requires in equilibrium that the electrochemical potential of any charge carrier crossing the interface be the same in the electrode and in the material next to it, may be expected to affect the surface dipole region even further. More discussion of this phenomenon, which may be characterized by a diffusion potential V_D which determines the strength of the dipole, has been given elsewhere[26].

Several interesting cases may be distinguished. With complete blocking of both positive and negative mobile charge carriers in the material, i.e. $(r_p, r_n) = (0,0)$, the surface charges may be considered to be specifically adsorbed according to an electrical adsorption isotherm which generally depends on both V_D and the applied p.d., V_a. If the surface charges are not completely blocked, then it seems sensible to conclude that there are no specifically adsorbed surface charges, and charge on the electrode alone balances that in the diffuse layer. Finally, if the diffuse layer charge is not blocked and that on the surface is, one would expect that in general any diffuse layer charge would be balanced both by specifically adsorbed charge and by charge on the electrode. The complete-blocking, $V_D \neq 0$ situation will be discussed further in the next section.

III. COMPLETE BLOCKING: QUASI-EQUILIBRIUM

Both positive and negative mobile species are taken to be completely blocked in the work considered in this section. Two situations have conventionally been examined: a single blocking electrode at $x = 0$ with the material of interest extending to $x = \infty$ in the right half space, and alternatively, two blocking electrodes at $x = 0$ and at $x = \ell$. A good experimental approximation to the first of these arrangements is to place a blocking working electrode at $x = 0$ and an indifferent or ohmic electrode at $x = \ell$ and ensure that ℓ is large compared to the bulk Debye length. Theoretical and, where possible, experimental curves of potential, field, and charge concentrations _versus_ x are instructive in such situations. Representative results are given in Refs. 20, 27-34, not all applying just to complete blocking. References 28, 29, and 31 include the substantial effect of intrinsic recombination on the curve shapes when one species of carrier is immobile. Space limitations preclude detailed discussion of such shapes here. A brief summary of some of the theoretical work through 1963 on static space charge distributions has been given elsewhere[35].

It is worthwhile to summarize expressions for diffuse-layer differential capacitance, C_d, in various blocking situations. For simplicity, all C_d results will be given only for the $z_n = z_p = 1$ case. In the present completely blocking case, C_d can be measured quasi-statically as $(\delta q/\delta V_a)$ or, when the system is linear, as the low-frequency limiting value of parallel ac capacitance. For a single blocking electrode with $V_D = 0$ and no specific adsorption, the intrinsic-conduction diffuse double layer differential capacitance is then given by[27]

$$C_d = (\varepsilon/4\pi L_D)\cosh(V_a^*/2) \ , \tag{16}$$

where $V_a^* \equiv V_a/(kT/e)$ and V_a is the applied potential difference. When both carriers are mobile, the appropriate Debye length reduces to

$$L_D = L_{D2} \equiv [\varepsilon kT/4\pi e^2(z_n{}^2 n_i + z_p{}^2 p_i)]^{1/2}$$
$$= [\varepsilon kT/8\pi e^2 c_i]^{1/2} \ , \tag{17}$$

where n_i and p_i are the bulk intrinsic concentrations, satisfying the electroneutrality condition $z_n n_i = z_p p_i$; in the present $z_n = z_p = 1$ case, $n_i = p_i \equiv c_i$. When say only the negative carriers are mobile,

$$L_D = L_{D1} \equiv [\varepsilon kT/4\pi e^2 c_i]^{1/2} \ . \tag{18}$$

Eq. (16) and subsequent C_d expressions hold adequately only as long as the diffuse layer charge carrier concentrations remain small compared to the concentration of intrinsic entities (atoms or ions) which make up the material. Grimley[36,37] has given a $V_D = 0$ treatment, applied specifically to Schottky defect carriers, where this restriction is relaxed. Nevertheless, Eq. (16), which includes no finite-size corrections, has usually been found adequate for liquid electrolyte experimental results[17].

The situation is somewhat more complicated in the two-electrode case. Assume again that $V_D = 0$ and there is no specific adsorption, so that $E = 0$ throughout the material before the application of V_a across the electrodes. Then[27]

$$C_d = (\varepsilon/8\pi L_{D2})\,\mathrm{ctnh}(M)\cosh(V_a^*/4) \ , \tag{19}$$

where

$$M \equiv \ell/2L_{D2} \ . \tag{20}$$

The two factors of two present here and not in Eq. (16) arise from the equal division of V_a across the two interphase regions and from the series combination of equal diffuse layer capacitance for these regions to give the total capacitance. Note that when $V_a \to 0$ and $(\ell/L_D) \to 0$, Eq. (19) reduces to $\varepsilon/4\pi\ell$, just the geometrical capacitance C_g of the material. Thus, the purely SCP part of C_d is

$$C_{SCP} = C_d - C_g = C_g[(M)\mathrm{ctnh}(M)\cosh(V_a^*/4) - 1] . \quad (21)$$

Again when $V_a \to 0$, it is clear that for $M \gtrsim 3$,

$$C_d \cong MC_g = \varepsilon/8\pi L_{D2} \equiv C_o , \quad (22)$$

just the ordinary unbiased capacitance of two diffuse double layers in series. This capacitance is intensive as far as ℓ-dependence is concerned, in accord with its approximate localization near the electrodes.

For either one or two electrodes, the situation is more complicated when one species of charge is immobile but bimolecular recombination between positive and negative charges can occur[28,29]. Note that it is an idealization to speak of applying V_a across completely blocking electrodes because of possible potential drops between the electrodes and the material studied. One can place the material to be investigated in an electric field, i.e., between capacitor plates, and calculate the applicable V_a from integration of the actual resulting field in the material from $x = 0$ to ℓ as in Eq. (7). Note that when $V_D = 0$, V_a must be zero when the electrode charge, q_m, is zero. The system may often be realized experimentally by placing very thin insulating layers between metallic electrodes and the material. If the capacitances of these layers are much greater than C_d for the material, then the V_a (across the material) which should be used in Eq. (19) will be well approximated by the actual p.d. applied across the electrodes. The case where the potential drop across such an insulating layer is not negligible has also been considered[28,38].

Let us now consider the possible presence of specifically adsorbed surface charge, q_s, with V_D not necessarily zero. Denote the total, integrated diffuse layer charge by q_d. For a single blocking electrode many Debye lengths away from any other electrode ($M \gg 1$), the relation $q_m + q_s + q_d = 0$ must hold. The differential capacitance may be calculated from $C_d \equiv dq_m/dV_a$ when the dependences of q_s and q_d on V_a are known. Although q_d should most properly be given by Grimley's expression[36] modified to include V_D and arbitrary valences, for illustrative simplicity I shall here use the conventional $z_n = z_p = 1$ diffuse layer result[26,27]

$$q_d = -A_1 \sinh[0.5(V_D^* + V_a^*)] , \quad (23)$$

where

$$A_1 \equiv (\varepsilon kT/2\pi e L_D) \equiv (4kT/e)C_o , \quad (24)$$

and $V_D^* \equiv V_D/(kT/e)$.

One next requires electrical adsorption isotherms for the individual components of $q_s \equiv q_s^+ + q_s^-$. For simplicity in writing equations, let us alternatively denote all positive-related (p and +) quantities with a subscript "1" and negative ones with a subscript "2". Assume that there are $N_s^+ \equiv N_{s1}$ and $N_s^- \equiv N_{s2}$ surface adsorption sites occupied by positive and negative charges, respectively. Then, $q_{sj} \equiv (-1)^{1+j}ez_jN_{sj}$, with j = 1, 2. Further assume that there is a fixed maximum number of adsorption sites for charge of each sign, N_{sm1} and N_{sm2}. Then the maximum surface charge components are $q_{smj} = (-1)^{1+j}ez_jN_{smj}$. Under these conditions the Langmuir isotherm is an appropriate choice[26] except that it takes no account of planar interaction between adsorbed entities, important when $N_s \equiv N_{s1} + N_{s2}$ begins to become an appreciable fraction of N_{sm1} or N_{sm2}. Nevertheless, on using the Langmuir isotherm for illustrative purposes, one may write[26]

$$(N_s^{\pm}/N_{sm}^{\pm}) = [1 + \exp(\Delta G^{\pm}/kT)]^{-1} , \tag{25}$$

where the ΔG's are electrochemical free energies of adsorption from the bulk, referenced to $(N_s^{\pm}/N_{sm}^{\pm}) = 0.5$ at $\Delta G^{\pm} = 0$.

Following conventional practice, one may separate each $\Delta G^{\pm}$ into a chemical part independent of applied potential, $\Delta G_o^{\pm}$, and a potential dependent part, $ez_p\lambda^+V_a$ or $-ez_n\lambda^-V_a$. The $\lambda^{\pm}$ quantities introduced here are usually of the order of unity or less and are included to account at least approximately for discreteness of charge effects[15,16]. Thus, $\lambda^{\pm}V_a$ are micropotentials rather than the macropotential, V_a. Separation of $\Delta G^{\pm}$ yields $\Delta G_j^* \equiv (\Delta G_j/kT) \equiv \Delta G_{oj}^* - (-1)^j(z_j\lambda_jV_a^*)$. Finally, let $\psi_o^* \equiv 0.5(V_D^* + V_a^*)$. Then the foregoing equations lead for $z_n = z_p = 1$ to

$$C_d = (2C_o)\cosh(\psi_o^*) + (e/4kT)\sum_{j=1}^{2}(-1)^{1+j}(\lambda_jq_{smj})\operatorname{sech}^2(\Delta G_j^*/2) . \tag{26}$$

A term equivalent, for $\lambda^{\pm} = 1$, to the first of the sech^2 expressions above has been cited by Pleskov[39] for the differential capacitance of electron surface states on a semiconductor surface. Note that ρ_j, defined as $e\lambda_j|q_{smj}|/8kTC_o \equiv (\lambda_j/2A_1)|q_{smj}|$, will usually be much larger than unity.

Raleigh, in unpublished work[40], has considered a simpler version of the present case, that which follows when the diffuse layer contribution is ignored, $N_{sm}^+ = N_{sm}^- \equiv N_{sm}$, $\Delta G_o^{\pm} = 0$, and

$\lambda\pm = 1$. Thus, when $V_a = 0$, $N_s^+ = N_s^- = 0.5N_{sm}$, a very special condition. Under these circumstances Eq. (26), rewritten for arbitrary z_n and z_p, reduces to just

$$C_d = (2\rho C_o)\operatorname{sech}^2(V_a^*/2) \ . \tag{27}$$

where $\rho \equiv (eN_{sm})(z_n + z_p)/2A_1$. The maximum value of the symmetrical C_d curve, $2\rho C_o = (e/4kT)[(eN_{sm})(z_n + z_p)]$, will generally be much larger than $2C_o$. Raleigh adjusts $(z_n + z_p)$ somewhat arbitrarily to yield agreement between $2\rho C_o$ and the peak of an experimental AgBr curve for C_d taken near the melting point of AgBr for a single blocking electrode situation[41,42]. He then finds good agreement between theory and experiment near the peak. Better agreement over a wider potential range could probably be achieved with inclusion of diffuse layer effects and with $\Delta G_o^{\pm}$ values non-zero and unequal. Raleigh takes the negative surface charges as Br^-; it seems more likely that silver ion vacancies, which have appreciably higher mobility, would predominate.

When $\Delta G_o^+ = \Delta G_o^-$, there is no equilibrium Frenkel space-charge double layer; instead, positive and negative adsorbed charges are of equal concentration when $V_a = 0$. It is thus clear that when $V_D \neq 0$, eV_D must be a function of $(\Delta G_o^+ - \Delta G_o^-)$. Thus, the V_D which appears in ψ_o^* is not independent of the values of $\Delta G_o^{\pm}$. This matter has been considered in some detail previously[26] for the case of $(\Delta G_o^{\pm}/kT) >> 1$, which allows the Langmuir/Fermi distributions of Eq. (25) to be reduced to Maxwell-Boltzmann distributions. An explicit linear relation between $(\Delta G_o^+ - \Delta G_o^-)$ and V_D and $\ell n|q_{sm}^+/q_{sm}^-|$ was found and used to calculate C_d curves for a two-blocking-electrode situation. Some of the results derived, particularly those for $\lambda^{\pm} \sim 0$, were qualitatively similar to the Raleigh AgBr experimental results, which show a central peak, unequal minima, and final rises. But in this two-electrode situation, most of this sort of behavior arose from unequal splitting for $V_D \neq 0$ of V_a between the two electrode regions, even for the $M >> 1$ case. The results were compared to an earlier, less accurate calculation[43] which erroneously assumed splitting of V_a into equal parts. Conditions are depicted very diagrammatically in Fig. 1 for the $V_D > 0$ case. There "e" and "s" denote electrode and surface regions. As V_a increases, one diffuse layer grows and the other decreases until finally the signs of the left-region diffuse charge and the right-region surface charge change, as shown in Fig. 1-c.

Even if Eq. (26), with appropriate values of $\Delta G_o^{\pm}$, $\lambda^{\pm}$, $N_{sm}^{\pm}$, and V_D, could fit Raleigh's AgBr data excellently, there remains an important stumbling block arising from the finite size of specifically adsorbed ions. Figure 1-d shows a crude picture of some ions specifically adsorbed on a metal electrode, say

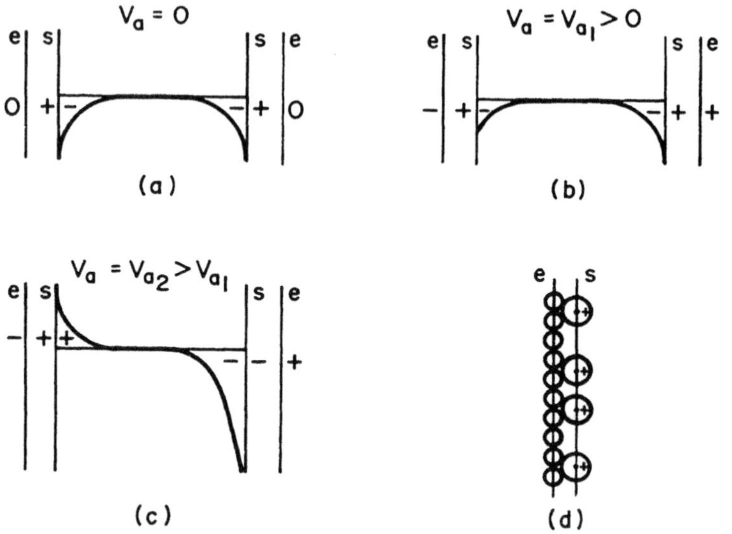

Fig. 1. (a) - (c). Signs of charge concentrations at electrodes (e), surfaces (s), and in diffuse layers for $V_D > 0$. (d). Specifically adsorbed charge at electrode surface.

platinum. The charge centroids of such ions are located at the inner Helmholtz plane[15], and the ion material between the center of charge of an ion and charges in the metal should have a low "dielectric constant," insofar as this concept is meaningful. One may expect an inner layer capacitance to be present associated with the charge-free region between the electrode and the inner Helmholtz plane. The approximate plane-parallel capacitor formed when charges in the metal and the adsorbed layer are smeared in their planes cannot be expected[44] to have an effective capacitance even as large as 100 $\mu F/cm^2$. But Raleigh finds a maximum capacitance of nearly 500 $\mu F/cm^2$ at $V_a = 0$. One would expect that the considerably smaller inner-region capacitance would be in series with any adsorption and diffuse layer capacitances and would dominate the combination. The solution to the problem of the missing inner layer capacitance is still unclear. It is possible, however, that it is essentially shorted out by partial charge transfer between the electrode and specifically adsorbed charges.

IV. STEADY-STATE AC RESULTS

I. Introduction

The most general situation of interest for a homogeneous material with a single species of mobile positive charge and a single species of mobile negative charge would involve two electrodes with different boundary conditions (r_p, r_n, and V_D/adsorption all different for the two electrodes), and extrinsic-intrinsic conduction with trapping and recombination. The applied p.d., V_a, would, in general, be made up of a static part V_{ao} and an alternating part V_{a1}, not necessarily small. For simplicity, and because this general situation has not yet been fully worked out, results for much simpler conditions will be discussed here.

Consider a homogeneous, isotropic material between two identical electrodes without V_D/adsorption. Take $V_{ao} = 0$ and consider that $(V_{a1})_{max} << kT/e$. Finally, only results for intrinsic conduction without trapping will be discussed, although a theory for extrinsic-intrinsic conduction without trapping has been given[22]. Most of the results following from the equations of Section II will be for zero recombination (full dissociation) of intrinsic carriers, i.e. $k_1 = k_2 = 0$, but some findings of current calculations which include intrinsic generation-recombination will be mentioned at the end of this section.

Before discussing detailed ac SCP results for a homogeneous material, a simplified SCP approach for inhomogeneous conditions will be examined in the next subsection and methods of data analysis examined in the following one.

2. Interfacial Polarisation

As its name implies, interfacial polarisation (IP) arises from mobile charges piling up at interfaces in a material. For example, the N-layer Maxwell-Wagner (MW) capacitor[45,46] consists of a series of layers, each of specified dielectric constant and conductivity, lying parallel to plane, parallel electrodes. Alternatively, one or more species of conducting particles of specific shapes may be considered to be dispersed in a carrier medium with different electrical properties[45-47].

Interfacial polarisation theory is simplified because no detailed account is included in it of mobilities, valences, recombination, or specific conditions at interfaces; instead everything is subsumed in the dielectric constant and conductivity assigned to each layer. More properly, the equations of Section II should be solved for each region with appropriate boundary conditions at each interface. Since this is usually impractical, it is interesting to compare results of the lumped-constant IP approach with those following from the detailed solution of the transport-continuity equations even for a homogeneous material.

The usual MW N-layer IP equivalent circuit made up of frequency-independent lumped-constant elements is shown at the

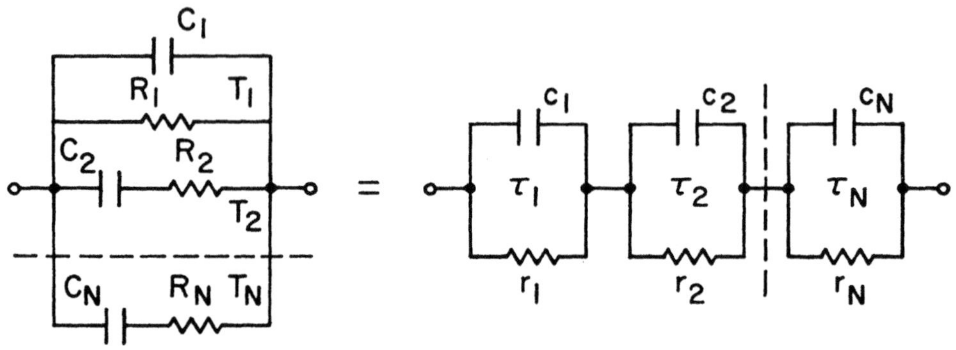

Fig. 2. Two equivalent circuits for interfacial polarisation.

right in Fig. 2. It is known as the Voigt model when used to describe mechanical viscoelasticity effects (with reinterpretation of the r's and c's) and describes a retardation situation[8,48]. Each parallel circuit represents the properties of a single layer. An alternate circuit, the Maxwell model, is shown at the left of Fig. 2. It conventionally describes a relaxation situation and is the usual model applied to describe a discrete or continuous ($N = \infty$) distribution of relaxation times for dielectric materials. As shown, these two circuits can have identical impedances at all frequencies if the relations between their elements are correctly chosen. It is possible to calculate the elements on the left directly from those on the right but the reverse solution cannot be carried out (for $N > 1$) explicitly and exactly. The following relations are applicable, however: $\prod_{\ell=1}^{N} T_\ell = \prod_{\ell=1}^{N} \tau_\ell$ and $R_1 = \sum_{\ell=1}^{N} r_\ell$, where the N time constants T_ℓ and τ_ℓ are defined in the figure. Note that C_1 in the left circuit should be identified with C_g, the geometric capacitance, and R_1 is the dc resistance of the system; it may sometimes be taken infinite to good approximation.

Maxwell-Wagner IP has frequently been invoked to explain observations of very large low-frequency apparent or effective dielectric constants (e.g., 49,50). It is often implied or stated that the high (apparent) dielectric constant is a true property of the material. Such is not the case; the error comes from calculating and using a properly intensive quantity, dielectric constant, in a situation where it is inapplicable. The problem is illustrated by the following simple example. Consider a two-layer MW situation with equal true dielectric constants, ε_o[4,6]. Assume that the layer thicknesses are d_1 and d_2, with $d = d_1 + d_2$ and $d_1 << d_2$, and take the resistivity of layer one much larger than that of layer two. Now if the effective low-frequency-limiting dielectric constant of the combination is calculated, one obtains $\varepsilon_{eff} \cong (d_2/d_1)\varepsilon_o >> \varepsilon_o$. But clearly this is not a true dielectric constant; it has been arrived at using d rather than the proper d_1 (<<d), and the resulting ε_{eff} is not an intensive property. In general, whenever a material is expected to be inhomogeneous and/or whenever space charge can occur near boundaries and electrodes it is wise initially not to calculate properly intensive quantities such as dielectric constant and conductivity, even with the appellation "apparent" or "effective," but instead to deal only with actual capacitances and conductances or specific (unit area) values of such quantities. Once the dependencies of frequency-independent circuit elements on electrode separation have been found, it may be possible to interpret some of them in terms of intensive bulk properties. It will certainly not be possible to do so, however, for all elements arising from interface-related processes.

Another way an incorrect or misleading large effective dielectric constant or capacitance can be obtained from even an N = 2 Voigt-model circuit is if experimental impedance data associated with such a circuit are interpreted in terms of a series capacitance and series resistance: $Z_T \equiv R_S + (i\omega C_S)^{-1}$, where Z_T is the total impedance. The N = 2 circuit leads to a frequency dependence of C_S of the following form: As ω decreases, C_S increases as ω^{-2} from a high-frequency value of c_1 (assuming $c_1 << c_2$), saturates over an appreciable frequency interval at $(c_1 + c_2)$, and then finally increases indefinitely again proportional to ω^{-2}. This rise, of course, should not be interpreted in terms of a dielectric constant increase; it appears because of inappropriate analysis of the circuit. For example, the total parallel admittance, $Y_T \equiv Z_T^{-1} \equiv G_P + i\omega C_P$, leads to a C_P which properly saturates at $(c_1 + c_2)$ as $\omega \to 0$.

Figure 3 shows two circuits which are electrically equivalent and arise from the SCP theory[22] to be discussed in IV-4. Here the dc discharge conductance is G_D, $G_\infty \equiv R_\infty^{-1} \equiv G_E + G_D$, and $G_{DN} \equiv G_D/G_\infty$, $G_{EN} \equiv G_E/G_\infty$. Note that the elements of $Y_i \equiv G_{iP} + i\omega C_{iP}$ and of $Z_i \equiv Y_i^{-1} \equiv R_i + (i\omega C_i)^{-1}$ are all frequency dependent in general. There are, however, two important conditions for which the circuit of Fig. 3 can be well approximated as an N = 2 type of circuit over an appreciable frequency range. Define Case A as that where in the range of interest C_i is essentially frequency independent and either R_i is as well, Case A_1, or, the usual case, A_2, $R_i << R_E \equiv G_E^{-1}$. Case B is that where over a given frequency range the frequency dependencies of the series-connected elements R_i and C_i are just those which lead to frequency independent values of the parallel elements, G_{iP} and C_{iP}. There are two Case B situations where the N = 2 circuit is well approximated. The first, of lesser interest, Case B_1,

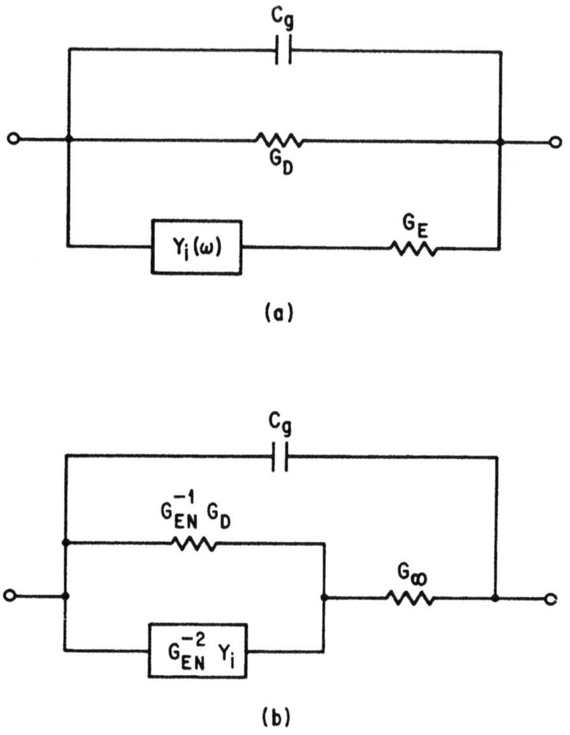

Fig. 3. Two exact equivalent circuits following from detailed theory[22] of charge motion in solids and liquids.

occurs when G_{iP} may be neglected, i.e. $\omega C_{iP} \gg G_{iP}$; it is then essentially equivalent to Case A_2. Case B_2, on the other hand, occurs when G_D may be neglected: $G_{iP}^{-1} + G_E^{-1} \ll G_D^{-1} \equiv R_D$. Then the circuit becomes like that of Fig. 3-b but with $G_{EN}^{-1}G_D$ replaced by G_{iP}, $G_{EN}^{-2}Y_i$ by C_{iP}, and R_∞ by R_E, essentially equal to R_∞ for this high dc resistance case. It can be shown[51] that such a circuit corresponds exactly to an N = 2 arrangement of the kind shown at the right of Fig. 2.

3. Data Analysis

Since even results for a homogeneous material with electrode effects and/or recombination can sometimes lead to a simple N = 2 parallel-series circuit, it is of interest to discuss how data associated with such a circuit might best be analyzed. The initial aim of analysis should be to determine the frequency-independent values, if any, of the capacitative and conductive elements of the equivalent circuit. Then, investigation of the dependence of these elements on temperature and electrode spacing, together with any other information known about the system, should help one decide to what degree the results arise from a MW IP or a homogeneous SCP system. Note that discussion of a $N = 2$ circuit is sufficient since $N = 2$ methods may be applied sequentially to pairs of a system exhibiting $N > 2$ circuit behavior.

In addition to an examination of the frequency response of series and parallel components of the circuit impedance/admittance, another approach may be particularly valuable. This analysis method involves plotting the real versus imaginary parts of some such complex quantity as admittance[52,53] or impedance[54-56] as parametric functions of frequency. Such Argand or "circle diagrams" have been used for many years in electrical engineering; when complex dielectric constant is the quantity considered, they are known as Cole-Cole plots[57].

There are four complex quantities worth investigating. Consider normalized forms denoted by a subscript "N", where resistances are normalized with R_∞ and capacitances with C_g. Let $Y_T \equiv Z_T^{-1}$ be the total admittance. Then we may write,

$$\mathcal{M} \equiv i\Omega Z_{TN} = C_{SN}^{-1} + i\Omega R_{SN} , \tag{28}$$

$$Z_{TN}^{*} \equiv R_{SN} + i(\Omega C_{SN})^{-1} , \tag{29}$$

$$Y_{TN} \equiv G_{PN} + i\Omega C_{PN} , \tag{30}$$

and

$$K_{DN} \equiv (Y_{TN}/i\Omega)^{*} = C_{PN} + i\Omega^{-1} G_{PN} , \tag{31}$$

where the asterisk denotes complex conjugation and $\Omega \equiv \omega C_g R_\infty$ as before. In the above, is the complex modulus of Hodge et al.[58], and K_{DN} is the conjugate of the complex effective dielectric constant in normalized form[59]. Note that $\mathcal{M}^{*} = K_{DN}^{-1}$. All the quantities defined above have positive imaginary parts. These parts are plotted on the positive y-axis versus the real parts of a given complex quantity on the positive x-axis.

What sort of shapes arise from these plots? Admittance-plane curves for several different circuits have been presented by Bauerle[53], and complex dielectric constant and complex conductivity-plane curves given for several other circuits by Grant[60]. For the $N = 2$ circuit, it turns out that $\mathcal{M}$ and Z_{TN}^{*} plots lead to two connected but well distinguishable semicircles with circle centers on the real axis provided the two time constants of the circuit are well separated. But as Hodge et al. have demonstrated[58], $\mathcal{M}$ shows good semicircle separation when the time constant difference arises primarily from a difference in resistances, e.g., $r_2/r_1 >> 1$, and Z^*_{TN} shows such separation when the difference comes from an appreciable difference between the two capacitances of the circuit. In all cases, both semicircles are present, but their sizes are very different in the above limiting cases. Generally, when one type of plot shows two distinguishable semicircles, only one will be apparent on the other. In the absence of further information, it is thus a good idea to construct both $\mathcal{M}$ and Z^*_{TN} plots initially. When two joined semicircles appear, all $N = 2$ circuit element values may be obtained from the real-axis intercepts and the frequencies at which the peaks of the semicircles occur.

The other two kinds of plots have been used more in the past but are generally not quite as instructive as the first two. For the $N = 2$ circuit, they never give two semicircles but, at best, lead to a single semicircle melding into a vertical line at its right end. Thus, the fact that there are two time constants present does not show up so clearly, and not all the circuit element values follow as readily from these plots. There thus seems to be little reason to use Y_{TN} and K_{DN} plots rather than $\mathcal{M}$ and Z^*_{TN} ones, at least in the initial stages of an investigation.

By restricting attention to the complex dielectric constant, Volger[46] concluded that in an N-layer MW situation, (N-1) relaxation times rather than N occur. As we have seen, this need not be the case. Bauerle[53] used an $N = 3$ circuit to analyze data on zirconia-yttria by admittance plane methods. He found only two connected semicircles but omitted any parallel capacitance across one of his three resistances. This omitted capacitance can be identified as C_g to good approximation. Since there is always some geometric capacitance in any real situation, it should properly not have been omitted. At sufficiently high frequencies, it leads to a semicircle in the Z^*_{TN} plane or a vertical line in the Y_{TN} plane. Bauerle's data did not extent to high enough frequencies to show this effect.

In almost all SCP cases of interest, the time constant $\tau_D \equiv R_\infty C_g$ will be appreciably shorter than other time constants present in the system. Furthermore, in most cases, C_g will also be much smaller than other capacitances present. Under these conditions, the parallel combination of C_g and R_∞ may be placed in series with the rest of the circuit to a high degree of approximation. For the special Cases A and B already discussed in connection with Fig. 3, the rest of the circuit may be well approximated by another resistance and capacitance in series, yielding an $N = 2$ Voigt-model circuit. Finally, whenever one wishes to fit impedance/admittance data to a theoretical model, it will be a good idea to do such fitting using generalized least squares, a technique which allows both the real and imaginary parts of the data to be statistically weighted in determining parameter estimates and their uncertainties[61].

4. Detailed Results for a Homogeneous Material

Over the years since 1952 there has been considerable work devoted to the solution of the equations of Section II for the small-signal ac situation with various simplifications and idealizations (e.g., 21, 22, 54-56, 59, 62-64). Here, I shall discuss recent work[22,55,56] which essentially subsumes most earlier results. Consider an intrinsic-conduction material, initially without recombination, and with identical plane, parallel electrodes. Further, V_D is taken zero, no specific adsorption effects are included, and the finite size of charge carriers is neglected. To define such a system as far as its normalized impedance is concerned, one may use the following parameters: $(r_p, r_n; \pi_m, \pi_z; 0, M)$. Here $\pi_m \equiv \mu_n/\mu_p$, $\pi_z \equiv z_n/z_p$, and the zero denotes intrinsic conduction. Because of symmetry $(r_n, r_p; \pi_m^{-1}, \pi_z^{-1}; 0, M)$ defines the same normalized situation.

It is found that the exact equivalent circuit of Fig. 3 can usually be well approximated by the N = 3 circuit of Fig. 4. Here $R_3 \equiv R_\infty$ and $C_3 \equiv C_g$; R_2 and C_2 can generally be taken frequency-independent, and R_1 and C_1 must be taken frequency-dependent in part of the frequency range[55,56]. Insofar as time constants may be defined for the three subcircuits, one usually finds $\tau_1 >> \tau_2 >> \tau_3 \equiv \tau_D$.

Figure 5 shows the simple frequency response for $C_{PN} \equiv C_P/C_g$ found with completely blocking, equal mobility, and equal valence

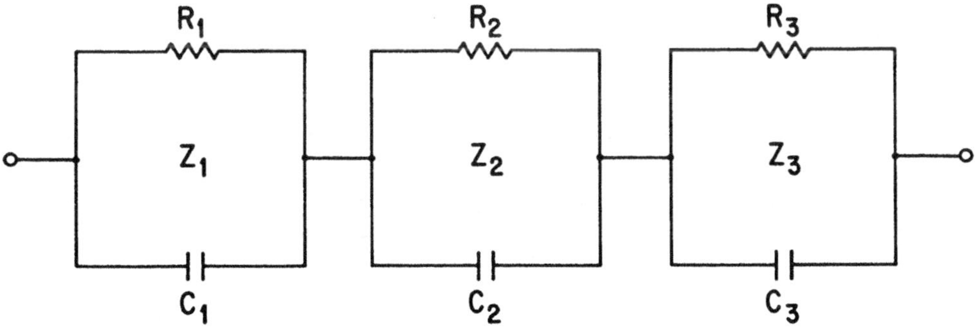

Fig. 4. Approximate equivalent circuit following from detailed SCP theory[22]. R_1 and C_1 are generally frequency dependent.

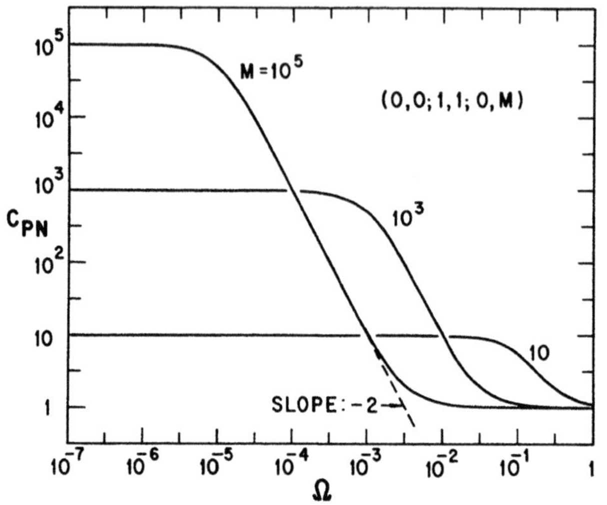

Fig. 5. Normalized parallel capacitance vs normalized frequency for complete blocking and several values of $M \equiv \ell/2L_{D2}$.

conditions. Here Fig. 4 reduces to an N = 2 Voigt model with $Z_1 = 0$, $R_2 = \infty$, and $C_2 = MC_g \equiv C_o$, just the ordinary diffuse layer capacitance for two layers in series. Somewhat similar results are obtained when $r_p = r_n \neq 0$[55,56]. Note that these results are an example of Case A_2^n reduction discussed in Section IV-2.

Figure 6 gives some results for another limiting case. Here $Z_2 = 0$, and, in general,

$$Z_{1N} \cong [\pi_m(i\Omega bM^2)^{\frac{1}{2}}\mathrm{ctnh}(i\Omega bM^2)^{\frac{1}{2}}]^{-1} , \tag{32}$$

where $b \equiv \delta_n\delta_p/\varepsilon_n\varepsilon_p$, $\delta_n \equiv (1 + \pi_z^{-1})^{-1}$, $\delta_p \equiv (1 + \pi_z)^{-1}$, $\varepsilon_n \equiv (1 + \pi_m^{-1})^{-1}$, and $\varepsilon_p \equiv (1 + \pi_m)^{-1}$. This approximate expression for Z_{1N} holds[56] when $r_p = 0$ for any π_m and π_z, not just the unity values of Fig. 6. Note that the Ω dependence in Fig. 6 involves appreciable regions with $\Omega^{-3/2}$ rather than the conventional Ω^{-2} behavior apparent in Fig. 5. In the intermediate frequency range, where the ctnh function is unity, Z_{1N} reduces to an ordinary diffusional Warburg impedance. Such an impedance must be modified as above, however, for a finite length of material, since as $\Omega \to 0$, there will eventually be less than one diffusion length available in ℓ.

For M >> 1, the usual case, the total low-frequency-limiting capacitance, C_{PO}, which follows from Eq. (32), is

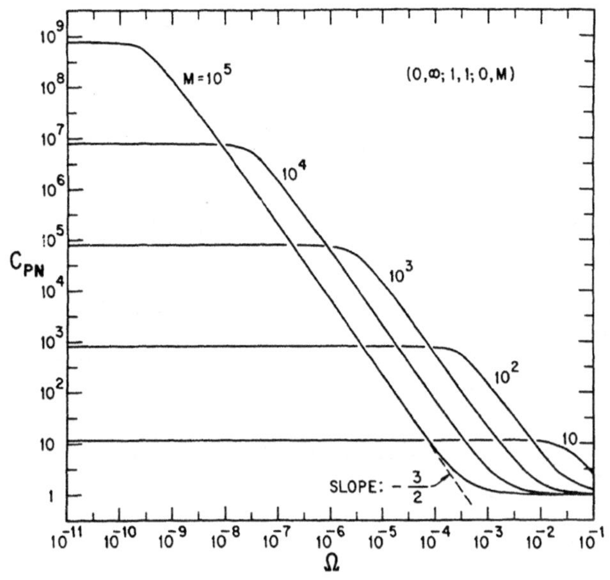

Fig. 6. Normalized parallel capacitance vs normalized frequency for $(r_p, r_n) = (0,\infty)$ boundary conditions and several M values.

$$C_{PO} \cong (\delta_n \delta_p M^2/3)C_g \cong (M/12)C_o , \qquad (33)$$

where the second form applies when $\pi_z = 1$. Notice that for large M, C_{PO}/C_o may greatly exceed unity. This large capacitance is connected with diffusion in a restricted space ($x = 0$ to ℓ) and is extensive, directly proportional to ℓ! It is of especial interest, both for its size and its genesis. Earlier work[59,63] shows conclusively that for $M >> 1$, C_{PO} is not primarily associated with stored space charge in the material when r_p and r_n are very different. It is a diffusional rather than a SCP capacitance, is sometimes called a pseudocapacitance, and is associated with the energy stored in the chemical reaction which occurs at the electrode when r_p or r_n is not zero[65]. When $r_p = r_n \neq 0$, on the other hand, there are still reactions present but no diffusion-related delay which can lead to diffusion capacitance[56].

Figure 7 shows some more complicated results which can appear when $0 \leq r_n \leq \infty$. Here $g_n \equiv 1 + (r_n/2)$. The different values of π_z are used to give greater separation between some of the curves, but the possible effect of π_z, which has of course a limited range, is much smaller than that of r_n and π_m. Note that Ω^{-m} regions occur with values of $m = 2$, 1.5, 1, and 0.5, all slopes which are frequently encountered experimentally. When $r_p > 0$ as

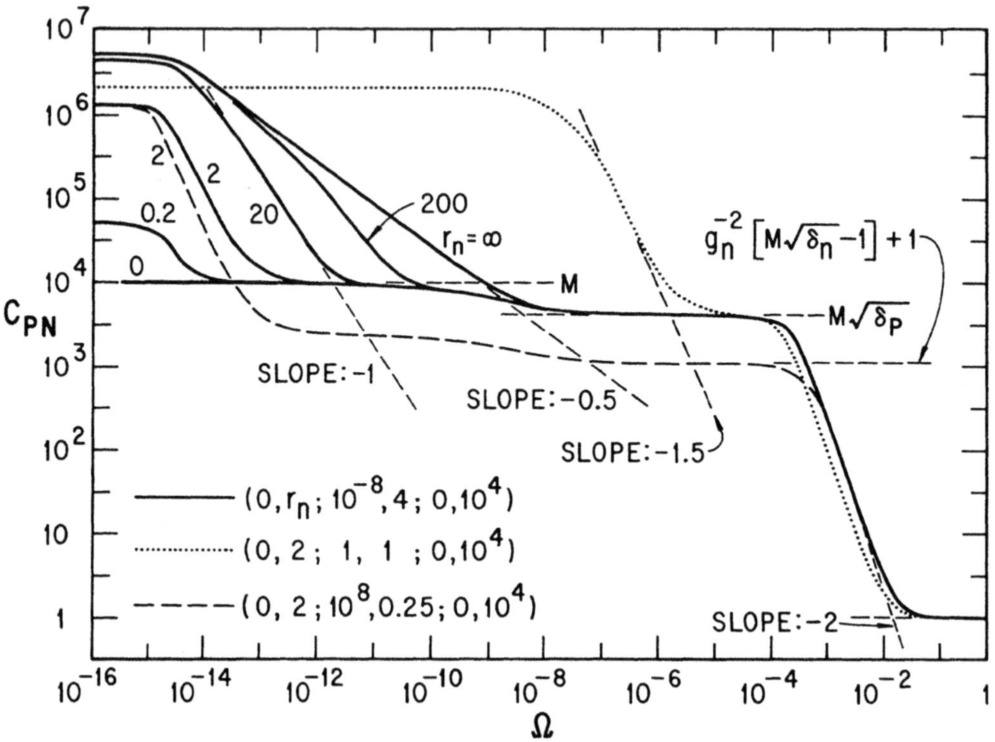

Fig. 7. Normalized parallel capacitance vs normalized frequency for several $(0,r_n)$ situations.

well, all intermediate slopes from m = 0 to 1.5 are also possible[55]. Generally, most measurable structure occurs in the curves when there is a large difference in mobilities and the more mobile species is much more strongly blocked than the other. Accurate capacitance measurements over a wide frequency range can be made in this $R_D >> R_E$ case but not when $R_E >> R_D$.

It has been shown[21,55,56,63,64] that when the dimensionless quantities r_p and/or r_n are neither zero nor infinite, they may be related to heterogeneous reaction rate constants, ξ_i. The result is $\xi_i \equiv (D_i/\ell)r_i$, where i = p or n. Since ξ_i is associated with an electrode reaction, it cannot depend on ℓ. Thus, both r_p and r_n must be proportional to ℓ when they are neither zero nor infinite and are associated with thermally activated reaction rates or surface recombination rates.

Figure 8 shows the general form of the Z_{TN}^* impedance plane plot following from the present theory[55,56]. The various arcs can be of any size relative to each other and all three do not necessarily show up simultaneously. Arc 1, which occurs at the

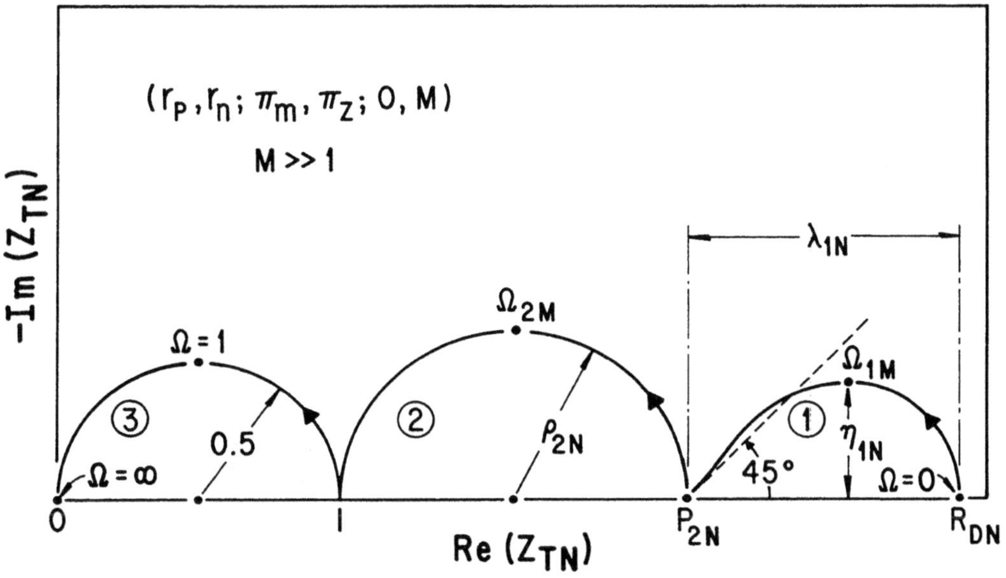

Fig. 8. Typical form of normalized impedance plane plot following from detailed SCP theory without recombination.

lowest frequencies, arises from the Z_1 of Fig. 4 and is associated with diffusion. The 45° region is that where conventional infinite-length Warburg response is well approximated. In the present normalized form, the size of arc 1 is independent of ℓ and depends primarily on π_m. With Z_T^* rather than Z_{TN}^* plotting, the arc 1 size will be directly proportional to ℓ since $R_\infty \propto \ell$.

The central arc arises from Z_2 and is associated with finite electrode reaction rates; it is usually very well approximated by a semicircle with $C_{2N} \cong M$ and $R_{2N} \cong 2/(\varepsilon_n r_n + \varepsilon_p r_p)$. When either $\varepsilon_n r_n$ or $\varepsilon_p r_p$ is infinite, $R_{2N} = 0$ and no reaction arc appears. When $r_n = r_p \geq 0$ or $r_p = 0$, $r_n << 1 + \pi_m$, arc 1 does not appear or may be too small to measure. This is an N = 2 situation, an illustration of Case A_2 discussed earlier. In normalized form, the size of arc 2 is inversely proportional to ℓ through r_n and/or r_p, but in unnormalized form it is essentially independent of ℓ and involves circuit elements associated only with the interphase-electrode region. Finally, arc 3, which again is an excellent semicircle, arises from bulk effects, involves R_∞ and C_g only, is independent of ℓ in normalized form with a height of 0.5, and its size is directly proportional to ℓ for unnormalized conditions. Note that its peak occurs at $\Omega \equiv \omega\tau_D = 1$.

Arc 2 is not always found to be a perfect semicircle with its center on the real axis, as demonstrated by Fig. 9. When $\pi_m \sim M$,

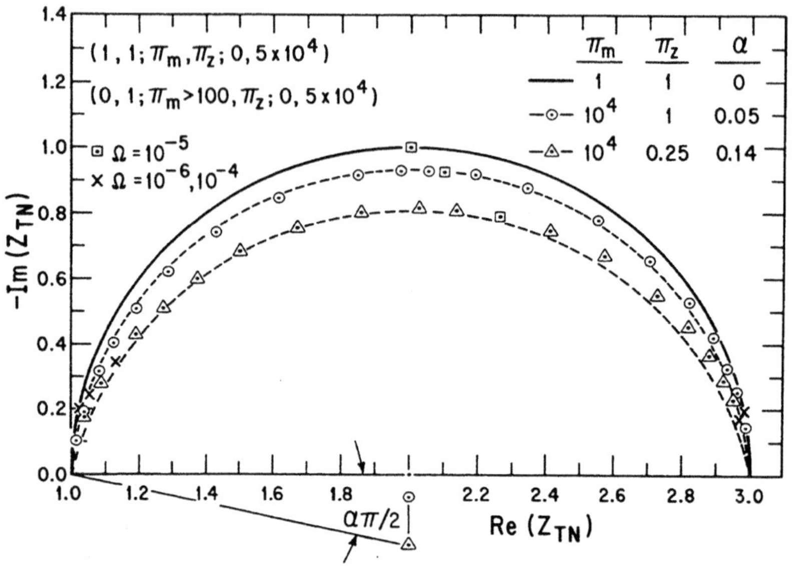

Fig. 9. Normalized impedance plane plot showing calculated arc 2, arising from slow electrode reaction, for several π_m, π_z values.

the resulting arc may be quite well approximated by a depressed semicircle, one with its center below the real axis. Many experimental data lead to depressed semicircles of this kind, although the depression is often greater than that apparent here, the maximum found thus far for the present theory. Figure 10 shows a situation where arc 1 is much larger than the other arcs and the effect of variable M is examined. Experiment frequently yields curves which approximate some of those shown in Fig. 10, although the available experimental frequency range usually does not allow all three arcs to be completely covered. Even coverage of parts of several arcs may allow analysis of the data to yield estimates of mobilities, boundary parameters, etc[56], however. In the usual M >> 1 case, it is generally wise to construct both Z_{TN}^* and Y_{TN} complex plane plots when initially analyzing data.

Up to this point the discussion has been limited to full dissociation results ($k_1 = k_2 = 0$). Some work has been published which deals with generation-recombination (GR) effects when only one charged species is mobile[66], but unpublished work in progress yields particularly interesting results in the $r_p \neq r_n$ two-mobile case (π_m neither zero nor infinite). Consider the M >> 1 situation with k_1 and k_2 non-zero. Two new parameters describing geminate GR are $\Lambda \equiv k_1/k_2c_i$ and $\xi_r \equiv (k_2c_i)^{-1}/\tau_D$, where $(k_2c_i)^{-1}$ is the effective recombination time. It turns

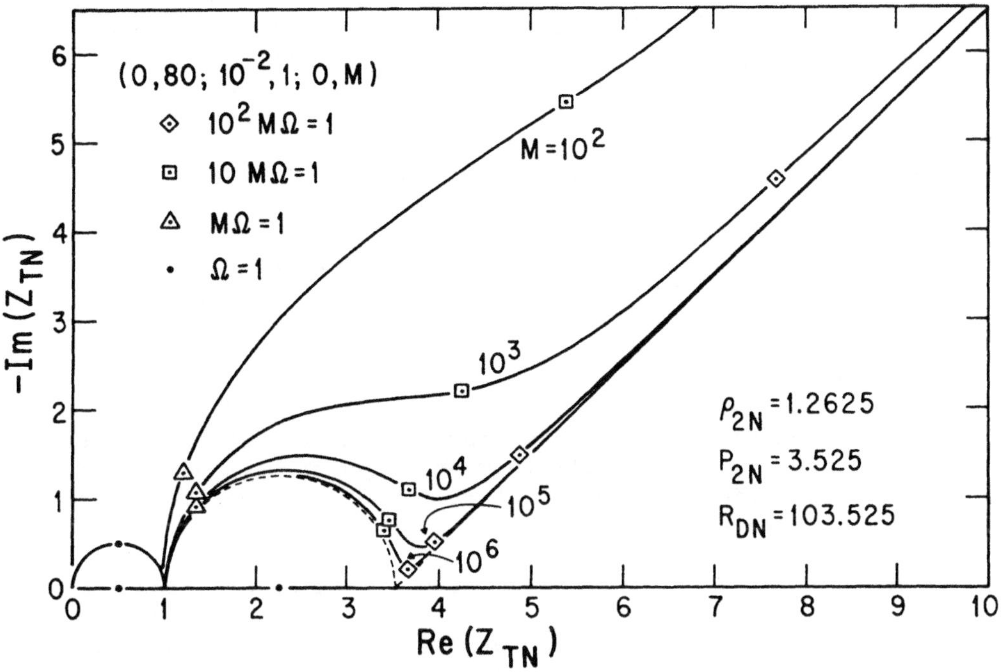

Fig. 10. Calculated normalized impedance plane plot for several values of M. Arc 1 not shown completely.

out for $\pi_z = 1$ that $(c_i/N) = \Lambda/(1 + \Lambda)$, where (c_i/N) is the dissociation ratio; thus for small dissociation $\Lambda \cong (c_i/N)$, and when $\Lambda \to \infty$, $c_i \to N$, full dissociation.

First, consider the low-frequency-limiting value of the parallel capacitance, $C_{P0} = C_g + C_{i0}$. In the general situation $(r_p, r_n; \pi_m, \pi_z; 0, M)$ take $r_p = 0$, $r_n \geq 1$, $\pi_m << 1$, $\pi_z = 1$, and $M >> 1$. Then one finds that the dominant part of C_{P0} is

$$C_{P0} \cong (M^2 C_g/12)(1 + 2\Lambda^{-1}) \ , \tag{34}$$

in agreement with Eq. (33) when $\Lambda \to \infty$. This expression may be rewritten as

$$C_{PN0} \cong (1/3)(\ell/4)^2[8\pi(z_e e)^2(2N - c_i)/\varepsilon kT] \ , \tag{35}$$

where z_e is the common value of z_n and z_p. For $c_i = N$, the capacitance involves noncombining positive and negative charges each of bulk concentration N, just as expected. But when $c_i << N$, the capacitance is twice as large! Note that it is then far larger than the value calculated conventionally with c_i replacing

$(2N - c_i)$. Evidently, not only does $c_i << N$ GR make it possible for all centers of bulk concentration N to contribute to C_{P0}, but it adds a further factor of two increase, probably related to the dynamics of GR itself. It is physically plausible to expect all originally neutral centers to contribute since the zero-frequency limit of the capacitance allows sufficient time for every neutral center to eventually dissociate and the resulting charges separate in the applied field.

GR also leads to interesting results in the non-zero Ω range. One often finds a new Z^*_{TN} complex-plane arc, located between arcs 1 and 2 or 2 and 3 of Fig. 8. There are then four arcs possible. An example where all four appear with reasonable separation is $(0, 10; 4x10^{-3}, 1; 0, 10^4)$ with $\Lambda = 10^{-4}$ and $\xi_r = 4x10^9$. The maximum values of $Im(Z^*_{TN})$ are found to be about 58, 50, 23, and 0.5, as Ω increases, for the diffusion, GR, reaction, and bulk arcs, respectively. Not all four arcs appear well separated under many conditions; in particular, the GR and reaction ones frequently meld together.

When $(r_p, r_n) = (0, \infty)$, there is no reaction arc, and the GR arc turns out to be a semicircle without depression when $\pi_m \xi_r \simeq 1$. This semicircle is an instance of Case B_2, discussed above. On the other hand when $\xi_r >> \pi_m^{-1}$, the GR arc may be of virtually the same shape as the finite-Warburg diffusion arc. The possibility of confusion between these two arcs thus arises but can be resolved from their different dependencies on length, ℓ. The size of the unnormalized GR arc is independent of ℓ, as it should be. Data which include even a part of the GR arc should allow the GR parameters to be estimated.

Finally, it is worth noting that when the Langevin[67,68] diffusion-limited theory for k_2 is used to evaluate ξ_r, one finds $\xi_r = 1$, not a well-known result. While it is, in fact, reasonable to expect that the shortest recombination time would be τ_D, when the mean free path of a charge carrier is large compared to the range of its screened Coulomb potential, one might expect that ξ_r could be much greater than unity. Lax[68] has discussed a situation where electrons are the major charge carriers and finds that the Langevin k_2 is much larger than experimentally observed values. To the degree that this may also be the case in solids with ionic/vacancy charge carriers, one needs to examine $\xi_r > 1$ effects there as well as those for $\xi_r = 1$.

REFERENCES

1. F. A. Kröger, The Chemistry of Imperfect Crystals, (North-Holland Publishing Company, Amsterdam, 1964), p. 880.
2. Ref. 1, pp. 879-908.

3. J. R. Macdonald, J. Appl. Phys. 34, 538 (1963).
4. N. Sinharay and B. Meltzer, Solid State Electronics 7, 125 (1964).
5. M. A. Lampert and P. Mark, Current Injection in Solids, (Academic Press, New York, 1970).
6. G. P. Owen and A. Charlesby, J. Phys. C: Solid State Phys. 7, L400 (1974).
7. J. R. Macdonald and M. K. Brachman, Rev. Mod. Phys. 28, 393 (1956).
8. J. R. Macdonald and C. A. Barlow, Jr., Rev. Mod. Phys. 35, 940 (1963).
9. C. Bucci and R. Fieschi, Phys. Rev. Lett. 12, 16 (1964); C. Bucci, R. Fieschi, and G. Guidi, Phys. Rev. 148, 816 (1966).
10. B. K. P. Scaife, J. Phys. D: Appl. Phys. 7, L171 (1974).
11. H. M. Gupta and R. J. Overstraeten, J. Phys. C: Solid State Phys. 7, 3560 (1974).
12. H. Scher and M. Lax, Phys. Rev. B7, 4491, 4502 (1973).
13. H. Scher and E. Montroll, Phys. Rev. B15, to be published.
14. C. Kittel, Introduction to Solid State Physics, 4th Ed., (John Wiley, New York, 1971), p. 270.
15. C. A. Barlow, Jr. and J. R. Macdonald, Adv. in Electrochem. and Electrochem. Eng. 6, 1 (1967).
16. W. R. Fawcett, J. Chem. Phys. 61, 3842 (1974).
17. M. J. Sparnaay, Rec. Trav. Chim. 77, 872 (1958); J. Electroanal. Chem. 37, 65 (1972).
18. G. Jaffé, Ann. Physik 16, 217, 249 (1933); Phys. Rev. 85, 354 (1952).
19. J. R. Macdonald, J. Appl. Phys., to be published.
20. J. R. Macdonald, Solid State Electronics 5, 11 (1962).
21. H. Chang and G. Jaffé, J. Chem. Phys. 20, 1071 (1952).
22. J. R. Macdonald, J. Chem. Phys. 58, 4982 (1973).
23. J. Frenkel, Kinetic Theory of Liquids (Oxford University Press, New York, 1946), p. 36.
24. K. L. Kliewer and J. S. Koehler, Phys. Rev. 140, A1226 (1965).
25. L. M. Slifkin, J. de Physique 34, C9-247 (1973).
26. J. R. Macdonald, J. Appl. Phys. 45, 73 (1974).
27. J. R. Macdonald, J. Chem. Phys. 22, 1317 (1954).
28. J. R. Macdonald, J. Chem. Phys. 29, 1346 (1958).
29. J. R. Macdonald, J. Chem. Phys. 30, 806 (1959).
30. P. M. Sutton, J. Am. Ceram. Soc. 47, 188 (1964).
31. P. M. Sutton, J. Am. Ceram. Soc. 47, 219 (1964).
32. J. W. Mayer et al., Phys. Rev. 137, A295 (1965).
33. M. Maitrot et al., Rev. de Phys. Appl. 6, 369 (1971).
34. R. P. Buck, J. Electroanal. Chem. 46, 1 (1973).
35. J. R. Macdonald, J. Chem. Phys. 40, 3735 (1964).
36. T. B. Grimley, Proc. Roy. Soc. (London) A201, 40 (1950).
37. D. O. Raleigh, Electroanal. Chem. 6, 87 (1973).
38. D. Miliotis and D. N. Yoon, J. Phys. Chem. Solids 30,

1241 (1969).
39. Yu V. Pleskov, Prog. in Surf. and Membrane Sci. 7, 57 (1973).
40. D. O. Raleigh, private communication.
41. D. O. Raleigh and H. R. Crowe, Solid State Comm. 8, 955 (1970).
42. D. O. Raleigh, J. Electrochem. Soc. 121, 633 (1974).
43. E. Fatuzzo and S. Coppo, J. Appl. Phys. 43, 1457 (1972).
44. R. D. Armstrong, J. Electroanal. Chem. 52, 43 (1974).
45. A. R. von Hippel, Dielectrics and Waves, (John Wiley, New York, 1954), pp. 228-234.
46. J. Volger, Prog. in Semiconductors 4, 207 (1960).
47. L. K. H. Van Beek, Physica 26, 66 (1960).
48. J. Schrama, Thesis, Univ. of Leiden (1957), pp. 32, 60.
49. I. S. Goldstein, J. Appl. Phys. 45, 2447 (1974).
50. H. H. Byer and L. G. Bobb, J. Appl. Phys. 45, 3739 (1974).
51. I. M. Novosel'skii et al., Soviet Electrochemistry 8, 546 (1972).
52. J. R. Macdonald, J. Chem. Phys. 23, 275 (1955).
53. J. E. Bauerle, J. Phys. Chem. Solids 30, 2657 (1969).
54. J. R. Macdonald, J. Electroanal. Chem. 32, 317 (1971).
55. J. R. Macdonald, J. Electroanal. Chem. 53, 1 (1974).
56. J. R. Macdonald, J. Chem. Phys. 61, 3977 (1974).
57. K. S. Cole and R. H. Cole, J. Chem. Phys. 9, 341 (1941).
58. I. M. Hodge et al., J. Electroanal. Chem. 58, 429 (1975).
59. J. R. Macdonald, J. Chem. Phys. 54, 2026 (1971).
60. F. A. Grant, J. Appl. Phys. 29, 76 (1958).
61. D. R. Powell and J. R. Macdonald, Computer J. 15, 148 (1972); 16, 51 (1973). Also, J. R. Macdonald, Am. J. Phys. 43, 372 (1975).
62. J. R. Macdonald, Phys. Rev. 92, 4 (1953).
63. R. J. Friauf, J. Chem. Phys. 22, 1329 (1954).
64. J. H. Beaumont and P. W. M. Jacobs, J. Phys. Chem. Solids 28, 657 (1967).
65. J. R. Macdonald, J. Appl. Phys. 44, 3455 (1973).
66. J. R. Macdonald, J. Phys. C: Solid State Phys. 7, L327 (1974); 8, L63 (correction) (1975).
67. P. Langevin, Ann. Chim. Phys. 28, 289, 433 (1903).
68. M. Lax, Phys. Rev. 119, 1502 (1960).

DISCUSSION

ARMSTRONG : In superionic conductors the Debye length is often less than a typical atomic size, which suggests that your theories are not directly applicable in this case.

MACDONALD : To some degree this conclusion is doubtless true. Instead of an ordinary diffuse double layer extending over many crystal planes being present, the double layer capacitance may be a sort of inner-layer capacitance associated with charge in the plane immediately next to the electrode. Although this would effect the d.l. capacitance which arises in my theory, it should not necessarily have a large effect on the reaction, diffusion and recombinaison terms in the theory. It seems likely that most of the results of my theory would still be applicable, perhaps with a very large value of $M(=e/2L_D)$ inserted in appropriate parts of the theory.

WAGNER : Would you please provide us with a physical picture of the capacitance and/or pseudocapacitance for a (0,0) completely blocking and (∞,0) partially blocking electrodes ?

MACDONALD : The (0,0) double layer capacitance is often taken to be a diffuse (space distributed) space charge capacitance arising from the competing effects of electric field and diffusion. It was originally discussed by Gouy and by Chapman. The (∞,0) or (0,∞) low-frequency limiting capacitance is more complicated. It arises when a Warburg impedance is limited because the associated diffusion length fills the entire space between electrodes. At sufficiently low frequencies this must happen for any finite electrode separation. This low-frequency-limiting capacitance is not a pure space charge capacitance but is associated with diffusion to an electrode, and, in the usual case, with electron transfer there.

VOÏNOV : When you represent an interface by a capacitor (double layer) in parallel with a resistance (electrode reaction) do you imply that in open-circuit the capacitor is not charged ?

MACDONALD : I only assume the capacitor is charged when some potential difference occurs across it. In either a.c. or d.c. experiments, potentials are present which determine the charge on a given capacitor and, sometimes, determine its magnitude as well.

BONANOS : In most experimental situations the system of electrodes, interfaces, bulk, grain boundary properties etc. are effectively connected in series. What is the relevance of the Voight and Maxwell models in such situations ?

MACDONALD : The Voigt and Maxwell models are entirely equivalent electrically for proper element choices. One will generally use the one which a) arises naturally from theory and/or b) affords the most transparent explanation of the various physical processes assumed to be present.

HENISCH : How is the contour of the built-in diffusion barrier modified by the induced displacement of charge carriers in the bulk ?

MACDONALD : The barrier depends on the potential difference between bulk and surface. In equilibrium this potential difference is just the diffusion potential V_D. In other cases, the appropriate potential is $(V_D+\psi_a)$ where ψ_a is the electrostatic potential difference. The expression for the contour of the barrier, i.e. $\psi(X)$ (where ψ is the mean electrostatic potential) as a function of X, is too complicated to give here. In principle, it should be possible to eliminate a barrier by the application of a potential difference opposite to V_D. But when one uses symmetric electrodes with a similar or identical barrier at each, the application of an external potential difference increases the barrier height at one electrode and decreases it at the other. Thus there will never be a condition where both barriers are simultaneously absent.

HEYNE : Can you clarify what exactly you mean by <u>un</u>coupled ? I understand that coupling by Poisson's equation is always present.

MACDONALD : In the supported electrolyte case a large concentration of a supporting solute is present. The positive and negative ions of this fully dissociated salt are usually assumed to be completely blocked. In addition the charges of interest (perhaps only partly blocked) are present at a much lower concentration. Therefore, the field gradient is primarily established by the high concentration species, and the low concentration ones are largely uncoupled. In other words, a charge in the concentration of the low-concentration negative species at a given location need not be compensated by a corresponding charge in the concentration of the low-concentration positive species.

RALEIGH : Could you comment on the relative nature of Warburg impedance in supported and unsupported electrolytic media ? The concept of a Warburg impedance in an unsupported medium may be confusing, since it was originally derived for field-free diffusion in supported electrolytes.

MACDONALD : The Warburg impedance arises at frequencies where diffusion to an electrode is the rate limiting step. It can therefore appear in either supported or unsupported situations, with appropriate differences arising from the different effects of Poisson equation charge coupling in the two cases. The original derivation of the Warburg impedance was for a single electrode (semi-infinite) situation. All realizable situations involve finite dimensions. Therefore, one would expect the finite length Warburg formula to apply in most cases rather then the original infinite length one.

POLARIZATION STUDIES ON SOLID STATE ELECTROLYTES

J. Bruce Wagner, Jr.

Materials Research Center and
Department of Materials Science and Engineering
Northwestern University, Evanston, Illinois 60201 (USA)

The purpose of this tutorial paper is to review the theory, the method of measurement and treatment of the data for experiments on polarized solid electrolytes according to the theory of C. Wagner and the developments of H. Rickert, K. Weiss, D. O. Raleigh, A. Joshi and others. Methods to obtain the total and electronic conductivities, transport number, number and mobility of electronic charge carriers and the double layer capacitance at the electrolyte-electrode interface are to be discussed. Examples will be chosen which exhibit good agreement with theory--primarily the silver and copper halides.

In addition to great theoretical insight, the power of C. Wagner's methods lie in the specification of the chemical potentials of each constituent of the material to be studied. Thus in the measurements to be discussed, the chemical potentials are considered to be fixed by coexistence of the electrolyte with a reversible electrode or by coexistence in a gaseous atmosphere.

A useful tool for screening electrolytes and for input data for determining transport numbers is the total electrical conductivity, σ, where $\sigma = \sigma_{ion} + \sigma_e$ and the subscripts ion and e refer to the ionic and electronic partial conductivities, respectively. The total conductivity can be written as

$$\sigma = \sum_i c_i \, |z_i| \, e\mu_i \qquad (1)$$

Where c_i denotes the concentration of the ith species of valence $|z_i|$, e is the electronic charge per particle, and μ_i is the drift mobility. The drift mobility is related to the

M. Kleitz and J. Dupuy (eds.), Electrode Processes in Solid State Ionics, 185–222.

absolute mobility, B_i, by $B_i = (\mu_i/|z_i|e)$. For electrons and electron holes, $|z_i| = 1$ and this term usually is omitted from the equations.

What is needed are methods to separate the partial ionic and partial electronic conductivities and hence the transport numbers. The transport number, t_i, where i refers to the ith species, is defined as the ratio of the current carried by the ith species, I_i, to the total current carried by all species,

$$t_i = I_i/\sum_i I_i \tag{2}$$

For a material under fixed chemical potentials, the ratio of the partial conductivity of the ith species, σ_i, to the total conductivity, σ, also yields the transport numbers as,

$$t_i = \sigma_i/\sum_i \sigma_i = \sigma_i/\sigma \ . \tag{3}$$

Methods to obtain t_i and σ_i will be described briefly below. First consider the measurement of the total conductivity. This is done most simply by measuring the a.c. conductivity on a sample (e.g. MX) of well defined chemical potentials and hence well defined stoichiometry. A two probe method may suffice. Both ions and electrons may move at frequencies of the order of 1000 Hz. The sample may be sandwiched between two electrodes consisting of the parent metal M (reversible to ions and electrons) in which case the activity of M is unity and that of X has the lowest possible value. To test the dependence of σ on the X/M ratio, a sample may be heated between or suspended by wrapped inert electrodes (electrodes blocking to ions but allowing transfer of electrons, e.g. platinum) and a given partial pressure of the metalloid, p_{X_2} passed over the sample. By varying p_{X_2} the dependence of σ on p_{X_2} may be determined.* In either case, the dependence of the a.c. conductivity on frequency should be tested [1]. Experimental details of these types of measurement and precautions necessary have been summarized by Rapp and Shores [1] and by Blumenthal and Seitz [2].

The classical method of determining the transport numbers in electrolytes is by the validity of Faraday's law. Tubandt

*If additional chemical potentials must be fixed, appropriate gas mixtures may be used to equilibrate the sample in accord with the Gibbs Phase Rule. A discussion of the techniques for handling gas mixtures (primarily for oxygen, sulfur, nitrogen and carbon chemical potentials) is available by Schwerdtferger and Turkdogan [3].

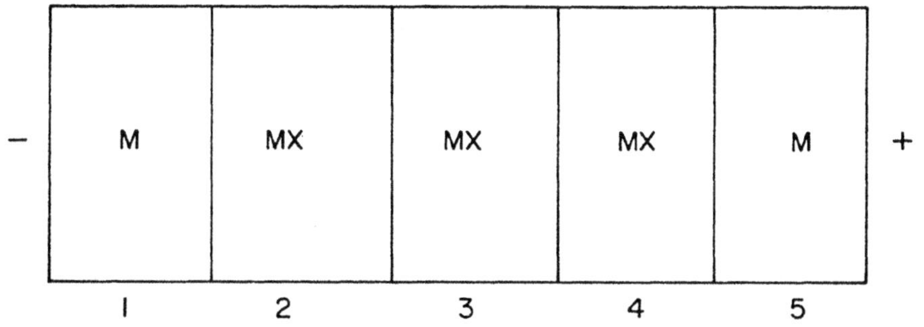

Fig. 1. Schematic representation of a Tubandt cell experiment.

utilized this method for solids as an analog to the Hittorf transport number method in aqueous solutions. In Tubandt's method three (or more) pellets of MX are sandwiched between reversible electrodes. The pellets are weighed prior to assembling into the configuration shown in Fig. 1. A known d.c. current for a known time is passed through the cell with the polarity as indicated. Afterwards the pellets are separated and reweighed. For example if pellet 5 loses weight and pellet 1 gains the same amount of weight in proportion to the number of equivalents passed, the transport number of M may be inferred as, $t_M^{+\nu}$ (in MX) = (equivalents of M gained by pellet 5) ÷ (total equivalents of current passed). Similar considerations apply to the migration of X. Details and examples are reported by Tubandt in Handbuch der Experimental Physik [4] and in the later article by Lidiard [5].

The transport number can also be obtained by measuring the open circuit potentials of a system under a gradient in chemical potential, e.g. a metal being oxidized to an oxidation product. This method was first described theoretically by C. Wagner and he applied the technique to silver sulfide being formed by reaction of silver and sulfur [6]. The potential is measured between the metal M and the outer surface of the compound, MX, coexisting with the gas phase (X_2). Local equilibrium must be established at the interfaces between M-MX and MX-X_2, i.e. a diffusion controlled reaction must exist for the method to work. A scale product is allowed to form on the metal to which one inert lead wire is connected. Then a second probe is placed in contact with the scale and the open circuit voltage obtained. The ratio of the measured open circuit potential E to that calculated from the free energy of formation, E_{theo}, [$\Delta G_f^o = nF\, E_{theo}$] yields the transport numbers as

$$(E/E_{theo}) = t_{ion} = (1 - t_e) \; . \qquad (4)$$

If the predominant migrating species, cation or anion, is known, e.g. from marker movement studies, the ionic transport number can be assigned to a particular species. Dravnieks and McDonald [7] utilized this method for a variety of electrolytes and electronic conductors.

A variation of this type of measurement is the measurement of the open circuit voltage of an electrolyte between two reference electrodes, e.g. $p_{X_2}^{(1)}$ and $p_{X_2}^{(2)}$, or between two solids which fix the chemical potentials, e.g. two different two-phase mixtures each of which fixes the chemical potential of a constituent at a different value [8].

The accuracy and precision of the Tubandt method is limited by the difficulty in separating each pellet from its neighbor (they sinter), by possible evaporation losses, and by errors due to weighing. The measurement of the open circuit voltage of an electrolyte (scale) under a chemical potential gradient is limited by the contacts to the scale and yield an average value unless special experiments are carried out to obtain the cell voltage as a function of chemical potential [1].

In any event the measurement of the electronic conductivity in a mixed conductor is very precisely and accurately carried out by using C. Wagner's asymmetric polarization cell method which is the primary subject of the present paper.

Consider a binary compound that is a mixed conductor, MX placed between an electrode reversible to ions and electrons and an electrode reversible to electrons but not ions. On application of a d.c. potential (below the decomposition potential of MX) with the positive pole on the ion blocking electrode, the ion current within MX may be suppressed and only a small electronic leakage current will flow at steady state. Initially both ions and electrons tend to migrate. This transient situation will be discussed later. Eventually, a depletion of ions will result at the MX-ion blocking electrode as no more ions are supplied at the positive

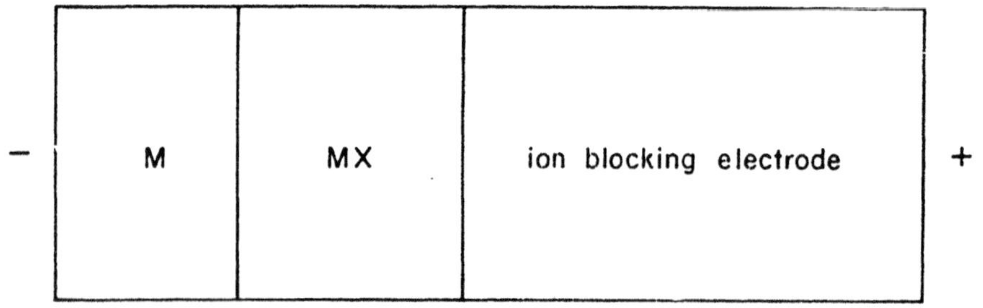

Fig. 2. Schematic representation of a d.c. polarization cell or C. Wagner asymmetric cell. M is a reversible electrode which exchanges both ions and electrons.

electrode and a steady state is achieved wherein the leakage current due to electrons, $\ominus$, or electron holes, $\oplus$, is the only conduction path. If these electronic charge carriers obey Boltzmann statistics, their mobilities are independent of concentration (ideal solution laws apply) and if the contact resistances at the electrode-electrolyte interfaces are negligible then the electronic leakage current, I, is given by [8]

$$I = I_{\ominus} + I_{\oplus} = (RTA/LF)\,\{\sigma_{\ominus}^{o}\,[1 - \exp(-EF/RT)] + \sigma_{\oplus}^{o}\,[\exp(+EF/RT) - 1]\} \qquad (5)$$

In the above equation, the following are defined :

- I = total electronic current at steady state
- $I_{\ominus}$ = electronic current due to electrons
- $I_{\oplus}$ = electronic current due to electron holes
- A = cross sectional area of the electrolyte assumed constant and the same as the interface area of the electrodes.
- L = thickness of the electrolyte
- $\sigma_{\ominus}^{o}$ = electronic conductivity due to electrons in MX equilibrated with M
- $\sigma_{\oplus}^{o}$ = electronic conductivity due to electron holes in MX equilibrated with M

Equation (5) can be evaluated by analyzing current-voltage plots. The relative contributions of electrons and of holes will vary with material and with the applied potential. Note that increasing values of E tend to increase the current due to electron holes and depress the current due to electrons. In some cases the partial electronic conductivities are so dissimilar that one of the two terms in Eq. (5) may be neglected. For example, CuCl in equilibrium with its parent metal, Cu, conducts electronically via electron holes ($\sigma_{\oplus}^{o} \gg \sigma_{\ominus}^{o}$). Cuprous chloride has been studied recently by A. V. Joshi [9,10] and current-voltage data for it are shown in Fig. 3.

The ternary compound, Ag_3SI, has been studied in a cell

Ag | Ag_3SI | graphite III

by Takahashi and O. Yamamoto (11). Ag_3SI conducts electronicly via electrons and a plot of Eq. (5) neglecting the term for electron hole conduction yielded a straight line from which $\sigma_{\ominus}^{o}$ was determined.

In many cases electronic conduction may occur with

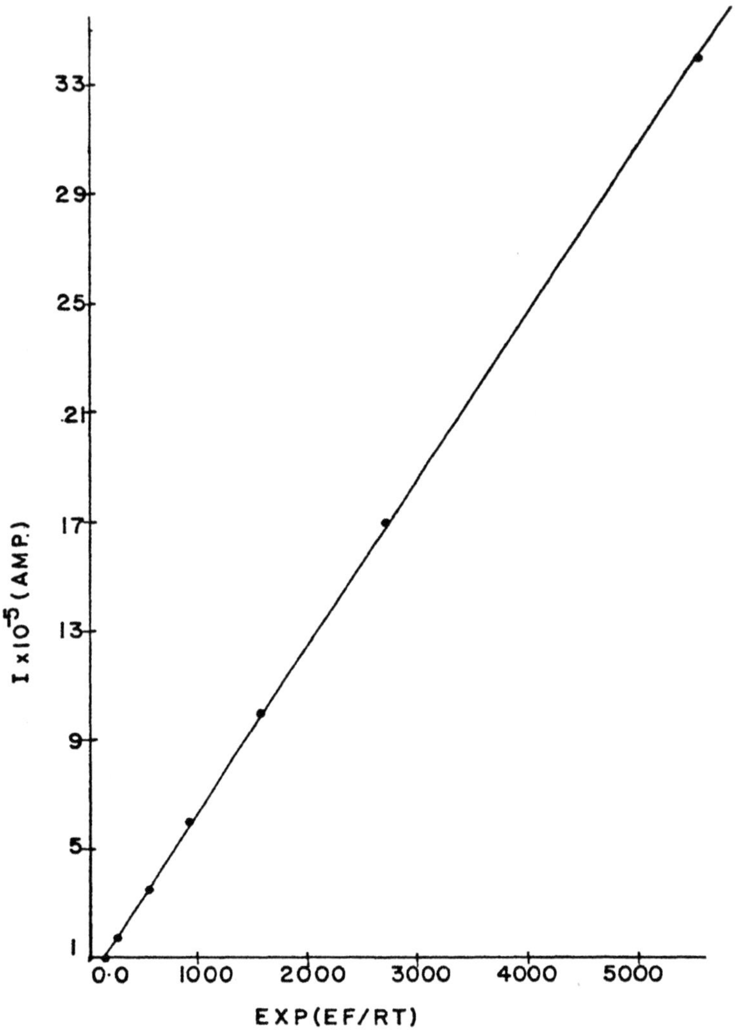

Fig. 3. Current-voltage data for the cell, Cu|CuCl|graphite +, at 383°C. After Joshi [9,10].

contributions from both electrons and electron holes and Eq. (5) cannot be simplified by neglecting one term. When, for example, $\sigma_{\ominus}^{o} > \sigma_{\oplus}^{o}$, but conduction is via both species, a plot of log I versus E may yield behavior shown schematically in Fig. 4.

Some examples of materials which conduct by both electrons and electron holes are the $CaO\text{-}ZrO_2$ electrolyte studied by Patterson et al. [12], the silver beta alumina electrolyte studied by Whittingham and Huggins [13] and silver chloride coexisting with silver as one electrode [14]. Under these conditions, it is

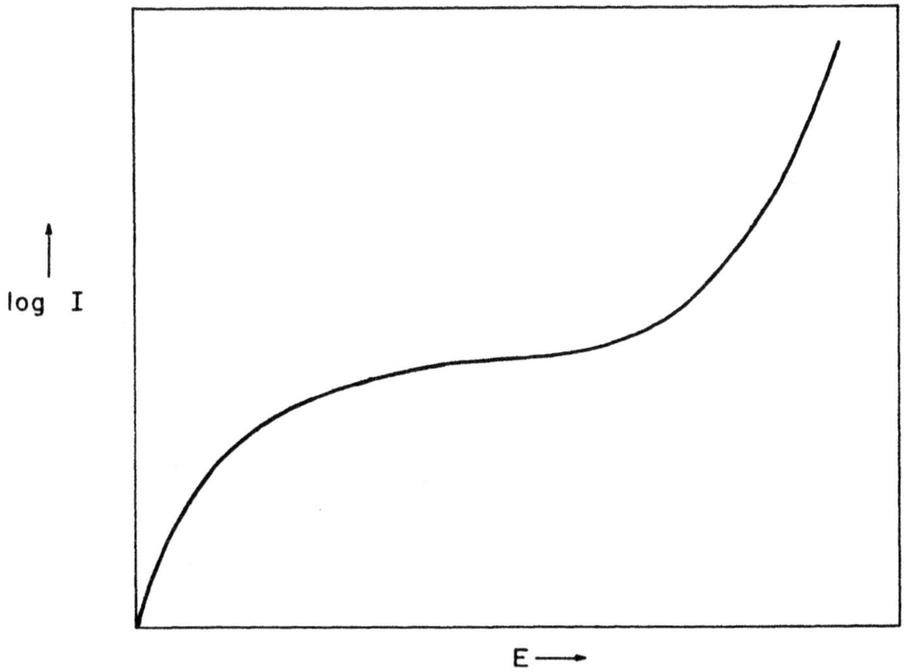

Fig. 4. Schematic diagram of log I versus E for the cell, M|MX|graphite, in which MX conducts by both electrons and electron holes, i.e., $\sigma_{ion} > \sigma_e$ and $\sigma_\ominus^{o} > \sigma_\oplus^{o}$.

expedient to rearrange Eq. (5) as follows:

$$I/[\exp(EF/RT) - 1] = RTA/LF\,[\sigma_\ominus^{o} \exp(-EF/RT) + \sigma_\oplus^{o}] \qquad (6)$$

A plot of the left hand term versus exp(-EF/RT) yields $\sigma_\ominus^{o}$ from the slope and $\sigma_\oplus^{o}$ from the intercept on the I/[exp(EF/RT) - 1] axis [12]. A similar expression holds for anion conductors [12].

Reversible electrodes may also include the metalloids as shown schematically in Fig. 5,

$+\,X_2$(gas) | MX | ion blocking electrode IV

For example the cell,

$\oplus$ graphite, Br_2(gas) | AgBr | graphite or Pt V

has been studied by D. O. Raleigh [15] and the cell,

$\oplus$ graphite, Cl_2(gas) | AgCl | graphite VI

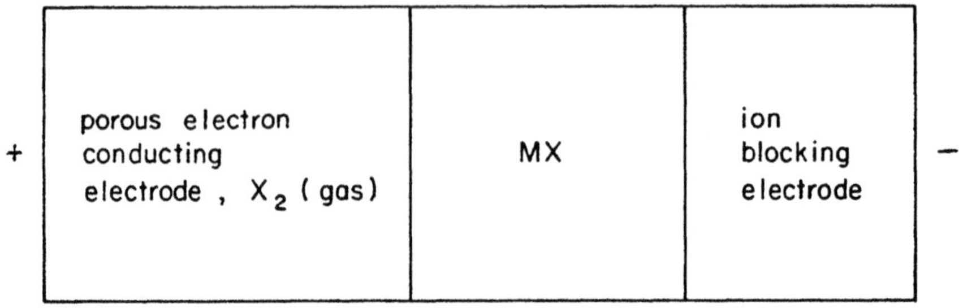

Fig. 5. Schematic diagram of a d.c. polarization cell involving the metalloid gaseous phase as a reversible electrode.

has been studied by van der Meulen and Kröger [14]. The advantage of this type of cell (cell IV) is that the activity of the metal can have a very low value. Therefore, the use of cell IV may allow one to study the electrolytes which exhibit primarily electron hole conduction for low values of the activity of M. The equation for this case is,

$$I = I_{\ominus} + I_{\oplus} = (RTA/LF)\{\sigma_{\oplus}^{o}\ [1 - \exp(-EF/RT)] + \sigma_{\ominus}^{o}\ [\exp(EF/RT) - 1]\} \tag{7}$$

Note for this case that increasing the applied potential to cell IV, decreases the concentration of electron holes and increases the concentration of electrons in contradistinction to the discussion for cell II.

A combination of the a.c. total conductivity and electronic conductivity derived from d.c. polarization measurements allows the determination of the electronic transference number, t_e, according to Eq. (3) under the assumption that $\sigma_{ion} \gg \sigma_e$ and hence $\sigma \cong \sigma_{ion}$. A plot of these data for CuCl is shown in Fig. 6. The data shown in Fig. 6 represent values of the conductivities of CuCl coexisting with copper. However, the electronic and ionic conductivities may both depend on the stoichiometry fixed by the chemical potentials of the constituents. Shown in Fig. 7 are two schematic graphs [18] for the behavior of CaF_2. The upper graph shows the isothermal variation of the defect concentration with partial pressure of fluorine (a Brouwer or Kröger-Vink diagram) and the lower one shows the corresponding behavior of the conductivities.

When such data are collected as a function of temperature and plotted as a three dimensional diagram, the electrolytic domains--regions in the σ_{ion}, p_{X_2}, T diagram wherein ionic conductivity predominates are shown. See Fig. 8.

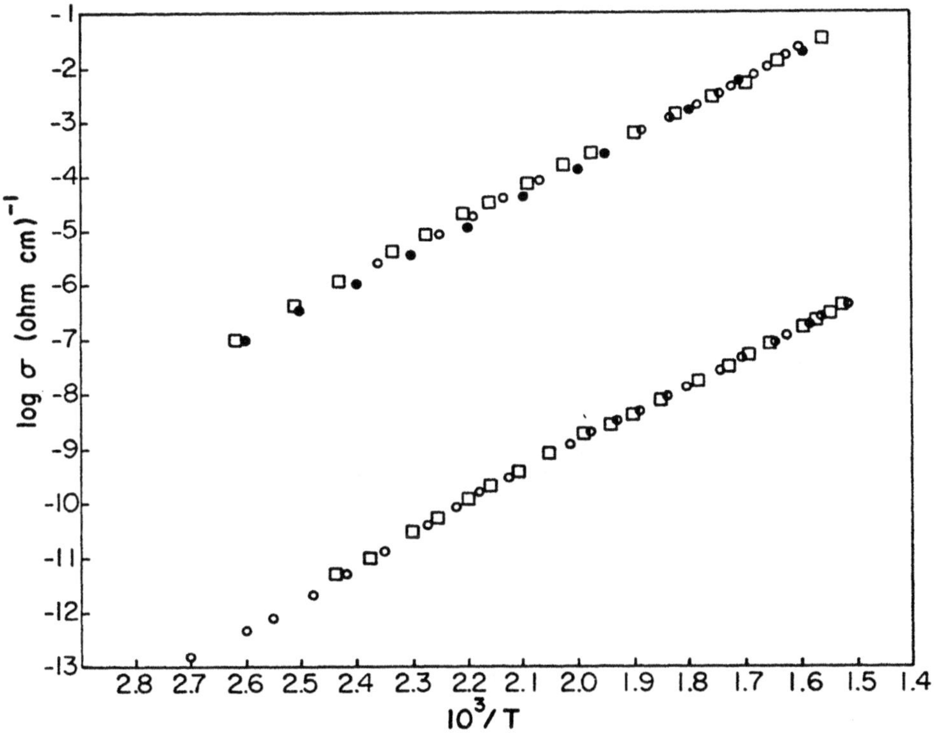

Fig. 6. Logarithm of the total conductivity ($\sigma \cong \sigma_{ion}$) upper curve and the electronic hole conductivity σ_+^o both for CuCl coexisting with copper. The open square and circles are the data of Joshi [9,10] and the closed circles those of Hsueh and Christy [16] and of Wagner and Wagner [17].

Such plots are extremely useful in choosing electrolytes for a given application.

In the above discussions, experiments were described which conform to the theory of C. Wagner [8]. In the following, some examples of phenomena which lead to deviations or to apparent deviations are discussed.

Recently Joshi and Liang [19] have proposed a criterion to check whether the condition, $\sigma_{ion} \gg \sigma_e$ is met. This may be written as,

$$\sigma_{ion} \gg \sigma_{\oplus}^o \exp(EF/RT) \qquad (8)$$

for electron hole conductors and hence,

$$E \ll (-\ 2.303\ RT/F) \log [\sigma_{\oplus}^o/\sigma_{ion}] \equiv E_c \qquad (9)$$

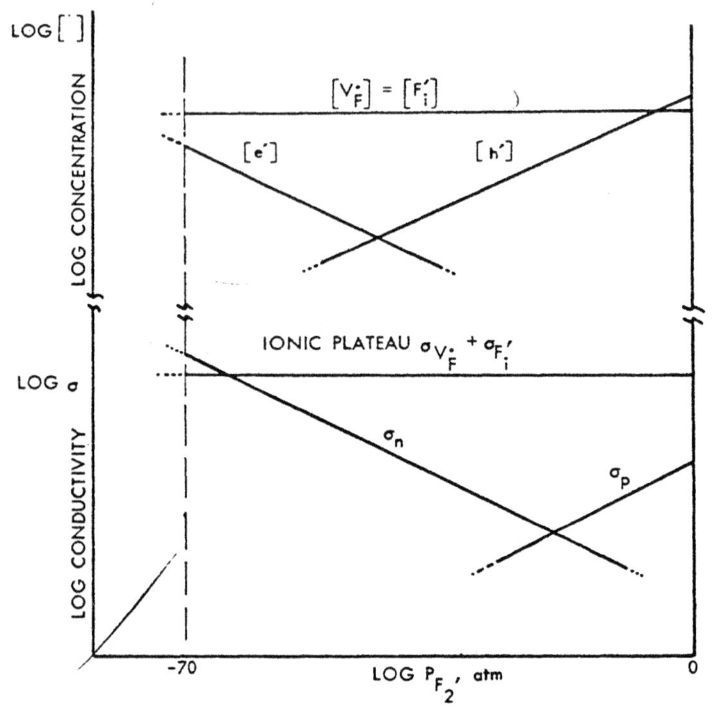

Fig. 7. Schematic diagrams of the log of defect concentrations and log of conductivities for CaF_2 equilibrated with various flourine pressures. Reprinted from Ref. [18] p. 498 by courtesy of Marcel Dekker, Inc.

where E_c is defined as the critical potential. According to these authors the C. Wagner theory is valid if the applied potential, E, is much less than critical potential, E_c. Joshi and Liang applied this test to PbF_2 using the cell,

- Pb | PbF_2 | graphite + . VII

As long as $E \ll E_c$, the experimental data were in accord with theory. They suggest [19] this may be one cause of deviations from the ideal slope, [F/2.303 RT], of log $I_{\oplus}$ versus E plots reported by other authors for $PbBr_2$ and PbF_2. C. Wagner's theory requires that the concentration of lattice defects>> electronic defects.

Recently T. Jow [20] has studied CuI in the cell

- Cu | CuI | graphite + VIII

Jow has found that log $I_{\oplus}$ vs E plots are linear at potentials less than the critical potential for T = 450° and 435°C (α phase) and at 384°C (β phase). On the other hand, in the γ phase, the total conductivity and electron hole conductivity approach one

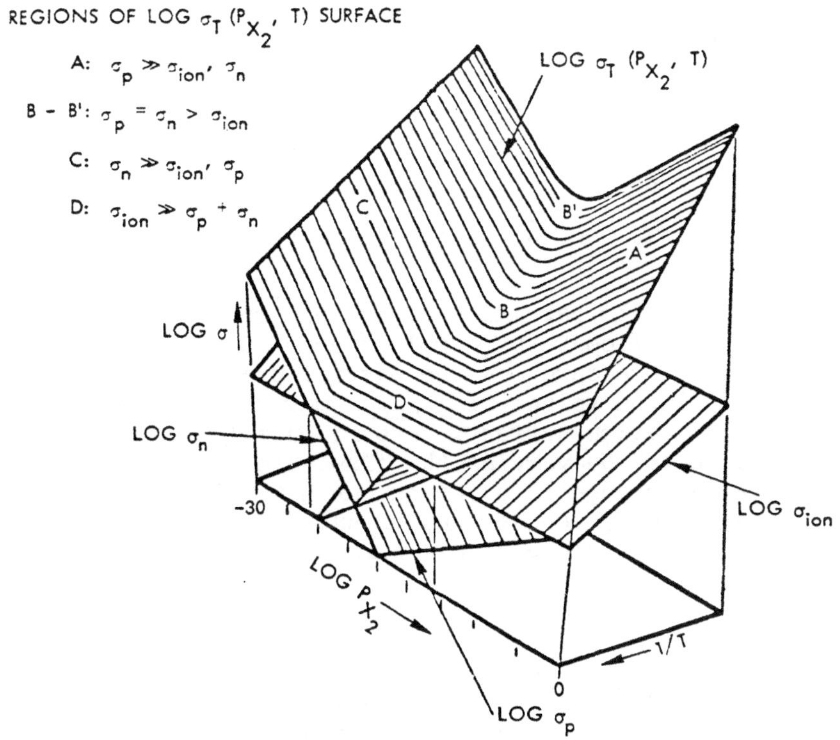

Fig. 8. Schematic diagram of the log σ surfaces over the log p_{X_2} - 1/T plane for the total conductivity partial electron and electron hole conductivities and ionic conductivity here denoted by σ_T, σ_n, σ_p and σ_{ion}, respectively. Reprinted from Ref. [18] p.501 by courtesy of Marcel Dekker, Inc.

another at temperatures below 220°C. Under these conditions, the log I_+ vs E plots appear to exhibit the theoretical slope, (F/2.303 RT). It may happen that the Cu|CuCl couple is behaving like a metal-semiconductor junction which also exhibits similar current-voltage characteristics. Further work is needed to examine this hypothesis.

Besides the usual problem of contamination, by impurities pre-existing in the crystal, they may enter from the gas phase, e.g. water vapor reacting with cuprous halides. Problems can arise from the decomposition of the electrolyte in a flowing inert gas stream. According to Wagner and Wagner [17] an upper limit for the decomposition current can be obtained from calculations of the partial pressures of X and of X_2 (assumed the major volatile species) for the reactions

$$MX = M + X(gas) \tag{10}$$

and

$$MX = M + 1/2\ X_2(\text{gas}) \tag{11}$$

at the inert electrode (positive pole in cell II) for a given value of the applied potential, E. This yields for the current due to the decomposition, I_{decomp},

$$I_{decomp} = I_X + I_{X_2} = F\,\dot{n}\,[p_X/P] + 2F\,\dot{n}\,(p_{X_2}/P) \tag{12}$$

where I_X and I_{X_2} are the partial currents due to the evolution of X and X_2, $\dot{n}$ is the flow rate in moles/sec of inert gas past the sample, P the total pressure, p_X the partial pressure of X and p_{X_2} that of X_2. Calculated values of I_{decomp} may be compared with the observed polarization current. Experimentally the decomposition current is tested by varying the flow rate, $\dot{n}$, over the sample and testing for any change in observed current. Similar considerations have been made by van der Meulen and Kröger [14] for AgCl. These authors also reported unstable currents for the cell

$$- \text{Ag} \mid \text{AgCl} \mid \text{C} + \qquad\qquad \text{IX}$$

which they suggested might be due to "bursts of ionic conduction released when a trapped dislocation breaks loose from a pinning point." The effect of stress on the current in a cell should be studied further because of the practical aspects of confining solid electrolytes either to obtain good contact resistance or to seal the electrolyte in a container for a battery application. Another contribution to the observed current could be leakage along the surface of the cell [14]. This is very difficult to eliminate. Some suggestions are provided by Blumenthal and Seitz [2]. Finally, the problem of contact resistance must be mentioned. For relatively plastic electrolytes, a small pressure against the cell is sufficient to cause good contact after an elevated temperature anneal. An experimental check is to use identical cells of the same cross sectional area but of varying electrolyte length. According to Eq. (5), a plot of I versus 1/L should yield a straight line for the same applied potential and temperature if contact resistance is negligible. Likewise a plot of the apparent value of the conductivity versus 1/L where L approaches infinity provides a graph which allows one to choose a length whereby the contact resistance is negligible. The use of a double cell involving samples of two lengths will be discussed below in connection with transient measurements.

Most studies have used the dimensions of the electrolyte as determined outside the cell to calculate the value of the cross sectional area, A. Better accuracy can be obtained [15] by determining the cell constant, G, where G = A/L in Eqs. (5) and

(7). If the total conductivity, σ, of the electrolyte has been independently determined then $G = A/L = R\sigma$ where R is the measured resistance of the electrolyte in the cell by a.c. measurements.

Next consider the phenomena which occur when a cell such as cell II is at steady state under an applied voltage, E_1, and a second voltage E_2, ($E_2 > E_1$) is suddenly applied to the cell. There is a transient response of the current until a new steady state is attained. During this transient period three phenomena may occur. These are:

(1) Charging of a double layer capacitance at the MX|ion blocking electrode, and

(2) Redistribution of electronic carriers within the electrolyte.

If the non-reversible electrode (the positive electrode in cell II) can dissolve a constituent of MX, e.g. M, then the third phenomenon may occur:

(3) Diffusion of M into the positive electrode until the electrode becomes equilibrated with the new activity of M fixed by the new potential, E_2.

Each of these effects can be studied separately provided special precautions are taken to eliminate the others. For example, diffusion into or out of the positive electrode [Cell II] can be made negligible by choosing a material which has virtually no solubility for the constituents of the electrolyte, MX. Under such conditions, when a cell such as cell II is at steady state under a potential, E_1, and abruptly the potential is changed to a new potential, E_2 where $E_2 > E_1$, the electrolyte responds. During this transient time, some metal ions in MX must be plated out on the cathode (the reversible electrode, M) in accord with the new activity of M imposed by the potential E_2. Likewise there will be a redistribution of electrons and of electron holes. C. Wagner [21] has shown that this redistribution time is due to the counter diffusion of ions and electronic carriers and is determined exclusively by mobility of the excess electronic carriers as minority constituents. For cell II, at steady state under E_1 and under E_2 respectively, the activity of M, a_M, is,

$$a_M = \exp(-E_1F/RT) \tag{13}$$

and

$$a_M = \exp(-E_2F/RT) \ . \tag{14}$$

The concentrations of electrons and of electron holes at steady state under E_1 are

$$c_{\ominus} = c_{\ominus}^{o} \exp(-E_1F/RT) \qquad (15)$$

and

$$c_{\oplus} = c_{\oplus}^{o} \exp(+E_1F/RT) \ . \qquad (16)$$

These become at the new steady state,

$$c_{\ominus} = c_{\ominus}^{o} \exp(-E_2F/RT) \qquad (17)$$

and

$$c_{\oplus} = c_{\oplus}^{o} \exp(+E_2F/RT) \qquad (18)$$

In Eqs. (15) - (18), $c_{\ominus}^{o}$ and $c_{\oplus}^{o}$ denote the concentrations of electrons and electron holes in MX coexisting with the reversible electrode [M in cell II]. Thus the contribution due to the redistribution of electronic carriers can be minimized by decreasing the total number of electronic carriers available for redistribution. This can be done by (1) lowering the temperature; (2) decreasing the volume of the electrolyte, i.e. decreasing the length, L, for a given cross sectional area, A; and (3) adjusting the applied potential. For an electrolyte which conducts via electrons, increasing the potential decreases the concentration of electronic carriers and conversely for an electrolyte which conducts via electron holes.

D. O. Raleigh [22] pioneered in the study of the double layer capacitance at the interface between an electrolyte and an inert electrode. He carefully chose experimental conditions to minimize diffusion into the inert electrode and minimize the number of electronic carriers in the electrolyte under study. For example, some of his studies [23] involved the cell,

- Ag | AgBr | inert electrode + X

in which the sample geometry was about 0.15 cm long (i.e., L was small) by about 1.3 cm^2 area, A. The temperatures used were low (244 and 292°C), and AgBr coexisting with metallic silver conducts predominantly via electrons [24] so the potentiostatic measurements on applying $E_1 \rightarrow E_2$ where $E_2 > E_1$ assured the concentration of electrons was low. Thus the charge transfer measurements yielded the double layer capacitance at the AgBr inert electrode (single crystalline Pt or polycrystalline Pt or graphite).

The decay of the capacitance should obey,

$$I_C(t) = I_o \exp(-t/RC) \qquad (19)$$

where $I_C(t)$ denotes the charging or discharging current, I_o the current at $t = 0$, t the time, R the a.c. resistance of the electrolyte and C the capacitance. Raleigh's data obeyed Eq. (19) but he reported that the transient charging current for cell X did not decay with single time constant but generally consisted of a mixture involving two time constants. However, the point of the present discussion is that the double layer capacitance could be studied with the virtual exclusion of contributions due to redistribution of electronic carriers and from diffusion of ions into the inert electrode.

Raleigh [25] also studied the potentiostatic transient response of cells of type V. Under these conditions, Eq. (7) applies. Silver bromide equilibrated with gaseous bromine conducts electronically via electron holes [26]. Under these conditions, the activity of silver at the blocking electrode is,

$$a_{Ag} = \exp\,[-(+\,E_{theo} - E)]\;F/RT \qquad (20)$$

where E_{theo} denotes the decomposition potential and E the applied potential with the positive pole on the reversible electrode. The partial pressure of bromine at the blocking electrode, p_{Br_2},

$$p_{Br_2} = p^a_{Br_2}\,\exp\,(-2\,EF/RT) \qquad (21)$$

where $p^a_{Br_2}$ denotes the experimentally applied pressure of bromine at the reversible electrode. See Fig. 5.

At voltages of a few tenths of a volt at low temperatures, there will be no discharge of bromine at the blocking electrode and hence no steady state ionic current involving bromine discharge [25]. Therefore a measurement of the potentiostatic transient response of cell V could be carried out. When the voltage was well below the decomposition voltage, the data indicated a charging of a double layer capacitance at the AgBr|Pt interface [25].

If the number of carriers is increased, then the contribution of their redistribution to the total charge transferred during a transient measurement must be considered.

The number of electronic carriers, N, assuming the predominance of either electrons or holes, in a cylindrical electrolyte at steady state is given by [27]

$$N = (1/2)AL\,[c^o - c(L)] \qquad (22)$$

where c^o denotes the concentration of carriers in number per cc at the reversible electrode, c(L) that at the inert electrode.

A linear gradient in concentration has been assumed for steady state. The concentrations of carriers may be related to the applied potentials by Eqs. (15) - (18). Thus, for example, for Ag|AgBr|Pt at steady state under an applied potential of E_1,

$$N_\ominus(E_1) = (1/2)AL\ [c_\ominus{}^o - c_\ominus{}^o \exp(-E_1F/RT)] \qquad (23)$$

and that for E_2 is

$$N_\ominus(E_2) = (1/2)AL\ [c_\ominus{}^o - c_\ominus{}^o \exp(-E_2F/RT)]\ . \qquad (24)$$

For $E_2 > E_1$, $N_\ominus(E_1) > N_\ominus(E_2)$. Therefore the total charge transferred, Q_e, due to redistribution of electrons is $[N_\ominus(E_1) - N_\ominus(E_2)]$ times the electronic charge, e,

$$Q_e = (1/2)e\ AL\ c_\ominus{}^o\ [\exp(-E_1F/RT) - \exp(-E_2F/RT)] \qquad (25)$$

On the other hand the charge transfer due to double layer capacitance is given by

$$Q_{dl} = A\ C_{dl}\ (E_2 - E_1)\ , \qquad (26)$$

and the total charge transfer is

$$Q = Q_e + Q_{dl}\ . \qquad (27)$$

K. Weiss [29] has developed a method for determining the mobility of charge carriers which is valid under conditions where $Q_e \gg Q_{dl}$, i.e. when double layer charging is negligible. The determination of the mobility is important in that the temperature dependence of mobility may provide insight into the conduction mechanism and moreover the mobility coupled with the conductivity allows a calculation of the concentration of charge carriers using Eq. (1). In Weiss' method, the cell such as cell II is polarized to steady state and the applied potential is abruptly removed and the voltage is allowed to decay to zero--a relaxation method. The re-equilibration of a crystal is determined by the diffusion coefficient of the electrons or electron holes [21]. The open circuit potential as a function of time, $\Delta V(t)$, for an electron conductor is according to Weiss [29],

$$\Delta V(t) = (RT/F)\ \ln\ [c_\ominus{}^o/c_\ominus(t)] \qquad (28)$$

where $c_\ominus(t)$ denotes the concentration of electrons as a function of time. The analogous expression for an electron hole conductor [9] is

$$\Delta V(t) = (RT/F)\ \ln\ [c_\oplus(t)/c_\oplus{}^o] \qquad (29)$$

where $c_\oplus(t)$ denotes the concentration of electron holes as a function of time. Consider the solution of Fick's second law for the electron hole case [9]. The concentration gradient in the sample of length L is constant, i.e. a linear change in concentration is assumed at steady state. See Eq. (22). Thus the concentration of electron holes at a distance, x, from the M|MX interface is,

$$c_\oplus(x) = c_\oplus^{\,o} + x/L\,[c_\oplus(x = L) - c_\oplus^{\,o}] \tag{30}$$

for t = 0 and $0 \leq x \leq L$. And also,

$$\left(\frac{\partial c_\oplus}{\partial x}\right)_{x=L} = 0 \qquad \text{for} \qquad t > 0 \tag{31}$$

With these boundary conditions, a particular solution to Fick's second law is

$$\left\{\frac{c_\oplus(x,t) - c_\oplus^{\,o}}{c_\oplus(x, t=0) - c_\oplus^{\,o}}\right\} = \frac{4}{\pi^2}\sum_{n=0}^{\infty}\frac{(-1)^n}{(2\,n+1)^2}\sin\left[\frac{\pi x(2n-1)}{2L}\right]\bullet \exp\left[-\frac{\pi^2(2\,n+1)^2\,D_\oplus t}{4L^2}\right] \tag{32}$$

where $D_\oplus$ is the effective diffusion coefficient of the electron holes in MX. Equation (32) reduces to

$$\left\{\frac{c_\oplus(x,t) - c_\oplus^{\,o}}{c_\oplus(x, t=0) - c_\oplus^{\,o}}\right\} \cong \frac{4}{\pi^2}\exp\left[-\frac{\pi^2\,D_\oplus t}{4L^2}\right] \tag{33}$$

for long times. Substituting Eq. (16) for $E_1 = E$, the applied potential at steady state and Eq. (29) into Eq. (33) yields,

$$D_\oplus t = -\frac{4L^2}{\pi^2}\ln\left[\frac{\pi^2}{4}\,\frac{\exp\left(\frac{\Delta V(t)F}{RT}\right) - 1}{\exp\,(EF/RT) - 1}\right]. \tag{34}$$

Equation (34) can be simplified when $\Delta V(t)F \gg RT$ as,

$$D_\oplus t = -\frac{4L^2}{\pi^2}\left[\frac{\Delta V(t)}{RT}\,F - \left(\frac{EF}{RT}\right) + \ln\left(\frac{\pi^2}{4}\right)\right] \tag{35}$$

Thus a plot of $\Delta V(t)$ versus time will yield the value of $D_\oplus$ from which the mobility may be calculated using the Nernst-Einstein equation,

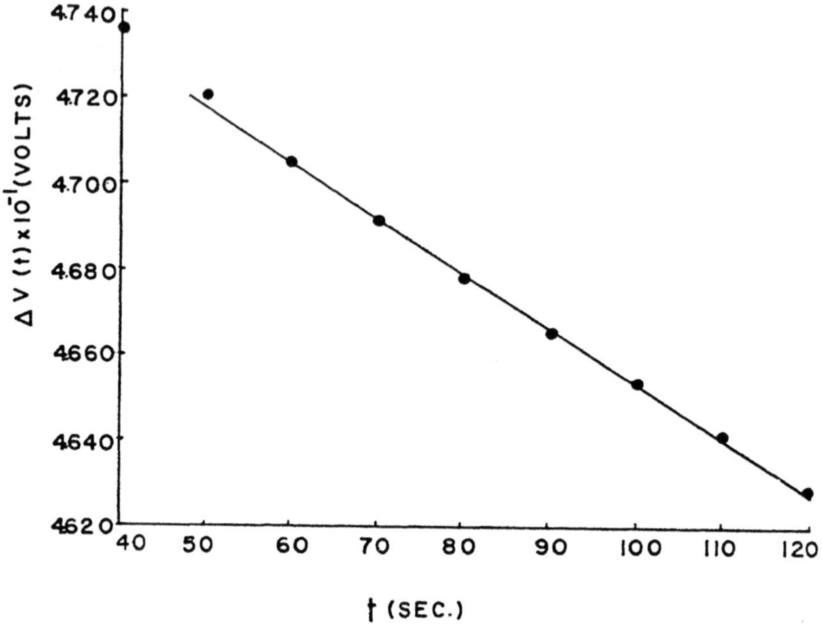

Fig. 9. Plot of $\Delta V(t)$ versus time for cuprous chloride ($L = 1.672$ cm) at 373°C. After Joshi [9].

$$D_{\oplus} = B_{\oplus} kT = \mu_{\oplus} RT/F \tag{36}$$

The analogous expression for the effective diffusion coefficient of electrons according to Weiss [29] is,

$$D_{\ominus} t \cong -\frac{4L^2}{\pi^2} \ln \left[\frac{\pi^2}{8} \bullet \left\{\frac{1 - \exp[-F\Delta V(t)/RT]}{1 - \exp[-EF/RT]}\right\}\right] . \tag{37}$$

A graph $\Delta V(t)$ versus t for CuCl in the cell,

$$- \text{Cu} \mid \text{CuCl} \mid \text{graphite} + \qquad \text{XI}$$

is shown in Fig. 9. Using Eq. (36), the mobility of electron holes was calculated.

Raleigh [27] cautioned against the use of Weiss' method unless the contribution due to charging of a double layer is eliminated or taken into account. Joshi [9] developed a double cell involving two similar cells of type II but differing only in the lengths of the electrolytes in order to separate the current due to charging from that due to electron hole redistribution. A schematic of this cell is shown in Fig. 10.

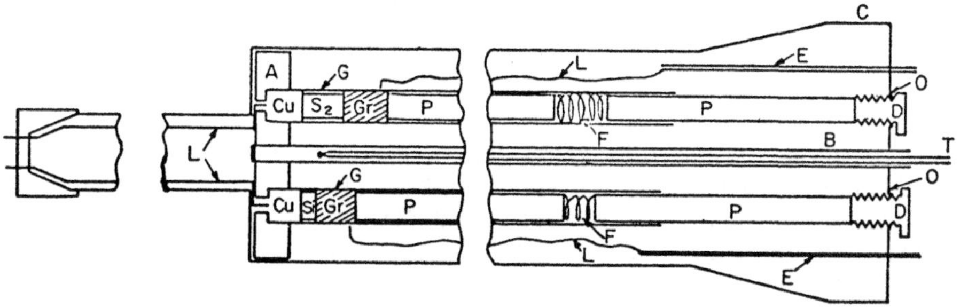

Fig. 10. Schematic diagram of a double cell consisting of two polarization cells identical except for differing lengths of the electrolytes [9]. In this diagram,

A - Lavite block	O - O-rings
B - Stainless steel block	P - Plungers
C - Teflon plus	T - Thermocouple
D - Stainless steel bolt	S_1 - Shorter sample of CuCl
E - Copper leads	S_2 - Longer sample of CuCl
F - Springs	Cu - Copper pellet
G - Glass tube cell holder	Gr - Graphite pellet
L - Platinum lead	

Joshi recognized that for a given sample size, double layer capacitance depended on the area of the sample not its length, [see Eq. (26)]. On the other hand, the redistribution of carriers depends on the number of carriers and hence the length of the sample. See Eq. (25). Following the suggestions of Raleigh [27] he developed a charge transfer experiment to separate these two effects. Again as an example, consider a double cell of type XI. Hence using Eq. (22) and Eqs. (16) and (18), the number of electron holes in a electron hole conductivity sample under applied potential, E_1 is

$$N_{\oplus} = 1/2\ AL\ c_{\oplus}{}^{o}\ [\exp(E_1F/RT) - 1] \tag{38}$$

and at steady state under E_2 is

$$N_{\oplus}' = 1/2\ AL\ c_{\oplus}{}^{o}\ [\exp(E_2F/RT) - 1] \tag{39}$$

Thus the total amount of charge transfer involved in changing the polarizing potential from E_1 to E_2 in analogy to Eq. (25) is

$$Q_e = 1/2\ e\ AL\ c_{\oplus}{}^{o}\ [\exp(E_2F/RT) - \exp(E_1F/RT)] \tag{40}$$

and the double layer charging is given by Eq. (26) so that using Eq. (27) the total charge transfer, Q_T, is given by

$$Q_T = (1/2)\, e\, AL\, c_\oplus{}^o\, [\exp(E_2F/RT) - \exp(E_1F/RT)] + A\, C_{dl}\, [E_2 - E_1] \quad (41)$$

for an electron hole conductor. Now if two samples, identical but of differing lengths, L_1 and L_2 where $L_1 > L_2$, are polarized, the contribution Q_{dl} may be separated from Q_e. This is possible if Q_{dl} is the same for both samples. Then for a sample of length, $(L_1 - L_2)$, one has

$$Q_e = (eA/2)(L_1 - L_2)\, c_\oplus{}^o\, [\exp(E_2F/RT) - \exp(E_1F/RT)] \ . \quad (42)$$

A similar expression holds for electron conductors.

Using the double cell, the value of Q_e for a sample of length L_1 (or L_2) may be subtracted from the total charge transfer Q_T for a single cell. Thus the value of Q_{dl} and C_{dl} may be obtained. Experimentally it is convenient to use $E_2 \gg E_1$ so that the first term of Eq. (42) is much larger than the second. Hence,

$$\ln Q_e = \ln \{(eA/2)(L_1 - L_2)\, c_\oplus{}^o\} + E_2F/RT \quad (43)$$

A plot of $\ln Q_e$ versus E_2 yields a slope of F/RT and the intercept yields directly the value of $c_\oplus{}^o$. If the value of $\sigma_\oplus{}^o$ is known from steady state measurements, see Eq. (5), then the mobility may be determined using Eq. (1).

Another method to determine the mobility is to measure the current as a function of time on application of a voltage step when the contributions of double layer charging and diffusion into the inert electrode are negligible. As an example, consider again a single cell of type XI in which a sample of CuCl of length L is polarized at steady state under a potential E_1. The concentration of electron holes at the Cu|CuCl interface is given by $c_\oplus{}^o$ and that at the CuCl|graphite interface by Eq. (16). Under steady state conditions, the concentration of electron holes as a function of distance x from the blocking electrode where x = 0 can be expressed as [10],

$$c_\oplus(x,0) = c_\oplus{}^o \exp(E_1F/RT) - x/L\, [c_\oplus{}^o \exp(E_1F/RT) - c_\oplus{}^o] \quad (44)$$

This is the initial condition for t = 0. When the applied potential is changed from E_1 to E_2, the new concentration of electron holes at the blocking electrode is given by Eq. (18) for x = 0, t > 0. For CuCl ($L \lesssim 1$ cm) at 200 - 300°C, the process occurs in one or two seconds and diffusion of electron holes in the electrolyte will occur. A particular solution to Fick's second law with these boundary conditions is,

$$c_{\oplus}(x,t) = \{c_{\oplus}^{o} \exp(E_2F/RT) \operatorname{erfc} x/(2\sqrt{D_{\oplus}t})$$
$$+ c_{\oplus}^{o} \exp(E_1F/RT) \operatorname{erf} x/(2\sqrt{D_{\oplus}t})$$
$$- [c_{\oplus}^{o} \exp(E_1F/RT) - c_{\oplus}^{o}]\, x/L\} \tag{45}$$

Equation (45) is valid for a semi-infinite rod, i.e. for short times in the experiment of Joshi [10] (~ 100 to 200 sec). Differentiation of Eq. (46) with respect to x yields

$$\left.\frac{\partial c_{\oplus}}{\partial x}\right|_{x=0} = \left\{- \frac{c_{\oplus}^{o} \exp(E_2F/RT)}{\sqrt{\pi D_{\oplus}t}} + \frac{c_{\oplus}^{o} \exp(E_1F/RT)}{\sqrt{\pi D_{\oplus}t}}\right.$$
$$\left. - 1/L\, [c_{\oplus}^{o} \exp(E_1F/RT) - c_{\oplus}^{o}]\right\} \tag{46}$$

The flux, $J_{\oplus}$, is

$$J_{\oplus} = - D_{\oplus} \left.\frac{\partial c_{\oplus}}{\partial x}\right|_{x=0} = I_{\oplus}/A \tag{47}$$

so that the current due to electron holes is,

$$I_{\oplus}(t) = - (eA\, D_{\oplus}) \left(\left.\frac{\partial c_{\oplus}}{\partial x}\right|_{x=0}\right)$$
$$+ \frac{eA\, D_{\oplus}}{\sqrt{\pi D_{\oplus}t}}\, c_{\oplus}^{o}\, [\exp(E_2F/RT - \exp(E_1F/RT)]$$
$$+ [eA\, D_{\oplus}c_{\oplus}^{o}/L]\, [\exp(E_1F/RT) - 1] \tag{48}$$

Substituting Eq. (36) and $\sigma_{\oplus}^{o} = c_{\oplus}^{o}\, e\, \mu_{\oplus}$ into Eq. (48) yields,

$$I_{\oplus}(t) = L/(\sqrt{\pi D_{\oplus}t})\, [I_{\oplus}(E_2) - I_{\oplus}(E_1)] + I_{\oplus}(E_2) \tag{49}$$

where $I_{\oplus}(E_2)$ and $I_{\oplus}(E_1)$ are the steady state currents given by the second term of Eq. (5) for E_2 and E_1, respectively. Thus a plot of $I_{\oplus}$ versus the square root of time for short times, should yield a straight line from which $D_{\oplus}$ can be obtained. Joshi [9,10] has carried out such experiments and results for this current-time analysis are shown in Fig. 11.

For thin samples and lower temperatures, such plots of

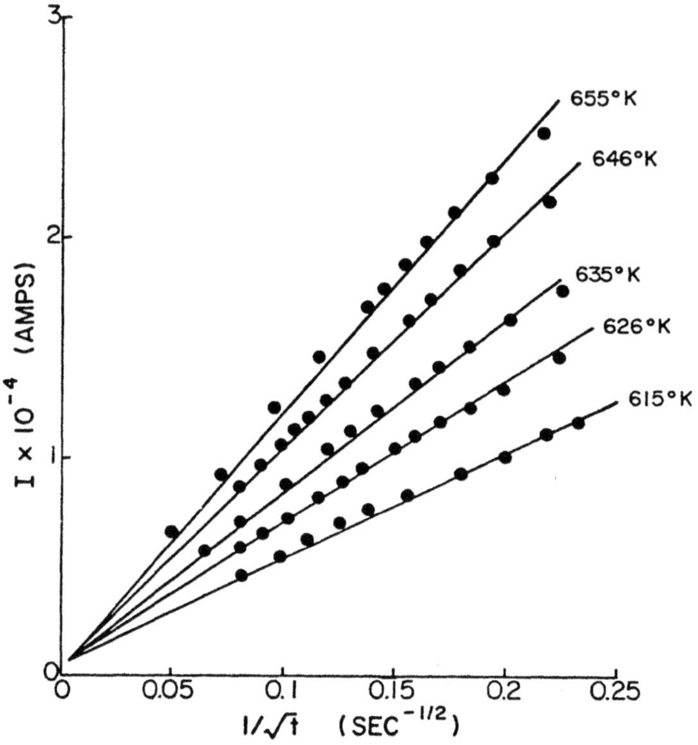

Fig. 11. Plots of $I_\oplus$ vs $1/\sqrt{t}$ for CuCl (L = 1.672 cm) for cell XI subjected to a potentiostatic change in voltage, E_1 to E_2. After Joshi [9,28].

$I_\oplus$ vs $1/\sqrt{t}$ may exhibit deviations from linearity because the contribution to double layer charging is no longer negligible. When a very thin sample is studied at lower temperatures, the double layer capacitance charging predominates. Under these conditions a plot of log $I_\oplus$ versus t yields a straight line according to Eq. (19). Results are shown in Fig. 12.

Raleigh [25] has studied the current-time response to potentiostatic step voltages applied to cell V and reported graphical resolution of the different components of the current-time plots which were non-linear for log $I_\oplus$ vs t. Recently Jow [20] has carried out similar tests on CuI in cell VIII. He has found the shape of the current-time plots can be qualitatively resolved into a double layer charging portion and a transient electronic hole current as is shown in the curves in Fig. 13. Observe that for a given thickness and temperature, both low and high polarizing potential steps yield a sharp rise in current at time t = 0^+ owing to a large contribution from double layer charging. The

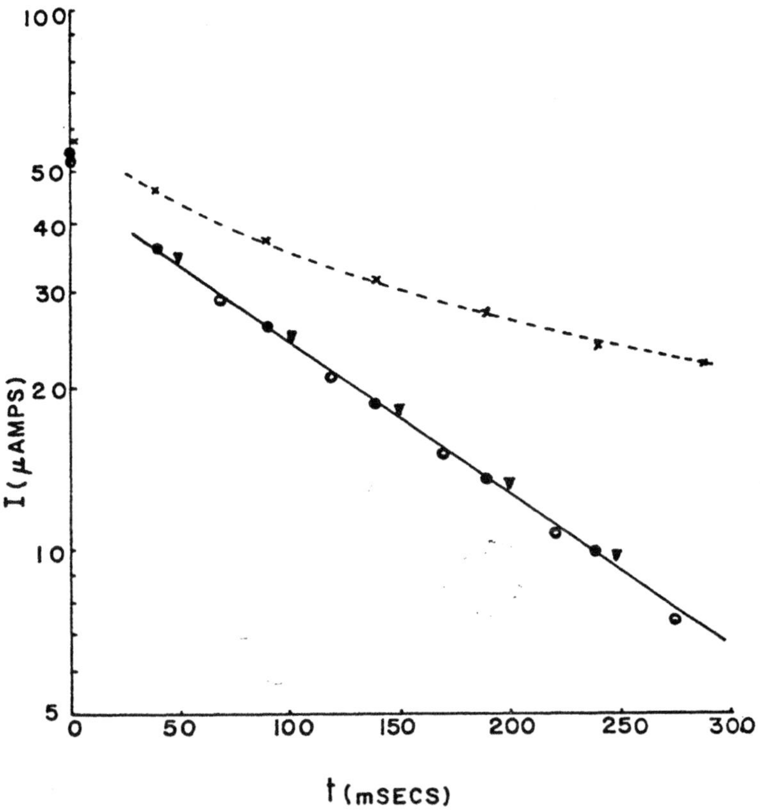

Fig. 12. Plots of log $I_{\oplus}$ versus t for a thin sample of CuCl ($L = 0.085$ cm) at 242°C. The voltage steps are:

Symbol	E_1(volts)		E_2(volts)
•	0.2476	-	0.2677
○	0.2678	-	0.2879
▽	0.2878	-	0.3080
x	0.4287	-	0.4489

Note that for the higher potentials the upper curve deviates from linearity. This is due to a finite contribution from the redistribution of electron holes [9].

transient hole current at steady states for E_1 and E_2 are given by the second term of Eq. (5). If the transient involves primarily the redistribution of electron holes, the shape of the current-time curve should follow an exponential curve as shown in the dotted line of Fig. 13a. As E_2 is increased, the concentration of electron holes increases and the relative contribution of double layer charging to the total process becomes smaller as is shown

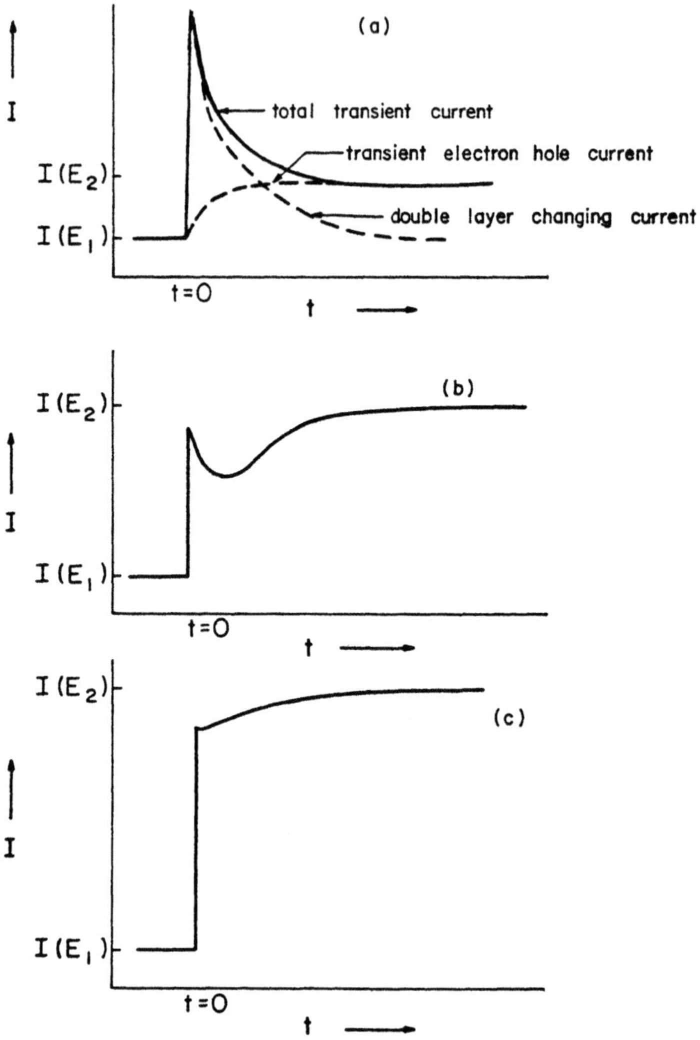

Fig. 13. Schematic plots of current versus time for Cu|CuI| graphite. The solid lines indicate the oscillographic trace obtained experimentally. The dashed curves represent the current due to double layer charging and to electron hole redistribution.

in Figs. 13b and c. Further work is needed to quantitatively separate these transient contributions.

Raleigh [23] has examined the behavior of cells of type X galvanostatically. The cell was initially polarized at a polarizing potential, E_1, where the activity of silver was low at the AgBr|inert electrode (typically -0.3 v, where $a_{Ag} \cong 0.0015$ at

244° and at 292°C). The cell was then switched to a constant current source which conducts Ag^+ ions to the blocking interface. The voltage was monitored as the dependent variable. Under these conditions Raleigh reports that the voltage-time expressions will be,

$$\Delta V(t) = E_1 + IR + (It/C_{dl}) \tag{50}$$

where I is the constant current, R the resistance of the electrolyte between the reversible and blocking electrode, t is the time and C_{dl} the double layer capacitance (assumed constant over the experiment). In this way the resulting voltage-time curve rises linearly with time until the decomposition voltage, E_{theo}, is approached at which time Faradaic processes--deposition of Ag on the blocking electrode occurred. Raleigh varied the blocking electrode and determined the silver coverage on the electrode surface by analysis of the galvanostatic voltage-time curves obtained as oscillographic traces.

If the inert electrode in cell II is replaced by one which can dissolve a component of MX, then diffusion into the electrode can be studied. When a cell is at steady state and suddenly a new value of the potential, E_2, is applied, the activity of M at the MX|positive electrode is given by Eq. (14). Thus the component of MX, assumed to be M in this example, will diffuse into the positive electrode in order to maintain the new activity imposed on it.

In order to carry out this type of diffusion study, the contribution from the current due to redistribution of electronic carriers and for charging or discharging the double layer capacitor it must be small and known. Furthermore, the "sink" electrode must be an electronic conductor, exhibit solubility for M (or X) and diffusion must be rapid. H. Rickert and coworkers [30,31,32] were among the first to use this technique to study oxygen diffusion in metals. An excellent review by Rickert [33] is available which outlines the theory and also experimental details. Actually, the polarization cells involving AgBr as electrolyte mentioned above are readily adaptable to diffusion studies and Raleigh [34, 35] carried out such studies using the cell,

Ag | AgBr | Au XII

i.e. the inert electrode, C or Pt, in cell X was replaced by gold as a sink for Ag.

As mentioned previously, AgBr, conducts electronicly, primarily via electrons when it coexists with Ag. Thus, although there will be some electronic leakage according to the first term of Eq. (5), the number of electrons will be small, Eq. (23), and the time for redistribution of carriers will be relatively short thus allowing the use of a potentiostatic technique. At the AgBr|Au interface, the activity of silver is given by Eq. (13) for E_1 and Eq. (14) for E_2. If c_0 and c_1 are the concentrations of silver corresponding to $a_{Ag}(E_1)$ and $a_{Ag}(E_2)$ respectively, the diffusion flux, J, in grams per cm^3 will be,

$$J = -D \left.\frac{\partial c}{\partial x}\right|_{x=0}$$

$$-D(c_1 - c_0) \bullet \left[\frac{d}{dx} \operatorname{erfc} \frac{x}{2\sqrt{Dt}}\right]_{x=0} = (c_1 - c_0)\left(\frac{D}{\pi t}\right)^{1/2} \qquad (51)$$

The current in the external circuit, I_D, is given by Faraday's law as

$$I_D = JAF \qquad (52)$$

where A is the cross sectional area (only unidimensional diffusion is considered in this case) and F is the Faraday constant. The current will continue until a new activity has been estiablished in the sink, the gold. The activity of silver as a function of concentration must be known. Corrections for the phase boundary motion must be made when necessary. Raleigh and Crowe applied small voltage steps to yield both in diffusion and out diffusion. Plots of the current versus the reciprocal of the square root of time were linear. However at infinite time, there was always a residual current. This residual current is due to the low level electronic leakage current in AgBr. At short times, below 10 sec at 394°C, there was a slight departure from linearity which Raleigh attributed to double layer charging.

Recently J. Goldman and J. B. Wagner, Jr. [36] carried out an analogous potentiostatic study utilizing the cell,

Cu | CuCl | Au . XIII

Because CuCl conducts via electron holes, if the length of CuCl is long compared to the gold, then the time to redistribute the electron holes may be rate determining [37] and the resulting diffusion current-time analysis will represent the effective diffusion coefficient of electron holes in CuCl rather than of copper in gold. A thin, 0.12 cm sample of CuCl was used and the contribution to the current due to diffusion of copper in gold, electronic leakage and double layer capacitances were separated as follows. The diffusion current is given by Eqs. (51) and (52) above.

The current due to the electron hole leakage is given by

$$I_{\oplus}(t) = -\frac{RTA}{LF}\sigma_{\oplus}^{o}[\exp(E_2F/RT) - 1] + \frac{2RTA}{LF}\sigma_{\oplus}^{o}[\exp(E_1F/RT) - \exp(E_2F/RT)] \bullet \sum_{n=0}^{\infty}\left[-\frac{n^2\pi^2 D_{\oplus}t}{L^2}\right]. \quad (53)$$

The current in Eq. (53) approaches a steady state value as t approaches infinity in which case Eq. (53) reduces to the second term of Eq. (5). The current due to charging a double layer capacitance is given by Eq. (19) where $I_o = E/R$ and E is the applied potential (E_2 in this case) and R the a.c. resistance of the electrolyte. The cell was subjected to voltage steps and the resulting current measured. In order to obtain the capacitance, the temperature was lowered to 228°C. This insured that diffusion of copper in gold was negligible and that the current due to electron hole distribution was small. Plots of logarithm of the current versus time yielded values of RC and the intercepts yielded $I_o = E/R$. The values of I_o were then plotted against E and the resulting linear plot yielded 1/R. Thereafter the value of RC was used at higher temperatures to insure that transient contributions due to double layer charging were negligible. This was accomplished by waiting a time equal to about 10 times RC_{dl} before the first data points were taken. It was assumed that C_{dl} was not dependent on temperature over the range studied.

The higher temperature (355 - 400°C) runs involved diffusion of copper into gold and data are shown in Fig. 14 and the resulting values of D(Cu → Au) are shown in Fig. 15.

From the intercepts of the curves in Fig. 14, values of the electronic leakage in CuCl were obtained. These were plotted against exp(EF/RT) in conformity with the second term of Eq. (5) and are shown in Fig. 16. The slopes yielded values of $\sigma_{\oplus}^{o}$ which were in excellent agreement with the data of Joshi [28] described earlier.

It should be mentioned that galvanostatic diffusion measurements may also be carried out as has been done by Pastorek and Rapp [39] and by Osterwald and Schwartzlose [40] for oxygen in copper. Recently, Rickert and coworkers have determined the chemical or effective diffusion coefficient in non-stoichiometric compounds. This involves, for example, replacing the gold sink in cell XII by a non-stoichiometric compound which conducts electronically. The cell used by Chu, Rickert and Weppner [32,33]

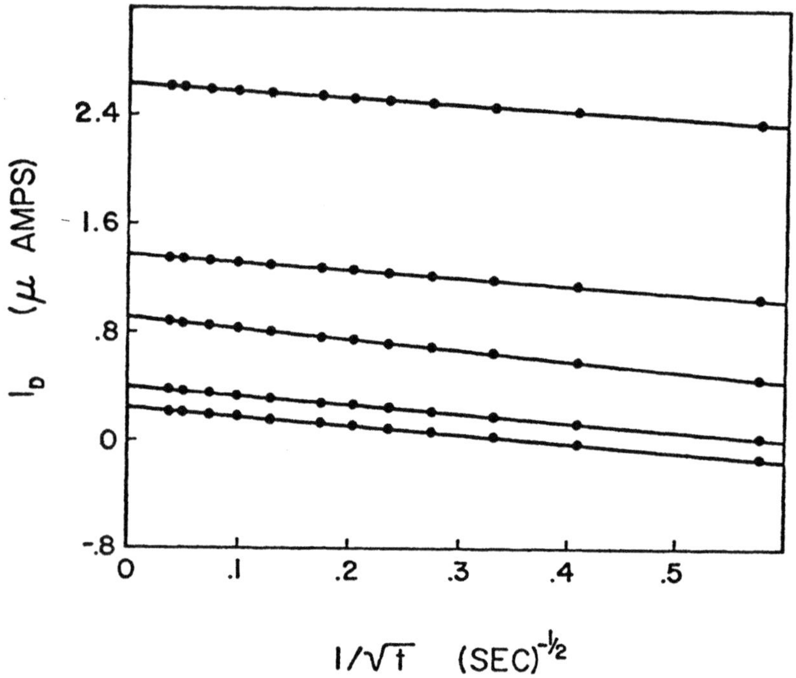

Fig. 14. Current versus $1/\sqrt{t}$ at 370°C for cell XIII. The slopes are linear indicating a diffusion process. See Eqs. (51) and (52). The voltages at infinite time given by the intercepts are due to the residual steady state electron hole leakage. After Goldman (36).

to study chemical diffusion in $Ag_{2+\delta}S$ was

$$Ag \mid AgI \mid Ag_{2+\delta}S \mid S(\ell) \qquad \text{XIII}$$

First measurements were made on $Ag_{2+\delta}S$ in equilibrium with sulfur. The silver sulfide was first brought into a stationary state involving a small concentration gradient of silver in silver sulfide by impressing a voltage on the cell. Then the circuit was interrupted and the voltage was measured as a function of time. Just as in the case of silver diffusing into gold, the concentration-voltage relations must be known prior to the experiment. These were carried out by solid state coulometric titrations earlier by C. Wagner [41]. The initial condition (steady state) for a sample of length L is,

$$c = c'' - (c'' - c')\, x/L \qquad \text{at } t = 0 \tag{54}$$

and the boundary conditions are

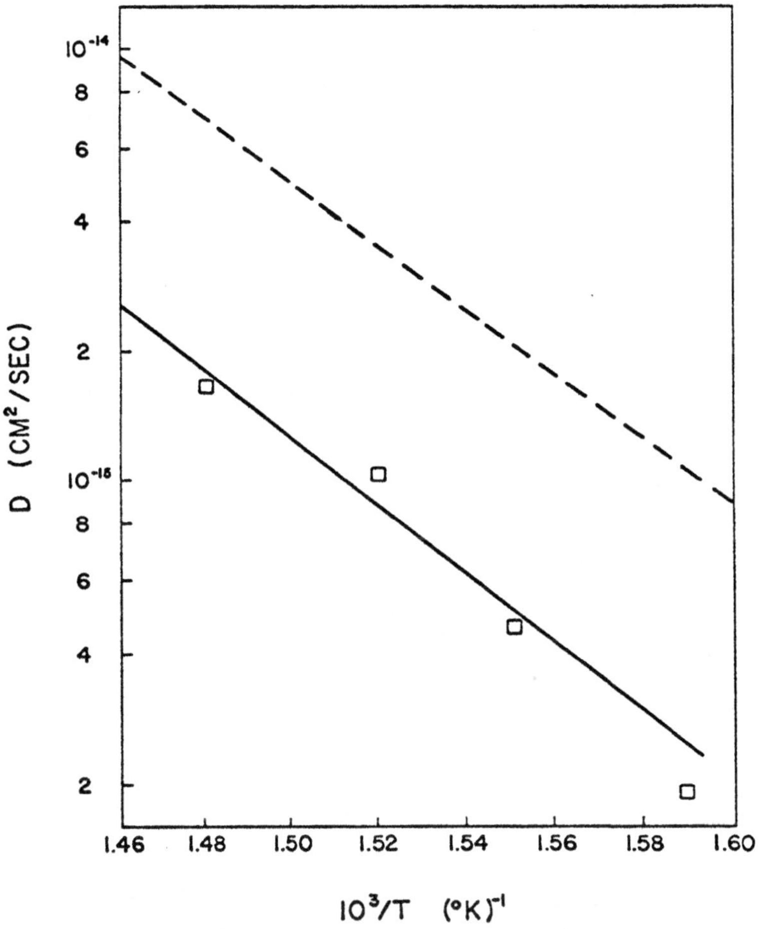

Fig. 15. Calculated diffusion coefficients for copper in single crystalline gold at 355°, 370°, 385° and 400°C. The dotted curve is extrapolated data of Vignes and Haeussler [38].

$$c = c' \qquad \text{for } x = L \text{ at } t \geq 0 \tag{55}$$

$$[\partial c/\partial x]_{x=0} = 0 \qquad \text{for } t > 0 \tag{56}$$

the solution for which yields,

$$[c - c''] = -\{2[c'' - c']/L\}\,[\sqrt{Dt}/\sqrt{\pi}] \tag{57}$$

for short times ($t \ll 4L^2/\pi D$). A plot of $\Delta V(t)$ versus time may be used to obtain a plot of $[c - c'']$ versus $\sqrt{t}$ according to Eq. (57)

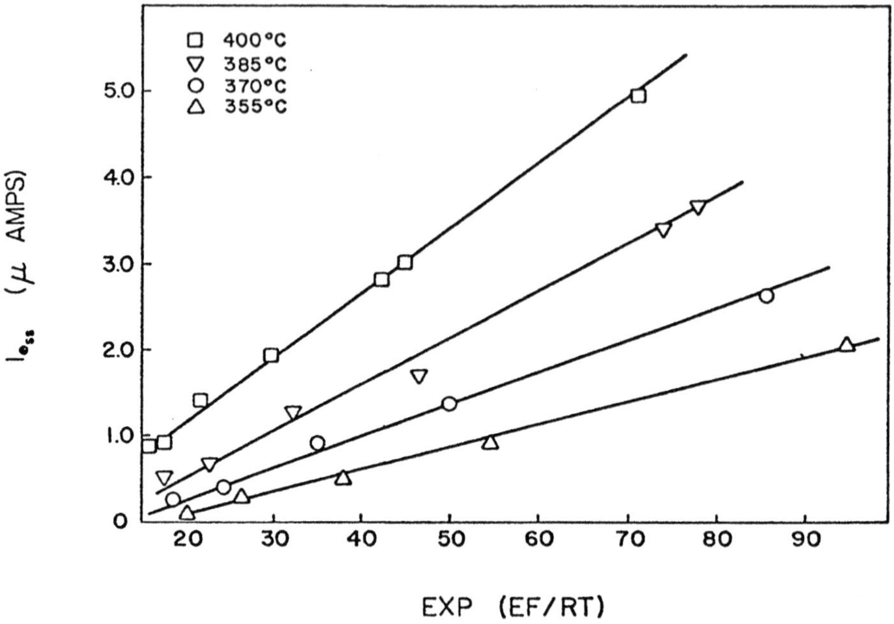

Fig. 16. Plots of the steady state leakage current in CuCl versus exp(EF/RT) for cell XIII. The data were taken from the intercepts of the diffusion current at infinite times. See Fig. 14. The slopes yield $\sigma_{\oplus}^{o}$ for CuCl and agree with the data of Joshi obtained by the charge transfer technique. After Goldman [36].

As mentioned earlier, the first diffusion measurements utilizing solid state electrolytes involved calcia or yttria stabilized zirconia to transport oxygen into a electrode material which exhibits solubility for oxygen. The cell is represented by;

$$p'_{O_2},\ Pt \mid \text{doped } ZrO_2 \mid Pt,\ p''_{O_2} \qquad \text{XIV}$$

where the effective oxygen pressure on the left hand side, p'_{O_2} might be provided by a gas mixture or by a two phase mixture such as Fe-FeO which yields a well-defined chemical potential of oxygen. The effective oxygen pressure on the right hand side is fixed by another reference state for the case of an equilibrium measurement in which case

$$\mu''_{O_2} - \mu'_{O_2} = 4\ EF \qquad (58)$$

and because $\mu_O = 1/2\ \mu_{O_2} = \mu_O^{o} + RT \ln a_O$, then the cell voltage is related to the corresponding activities of oxygen by

$$E = (RT/2F) \ln [a_O''/a_O'] \qquad (59)$$

For a diffusion measurement, oxygen may be transported from left to right through the electrolyte and into a sink potentiostatically just as described for the cells involving AgBr and CuCl as electrolytes.

The solution to Fick's second law for unidimensional diffusion [33] of oxygen into a metal such as silver is

$$I = 2\ F\ c_O \sqrt{D/\pi t} \qquad (60)$$

where c_O denotes the initial concentration. Hence a plot of I versus $1/\sqrt{t}$ yields a straight line. This technique has also been applied to the determination of chemical diffusion in semiconductors, i.e. the metallic electrode acting as a sink for oxygen has been replaced by a non-stoichiometric compound which is an electronic conductor as,

$$p_{O_2}, Pt(1)\ |\ \text{doped } ZrO_2\ |\ Fe_{1-x}O\ |\ Pt(2), N_2 \qquad XV$$

Application of Fick's second law to this situation yields, [32,33]

$$I = \sqrt{D}\ Q/L \sqrt{\pi t} \qquad (61)$$

for a sample of wüstite of cross sectional area, A and length L in which the time $\ll L^2/D$. Here Q denotes the total charge transported, D is the interdiffusion or chemical diffusion coefficient in wüstite and c' is the initial uniform concentration of metal in the wüstite. This can be adjusted by coulometric titration before the diffusion measurements are begun. c'' is the concentration at the interface for $t > 0$ as imposed by the new potential on the cell. For long times,

$$\log I = \log \left\{\frac{4\ Ae\ D\ [c'' - c']}{L}\right\} - [1.071/L^2]\ Dt \qquad (62)$$

A plot of log I versus t yields D directly. In this case the emf versus concentration data are not needed. A major advantage of this method is that the slow step involving the dissociation of gases on wüstite in the analogous thermogravimetric relaxation method is avoided (42).

SUMMARY

Methods to measure the transport number of electronic carriers using the C. Wagner asymmetric cell were reviewed. Transient measurements involving the potentiostatic application of $E_1 \rightarrow E_2$ were shown to yield three contributions to the observed current. These were (1) charging or discharging of a double layer capacitor, (2) redistribution of electronic carriers within the electrolyte and (3) diffusion of a constituent of the electrolyte into the positive electrode. Methods to separate and study each were presented.

Acknowledgments

This work was supported by Grant DAHC 15 73 G19 from the Advanced Research Projects Agency through the Materials Research Center of Northwestern University.

References

1. R. A. Rapp and D. A. Shores, "Solid Electrolyte Cells," Chapter 6C in Physiochemical Measurements in Metals Research, Volume IV, part 2, edited by R. A. Rapp, John Wiley and Sons, New York (1970).
2. R. N. Blumenthal and M. A. Seitz, "Experimental Techniques," Chapter 2 in Electrical Conductivity in Ceramics and Glass, Part A, edited by N. M. Tallan, Marcel Dekker, Inc., New York (1974).
3. K. Schwerdtfeger and E. T. Turkdogan, "Equilibria and Transport Phenomena Involving Gas Mixtures and Condensed Phases," Chapter 4B in Physiochemical Measurements in Metals Research, Volume IV, part 1, edited by R.A. Rapp, John Wiley and Sons, New York (1970).
4. C. Tubandt, Handbuch der Experimentalphysik, edited by W. Wien and F. Harms, Vol. 21, Part 1, p. 383, Akademische Verlagsgesellschaft, Leipzig (1932).
5. A. B. Lidiard, Handbuch der Experimental Physik, Vol. XX, p. 246, Springer-Verlag, Berlin (1957).
6. C. Wagner, Z. physik Chemie B21, 42 (1933).
7. A. Dravnieks and H. J. McDonald, J. Electrochem. Soc. 93, 179 (1948).
8. C. Wagner, Proc. Seventh Meeting C.I.T.C.E. Lindau (1955) p. 361, Butterworths Publications, London (1957).
9. A. V. Joshi, Ph.D. Thesis, Northwestern University (1972).
10. A. V. Joshi and J. B. Wagner, Jr., J. Electrochem. Soc. 122, 1071 (1975).
11. T. Takahashi and O. Yamamoto, Electrochim. Acta 11, 779 (1966).

12. J. W. Patterson, E. C. Bogren and R. A. Rapp, J. Electrochem. Soc. 114, 752 (1967).

13. M. S. Whittingham and R. A. Huggins, J. Electrochem. Soc. 118, 1 (1971).

14. Y. J. van der Meulen and F. A. Kröger, J. Electrochem. Soc. 117, 69 (1970).

15. D. O. Raleigh, J. Phys. Chem. Solids 26, 329 (1965).

16. V. W. Hsueh and R. W. Christy, J. Chem. Phys. 39, 3519 (1963).

17. J. B. Wagner, Jr. and C. Wagner, J. Electrochem. Soc. 104, 500 (1957).

18. J. W. Patterson, "Ionic Conductivity and Electrochemistry of Crystalline Ceramics," in Electrical Conductivity in Ceramics and Glass, part B, p. 453, edited by N. M. Tallan, M. Dekker Inc., New York (1974).

19. A. V. Joshi and C. C. Liang, Phys. Chem. Solids 36, 927 (1975).

20. T. Jow, Ph.D. Thesis research in progress, Northwestern University (1975).

21. C. Wagner, J. Phys. Chem. Solids 29, 1925 (1968).

22. D. O. Raleigh, "Electrode Processes in Solid-Electrolyte Systems," and his references therein, in Advances in Electroanalytical Chemistry, Vol. 6, edited by A. J. Bard, M. Dekker, New York (1973).

23. D. O. Raleigh, J. Phys. Chem. 71, 1785 (1967).

24. B. Ilschner, J. Chem. Phys. 28, 1109 (1958).

25. D. O. Raleigh, J. Phys. Chem. 70, 680 (1966).

26. C. Wagner, Z. phys. Chem. B32, 447 (1936).

27. D. O. Raleigh, Z. phys. Chem. [N.F.] 63, 319 (1969).

28. A. V. Joshi, "Electrochemical Studies on Single Crystalline Cuprous Chloride as a Solid State Electrolyte," in Fast Ion Transport in Solids, edited by W. van Gool, p. 173, Amsterdam-North Holland Publishing Co. (1973).

29. K. Weiss, Z. physik Chem. [N.F.] 59, 242 (1968).

30. H. Rickert and R. Steiner, Z. physik. Chem. [N.F.] 49, 127 (1966).

31. H. Rickert and A. A. El Miligy, Z. Metallkunde 59, 635 (1968).

32. H. F. Chu, H. Rickert and W. Weppner, "Electrochemical Investigations of Chemical Diffusion in Wüstite and Silver-Sulfide," in Fast Ion Transport in Solids, edited by W. van Gool, p. 181, Amsterdam-North Holland Pub. Co. (1973).

33. H. Rickert, Einführung in die Elektrochemie fester Stoffe, Springer-Verlag, Berlin (1973).

34. D. O. Raleigh, J. Electrochem. Soc. 114, 493 (1967).

35. D. O. Raleigh and H. R. Crowe, J. Electrochem. Soc. 116, 40 (1969).

36. J. Goldman and J. B. Wagner, Jr., J. Electrochem. Soc. 121, 1318 (1974).

37. J. B. Wagner, Jr., "Electronic Conductivity, Mobility and Double Layer Capacity in Solid State Electrolytes," in Fast Ion Transport in Solids, edited by W. van Gool, p. 489, Amsterdam-North Holland Pub. Co. (1973).

38. A. Vignes and J. Haeussler, Compt. Rend. 263, Series C, 1504 (1966).

39. R. L. Pastorek and R. A. Rapp, Trans. Met. Soc. AIME 245, 1711 (1969).

40. J. Osterwald and G. Schwartzlose, Z. physik Chem. [N.F.] 62, 119 (1968).

41. C. Wagner, J. Chem. Phys. 21, 1819 (1953).

42. L. W. Laub and J. B. Wagner, Jr., Oxidation of Metals 7, 1 (1973).

DISCUSSION

Den ENGELSEN : Professor Wagner distinguished two mechanisms in explaining the transient phenomena upon jump-biassing a dc polarization cell. These two mechanisms are (dis) charging of the double layer capacity and redistribution of electronic carriers. From an experimental point of view, this description is satisfactory. However, in my opinion, these two mechanisms are related each other. The question now is how ?

WAGNER : When a voltage step is applied, both diffusion of charge carriers (redistribution) and (dis) charging of the double layer may occur simultaneously. See fig.13. So far, we have not quantitatively separated the two contributions during a transient although the total charge transfered may be separated at the end of the experiment by methods described in the paper. The initial current transient is dominated by the double layer charging and the rate of approach to the new steady state is dominated by the redistribution of electronic carriers. The latter rate is the effective migration of electronic carriers accompanied by the ions produced by the new non-stoichiometry which is fixed by the new externally applied potential. The double layer charging for CuCl is observed at the lower applied potentials and conversely for redistribution.

HARTMANN : In addition to what you mentioned about chemical diffusion in silver-sulfide, we extended our measurements to a symmetrical cell with silver/silver iodide and two Pt-probes on each side of a long sample of Ag_2S or Ag_2Se which allowed us to establish a potential on each side and measure the EMF on each side independently from a flow of current. The relaxation of a silver concentration gradient recorded by EMF was used to measure $\tilde{D}$ as a function of deviation from ideal stoichiometry. For $Ag_{2+\delta}S$ at 200°C the values of $\tilde{D}$ are about 0.08 cm^2 sec^{-1} at equilibrium with silver and 0.25 cm^2 sec^{-1} near ideal stoichiometry. The consistency of the measurement is shown with the good agreement of the measured $\tilde{D}$ values with those calculated from Darken-Wagner equation. D_K is obtained from conductivity data and the thermodynamic factor calculated from the slope of the electrochemical titration curve.
Ref. B. Hartmann, H. Rickert, W. Schendler, Electrochim. Acta (In press).

DUDLEY : The two electrode-voltage-step method of measuring diffusion electrochemically has the disadvantage that the actual electrochemical potential at the interface is changed by the voltage drop across the bulk electrolyte resistance which varies with time. If instead a constant current is used, the IR contribution to the measured cell over potential is constant with time and can be eliminated from the calculation.

WAGNER : Both constant voltage and constant current methods have been used by other investigators.

RALEIGH : I wish to make the point that if you want to use a solid-electrolyte cell to measure diffusion coefficients in this way, you don't necessarily have to measure the double layer capacitance and electron redistribution characteristics in the electrolyte. When you apply a d.c. bias that fixes a significant activity of the diffusant at the substrate surface, double-layer charging should be completed fairly soon. If you plot the cell current versus $t^{-1/2}$ and get a significant linear range, it can be either diffusion into the electrode or electronic redistribution in the electrolyte. If you reduce the electrolyte thickness and the $t^{-1/2}$ slope is unchanged you conclude it is diffusion into the electrode. If the diffusion slope is altered, you see if you can reduce the electrolyte thickness until it is no longer altered. An experimenter who is primarily interested in diffusion can then see empirically whether he has a valid apparatus for studying this diffusion.

ETSELL : In your list of assumptions pertaining to the d.c. polarization technique, should not the fact that it is assumed that electronic conductivity is unaffected by the space charge built up at the irreversible electrode be included ?

WAGNER : Yes, any space charge effects are neglected in what I discussed.

HENISCH : One of the systems with asymmetric contacts discussed involved two carrier transport. In the ordinary way, this is governed by two current equations in terms of field and diffusion, two continuity relationships (incl. recombination), Poisson's equation, and boundary conditions for field, potential and carrier concentrations. These transport equations cannot be explicitly solved without simplifying assumptions. It would be interesting to see what assumptions can reasonably be made to render the problem algebraically tractable. The equations given appeared to envisage diffusion without space charge. How do (a) the local field,

(b) the local carrier concentrations, (c) local charge density and (d) the local potential vary with distance ? What provision is there for current continuity at the electrodes ? If the steady state (in the presence of the field) involves an ionic concentration gradient, how is the bulk conductivity defined.

WAGNER : The diagram which I projected for an asymmetric cell, polarized under E at steady state was as follows :

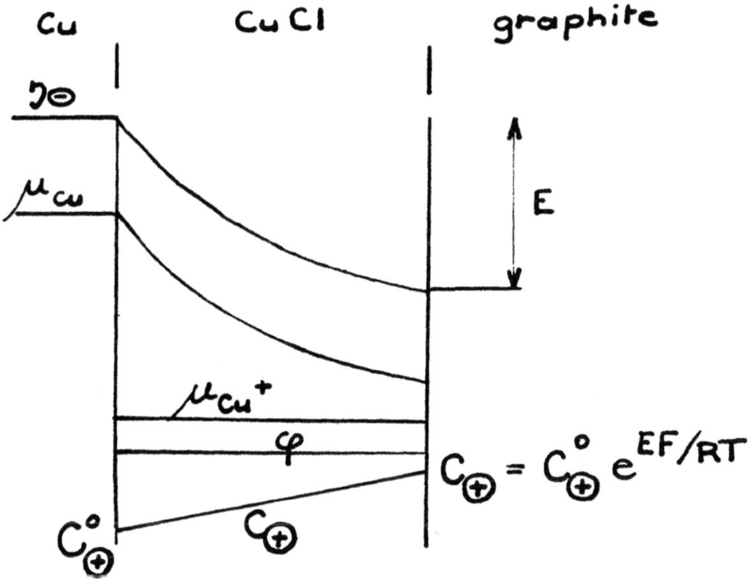

where $\eta_\ominus$ denoted the electrochemical potential of the electrons and in another terminology is known as the Fermi energy or Fermi level. In addition, the following symbols were introduced :

μ_{Cu} = chemical potential of copper

μ_{Cu^+} = chemical potential of copper ions

$C_\oplus$ = concentration of electron holes

$C_\oplus^\circ$ = concentration of electron holes at the Cu/CuCl interface

$j_\oplus$ = flux of electron holes = $I_\oplus/A$

E = applied d.c. potential less than the decomposition potential of CuCl

$D_\oplus$ = diffusion coefficient of electron holes related by the Nernst-Einstein equation to the mobility,

$$D_\oplus = \frac{\mu_\oplus}{e} kT$$

The local field, ϕ is virtually constant, $(\partial\phi/\partial x) \cong 0$, and the concentration of electron holes is linear across the electrolyte for the unidimensional case. The applied potential, E, is used solely to change the concentration of electron holes at the graphite interface. The concentration is fixed ($C^{o}_{\oplus}$) at the Cu/CuCl interface, i.e., local equilibrium is assumed to obtained at the reversible electrode.

The local charge density for this unidimensional case is linear so this flux of electron holes "leaking" through the predominantly ionic conductor is :

$$j_{\oplus} = \frac{I_{\oplus}}{A} = e\, D_{\oplus} \frac{\partial C_{\oplus}}{\partial x} = \text{constant}$$

(Note that the holes move in the opposite direction as electrons, $-(\partial n_{\ominus}/\partial x)$)

The steady state does not involve the movement of ions. These are blocked by the ion blocking electrode, graphite. Moreover, the concentration of species available for migration (interstitial copper and copper vacancies in the case of CuCl) is much larger than that of the electron holes. Consequently, on application of E, the relative concentration of electron holes is greatly affected while the concentration of copper ions is virtually unaffected. Hence $(\partial \mu_{Cu^{+}}/\partial x) \simeq 0$ and because no ions move, $(\partial \phi/\partial x) \simeq 0$ as shown in the diagram. Further details are available in reference 8.

ELECTRODE REACTIONS AT ELECTRODE-SOLID ELECTROLYTE INTERFACES. USE OF ELECTROANALYTICAL TECHNIQUES.

Jacques VEDEL

Laboratoire d'Electrochimie Analytique et Appliquée,
E.N.S.C.P., 11 rue Pierre et Marie Curie
75231-PARIS CEDEX 05

INTRODUCTION

The purpose of this chapter is to present a basic introduction to some electrochemical techniques which may be used in the study of exchange reactions at the interfaces between metallic electrodes and solid electrolytes. In fact, despite the great number of the described techniques, all of them are based on a few principles, which will be exposed and illustrated by examples drawn from published data.

When electric current folws through a galvanic cell :

Metal | electrolyte | Metal

the conduction mode changes at the interfaces between electrodes and electrolyte : in metal, charge carriers are electrons, and in electrolyte, they are ions. A charge transfer reaction :

$$O + ne \rightleftarrows R \tag{1}$$

occurs at the interface.

If, for instance, the oxidizing species O is consumed at the electrode surface (reduction reaction), its concentration locally decreases. A concentration gradient appears,

M. Kleitz and J. Dupuy (eds.), Electrode Processes in Solid State Ionics, 223–259.

and O begin to diffuse towards the electrode. Simultaneously, concentration of reducing species R increases and R diffuses away from the electrode. Furthermore, R and O may be involved in chemical reactions, the equilibrium state of which being perturbed by electrolysis. Thus, supply of O may be a function of a chemical reaction rate.

Hence, the overall rate of an electrochemical reaction does not depends only on the charge transfer kinetics. Seven steps have been considered |1|(Fig. 1).

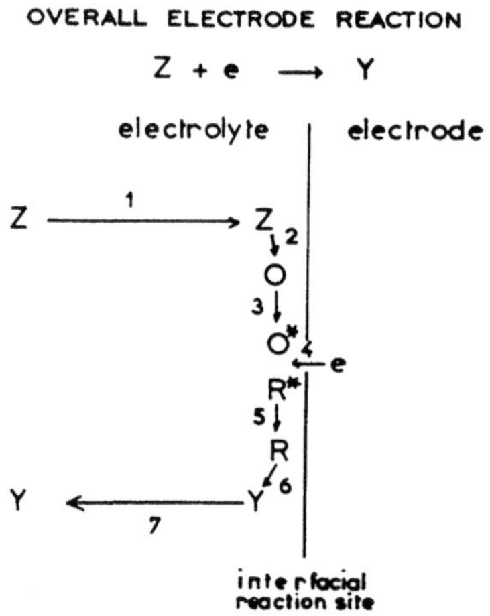

Fig.1. General schema of an electrochemical reaction.

- Step 1 : the reactant Z is brought close to the reaction interface by a mass transport process.
- Step 2 : A chemical reaction takes place, which transforms the reactant into an electroactive form O.
- Step 3 : A structural rearrangement occurs, yielding the species that takes place in the actual charge transfer reaction.
- Step 4 : The electron transfer occurs.
- Step 5 : A structural reorganization occurs, giving the stable product of charge transfer reaction R.
- Step 6 : A chemical reaction transforms the formed product.
- Step 7 : This species Y diffuses from the reaction site.

Oxidation reaction follows the reverse path. The slowest step is rate determining, allowing the separate study of each step of the reaction: when the cell is excited in a known way, the response to the excitation depends on the rate determining step, that is mass transfer controlled reaction if steps 1 or 7 are r.d.s., charge transfer controlled reaction (steps 3,4 and 5) or reaction controlled by coupled chemical reaction when steps 2 or 6 are the r.d.s.

Thus, influence of each component of the overall reaction will be examined successively, beginning with charge-transfer kinetics and following with mass transport control and influence of coupled chemical reaction.

CHARGE TRANSFER KINETICS

In the general scheme of electrochemical reaction (fig. 1), charge transfer kinetics occurs during steps 3, 4 and 5. Initially, oxidizing species is on the stable form O, and electron is in the electrode, at the potential ϕ^{el} (determined vs the electrolyte potential). During step 3, O undergoes some structural modification, allowing the transfer of electron from one energy level of the metal to a level of same energy in O. Electron-transfer occurs during step 4. O^* is transformed into R^*, that, during step 5, gives the stable form R. These transformations may be presented in a diagram energy-reaction coordinate |2| (Fig. 2). During reduction reaction, the energy of the system O + e first increases (curve 1). On point A, it intersects the curve representing the energy variation of $R \rightarrow R^*$ (curve 2), and the exchange $O^* + e \rightarrow R^*$ occurs. Activation energy for reduction reaction is $W_R^{\neq}$, and is $W_O^{\neq}$ for oxidation reaction. In $W_R^{\neq}$ is included the energy variation of the electron, which goes from the electrode potential to the electrolyte potential $\phi^{s.e.}=0$.

Let us modify the difference of potential between electrode and electrolyte from a value $\Delta\phi$. The energy of final state does not vary, whereas the initial state energy is modified from the quantity $-F\Delta\phi$. The conditions for the reaction are the same as before (curve 3), however the summit of the energy curve only increases from a value $-\alpha F\Delta\phi$, because the shape of the energy curve. The quantity α is called transfer coefficient. The activation energies become, respectively :

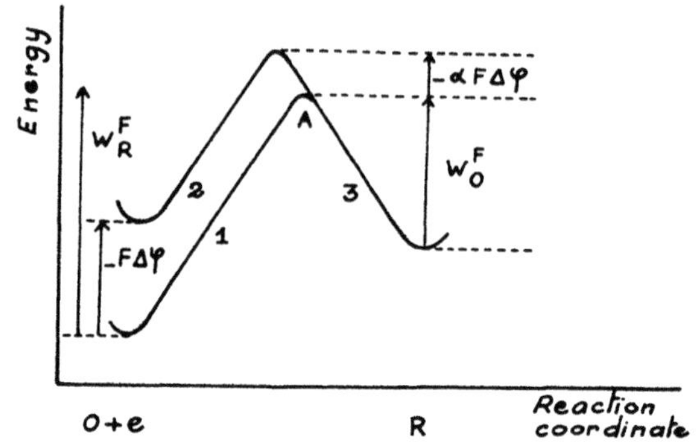

Fig.2. Diagram Energy-reaction coordinate for the reaction O + ne → R.

$$W_R^{\neq\,'} = W_R^{\neq} + (1 - \alpha)\Delta\phi$$
$$W_O^{F\,'} = W_O^{F} - \alpha\,\Delta\,\phi$$

Now the probabilities of transfer are given by the formulas :

$$P_R \propto \exp\left(-\frac{W_R^{\neq} + (1-\alpha)F\Delta\phi}{RT}\right)$$
$$P_O \propto \exp\left(-\frac{W_O^{\neq} - \alpha F\Delta\phi}{RT}\right)$$

and electrolytic currents are then :

$$I_{cat} = -FS\, C_o^{el}\ P_R = -FS\, C_o^{el}\ k_{cat}\ \exp(-(1-\alpha)F\Delta\phi/RT)$$
$$I_{an} = +\, FS\, C_R^{el}\ P_O = FS\, C_R^{el}\ \ _{an}\ \exp(\alpha F\Delta k/RT)$$

where S is the electrode area C_O^{el} and C_R^{el} the concentration of O and R, resp., on the interfacial reaction site and k_{cat} and k_{an}, cathodic and anodic reaction rate constants. (Conventionnally, cathodic currents, corresponding to a reduction reaction, are plotted as negative, and anodic currents as positive). Usually the potential parameter is the overvoltage η defined as the difference between the actual electrode potential and the equilibrium one (i.e., when no current flows through the interface). At equilibrium, cathodic and anodic currents are equals, the concentration on the electrode are equals to the equilibrium concentration C^{eq} and one obtains :

$$FS\ K_{cat} C_O^{eq} = FS\ K_{an}\ C_R^{el} = I_o$$

where I_o is the exchange current of the charge transfer reaction. Then, the total current flowing through the interface is :

$$I = I_{an} + I_{cat} = I_o \quad \frac{C_R^{el}}{C_R^{eq}} \exp(\alpha f \eta) - \frac{C_O^{eq}}{C_O^{eq}} \exp(-(1-\alpha) f \eta) \qquad (2)$$

$f = F/RT$

Expression (2) is the Butler-Volmer relation in which two quantities characterize the charge transfer kinetics :

- the transfer coefficient α, related to the charge transfer conditions, which affects only the symmetry of the I - η curves. Frequently, its value is about equal to .5.

- the exchange current I_O , related to structural modifications of red-ox species. Fig. 3 shows the influence of exchange current value on the variation of total current with overvoltage. When I_O has a high value, a small variation of the overvoltage leads to high current intensities. The Red-ox system is called reversible. When I_O is small, exponential terms have to be important to observe a current. The system is called irreversible. This corresponds to two limiting cases for the Butler-Volmer relation :

  . For reversible systems, the observed current values are always much less than I_O, and one has :

$$\frac{I}{I_O} = \frac{C_R^{el}}{C_R^{eq}} \exp(\alpha f \eta) - \frac{C_O^{el}}{C_O^{eq}} \exp(-(1-\alpha) f \eta) \simeq 0$$

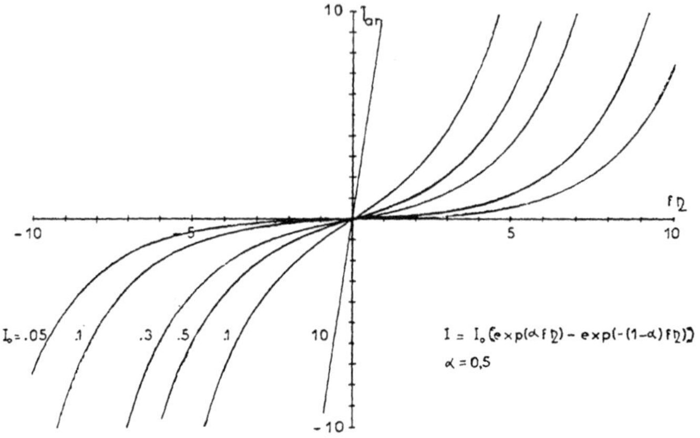

Fig.3. Influence of the exchange current value on the I - η curves.

$$\eta = (1/f)\ln C_R^{eq}/C_O^{eq} + (1/f)\ln C_O^{el}/C_R^{el} \tag{3}$$

or, by considering the normal potential E° of the red ox system :

$$E = E° + 1/f \ln C_O^{el}/C_R^{el} \tag{3'}$$

For reversible systems, the Nernst relation is valid for the concentrations on the electrode.

. For irreversible system : the observed current has only noticeable value for high values of one of the exponential terms. Then, the other one becomes negligible and one obtains the Tafel relations :

$$I \simeq I_{an} = I_0 \frac{C_R^{el}}{C_R^{eq}} \exp(\alpha f \eta) \tag{4}$$

for anodic currents, and

$$I \simeq I_{cat} = -I_0 \frac{C_O^{el}}{C_R^{eq}} \exp(-(1-\alpha)f\eta) \qquad (4')$$

for cathodic currents.

When charge transfer is the rate determining step the concentration ratio are equal to unity, and equations (4) and (4') become :

$$\eta_{an} = -\frac{1}{\alpha f} I_o + (1/\alpha f)\ln I_{an} \qquad (5)$$

$$\eta_{cat} = (1/(1-\alpha)f)\ \ln(-I_o) - (1/(1-\alpha\ )f)\ln I_{cat} \qquad (5')$$

These results will be applied to two kinds of problems encountered in electroanalytical studies i.e., the choice of a reference electrode and the study of electrochemical reactions by means of using the Tafel relations.

CHOICE OF A REFERENCE ELECTRODE

On derivation of the Butler-Volmer relation (eq.2), the potential of the working electrode (i.e., on which electrochemical reaction is studied) was referred to the bulk electrolyte potential. In fact, this potential is measured versus the potential of another electrode (the reference electrode) which must remain constant, even if another electrochemical reaction takes place on it. Reference electrodes are used in two or three electrodes systems (fig.4).

In two electrodes systems (fig. 4-a) a power supply (PS) impresses a potential difference, or a current, between the working electrode WE and the reference electrode RE. The difference between the voltage drop V, corresponding to a current I flowing through the cell and its equilibrium value V_e may be divided into three terms : the overvoltages on the working and reference electrodes, η^{WE} and η^{RE}, and the ohmic drop RI :

$$V - V_{eq} = \eta^{WE} + RI - \eta^{RE}$$

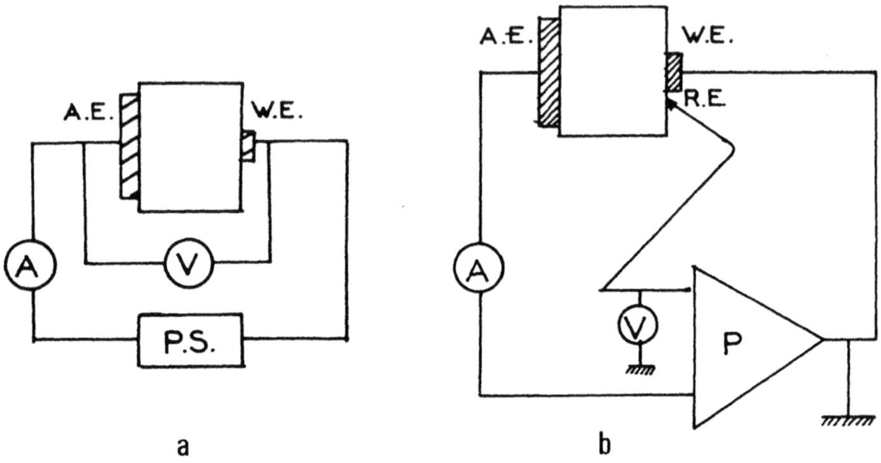

Fig.4. Use of reference electrodes : two (a) and three(b) electrode systems.

The measured quantity $V-V_{eq}$ is related to the overvoltage η^{WE} if (1) the ohmic drop is known (i.e., if the resistance R is determined by another technique) and (2) the overvoltage at the reference electrode is as small as possible and does not vary appreciably with I. The latter condition is realized if :

1. The current density on the reference electrode is as small as possible, i.e. the area of the reference electrode is much greater than the working electrode one.
2. The electrochemical reaction on the R.E. is reversible for that a small variation of η^{RE} may correspond to any variation of I.
3. The quantities of electroactive species reacting on the RE are sufficiently high to avoid concentration variation during electrolysis and to maintain constant the ratio C_o^{eq}/C_o^{el} (eq.3').

Remark : The ohmic drop has been evaluated by switching off the polarization. Then, the R.I. component vanishes immediately, whereas the overvoltage component decreases more slowly with time. This method seems to be right when mass transfer is the r.d.s., because the disappearance of overvoltage is then related to diffusion of the electroactive species.

In the three electrodes system (fig. 4.b) a potentiostat P monitors the potential of the working electrode versus the reference electrode by acting upon the current flowing between the working electrode and a third one, the auxiliary electrode. The potentiostat has a high input impedance and practically no current flows in the reference electrode circuit. The overvoltage on the reference electrode is eliminated. The ohmic dorp problem remains, because R.E. generally lies between WE and A.E., where the ohmic resistance of the electrolyte causes a potential gradient. PIZZINI and coworkers |4| studied the influence of the working electrode/auxiliary electrode configuration on the response of cells working at high temperatures (ceramic oxide electrolyte), using model electrochemical cells. Cells were simulated using copper electrodes pressed against conductive paper. The authors gave relations allowing the calculation of ohmic drop value for various cell configurations, knowing their geometrical parameters and the resistance of the electrolyte.

STUDY OF ELECTROCHEMICAL REACTIONS USING TAFEL PLOTS

Charge transfer kinetics may be investigated using expressions (4) or (5) if the charge transfer reaction is the rate determining step, that is if experimental conditions are such as $C_i^{el} = C_i^{eq}$. Such conditions are fulfilled if, for exemple the concentrations of electroactive species are sufficiently important so that their variations may be negligible, or as it will be shown later if electroactive species is the mobile ion in an electrolyte with only one mobile ion. Two examples will be presented, corresponding to the transfer of one and of two electrons.

1. Kinetics of the silver-silver rubidium iodide electrode

ARMSTRONG, DICKINSON, THIRSK and WHITFIELD |4| studied the kinetics of the $Ag/RbAg_4I_5$ electrode. $RbAg_4I_5$ is an ionic conductor with the silver ion the mobile species. A typical current voltage curve obtained on anodic oxidation is shown in fig. 5, together with the curve corrected for ohmic drop. It was observed that the magnitude of the current varied from one experiment to another, probably because of the variation of the contact area between electrode and electrolyte. For sufficiently high values of the overvoltage, Tafel plot is linear (fig. 5b) and the slope has been found equal to .2 V.

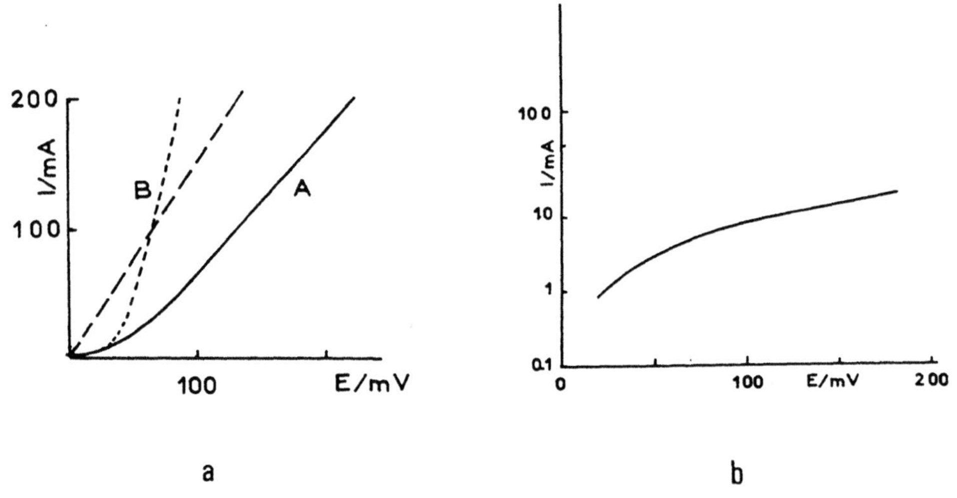

Fig. 5. Silver oxidation at the interface $Ag/RbAg_4I_5$ (ref. 4)

For the oxidation reaction $Ag - e \rightarrow Ag^+$, it is assumed that an ion crosses the interface between electrode and electrolyte. As in the preceedingly described case, an activation energy, depending on the electrode potential, is necessary and the current-potential relation is given by the Butler Volmer relation :

$$I = I_o \left| \frac{C^{el}_{Ag}}{C^{eq}_{Ag}} \exp(\alpha f \eta) - \frac{C_{Ag^+}}{C^{eq}_{Ag^+}} \exp(-(1-\alpha) f \eta) \right| \tag{6}$$

The authors supposed first the silver activity was always equal to unity, and also, the ratio C^{el}_{Ag}/C^{eq}_{Ag}. For high values of the overvoltage, relation (2) becomes :

$$I = I_o \exp(\alpha f \eta)$$

and the α value, resulting from the mechanism is about 0.3, and the observed results are consistent with the hypothesis of a limiting charge transfer kinetics. However, reactions involving the anodic oxidation of solid electrodes lead to overvoltage-current relations which look like the Butler Volmer relation. For instance, if surface diffusion of ad-atoms is rate-determining the I-η relation is (ref. 5, p. 290) :

$$I = (KI_o)^{1/2} \exp(\alpha f \eta/2) - \exp(-(1-\alpha/2) f \eta)$$

where the factor K depends on the diffusion conditions on the metallic surface, so that the observed behaviour could also be explained by this mechanism with $\alpha = 0.6$.

2. Kinetics of the O_2,Pt, Calcia stabilized zirconia electrode

FILAYEV, KARPACHEV and PAL'GUEV |6| studied the above electrode at 920 and 1020°C by determination of the I - η curves corresponding to the anodic oxidation reaction. They found that at the two mentioned temperatures the experimental Tafel slope were equals to (0.65 x 2.3xRT/F)V.

The overall electrode reaction is :

$$O^{2-} - 2e \rightarrow O \tag{7}$$

but the authors assumed that the oxygen ion loses its two electrons consecutively in a two-step reaction :

$$O^{2-} - e \rightarrow O^- \tag{8-1}$$

$$O^- - e \rightarrow O$$

The peroxide ion O^- appears only on the electrode surface, as a reaction intermediate, and is transformed either in oxygen or oxide ion. For a given overvoltage value, η, the total current I is caused for one half by reaction (8-1) and for the second one, by reaction (8-2). If I_{01} and I_{02} are the exchange currents and α_1 and α_2 the transfer coefficient for the two reactions, application of equation 2 leads to :

$$I/2 = I_{01}\left| C^{el}_{O^{2-}}/C^{eq}_{O^{2-}} \exp(\alpha_1 f\eta) - (C^{el}_{O}/C^{eq}_{O^-}) \exp(-(1-\alpha_1) f\eta)\right|$$

$$I/2 = I_{02}\left|(C^{el}_{O^-}/C^{eq}_{O^-}) \exp(\alpha_2 f\eta) - (C^{el}_{O}/C^{eq}_{O}) \exp(-(1-\alpha_2) f\eta)\right|$$

Eliminating the ratio $(C^{el}_{O^-}/C^{eq}_{O^-})$ gives

$$I = 2I_{01}I_{02}\frac{(C^{el}_{O^{2-}}/C^{eq}_{O^{2-}}) \exp\left|(\alpha_1+\alpha_2) f\eta\right| - (C^{el}_{O}/C^{eq}_{O}) \exp\left|-(2-\alpha_1-\alpha_2) f\eta\right|}{I_{02}\exp(\alpha_2 f\eta) + I_{01}\exp(-(1-\alpha_1) f\eta)} \quad (9)$$

As it will be shown later, the oxide ion being the mobile ion of the electrolyte, $C^{el}_{O^{2-}} = C^{eq}_{O^-}$

If the exchange currents I_{01} and I_{02} have approximately equal values, expression (9) gives for high overvoltages :

$$\log I = \log (2I_o) + (2.3/\alpha_1 f) \log I$$

to which corresponds the value 1.54 for the transfer coefficient α_1. Inversely, if $I_{01} >> I_{02}$, that is if the first oxidation reaction is much faster than the second one, expression (9) gives :

$$\log I = \log(2I_0) + (2.3/(1+\alpha_2) f) \log I$$

with $\alpha_2 = 0.54$.

Thus, the observed value of the Tafel line slope does not contradict the hypothesis of a slow charge transfer reaction. Nevertheless, the authors mentioned that the same value may also be obtained in the case when the discharge step is not rate-determining (for example if adsorbed atoms are involved in a charge transfer reaction).

Thus, no specific conclusion with regard to the reaction mechanism can be drawn from the use of Tafel's plots, that constitute only one element in the elucidation of electrochemical reaction.

MASS TRANSPORT PROCESS

Introduction

A species can be electrochemically oxidized or reduced if brought to the vicinity of the electrode surface. For high values of the overvoltage or for a reversible reaction, the rate of charge transfer is much larger than the rate of mass-transfer process. Under these conditions, the current is controlled by the latter process, and related to the flux of electroactive species. At the interface where the following electrochemical reaction occurs :

$$\nu O + ne \rightarrow \nu' R$$

there is no charge accumulation. Thus, relation between the electric current density j and the flux of material is

$$j = (nF/\nu)\,(J_O)_{x=0} \qquad (10)$$

where $(J_O)_{x=0}$ represents the flux of O species at the electrode surface. Generally, the flux J_i of a species of charge Z_i is related to the electrochemical potential gradient $\nabla\tilde{\mu}_i$ by :

$$J_i = -\frac{C_i D_i}{RT}\nabla\tilde{\mu}_i = -\frac{\sigma_i}{(Z_i F)^2}\nabla\tilde{\mu}_i \qquad (11)$$

where D_i = diffusion coefficient

σ_i = partial conductivity

$\tilde{\mu}_i = \mu_i + Z_i F \phi$

μ = chemical potenital of i species

ϕ = electrical potential of the phase.

If species i has an ideal behaviour, expression (11) becomes :

$$J_i = -D_i\left(\frac{dC_i}{dx} + \frac{C_i Z_i F}{RT}\frac{d\phi}{dx}\right) \qquad (12)$$

for unidimensional diffusion.

(1) If species i is not charged, $Z_i = 0$ and

$$J_i = -D_i \frac{dC_i}{dx} \qquad (13)$$

The flux of a non charged species is given by the Fick's first law.

(2) If the solid electrolyte has only one mobile ion, and if species i is this ion (for ex., i is Ag^+ ion in $RbAg_4I_5$, or O^{2-} in $ZrO_2.CaO$) ; on each point of the electrolyte, the electroneutrality condition is observed. The concentration of i is constant and equal to the concentration of the fixed charges, $dC_i/dx = 0$. The concentration of the mobile ion on the electrode surface is always equal to its concentration in the bulk of the electrolyte.

Ex : $RbAg_4I_5$: $i = Ag^+$

the iodine reduction reaction is written as

$$I + Ag^+ + e \rightarrow AgI\downarrow$$

only iodine atom I are diffusing species

$ZrO_2.CaO$: $i = O^{2-}$

the oxygen reduction reaction is

$$O + ne \rightarrow O^{2-}$$

O is the diffusing species.

(3) Species i is any ion in the electrolyte. Equation (11) holds. The flux J_i may be divided into two parts, the first one corresponding to a diffusion flux $J_{i,d}$ and the second one to a migration flux $J_{i,m}$

$$J_i = J_{i,d} + J_{i,m}$$

$$J_{i,d} = -Di(dc/dx) \; ; \; J_{i,m} = -(D_iC_iF/RT)(d\phi/dx)$$

$J_{i,m}$ is related to the total current density j_t flowing through the electrolyte by the mean of its transport number t_i

$$J_{m,i} = (t_i/Z_iF)j_t$$

and t_i is related to the mobilities (u_j) and concentrations C_j of other moving ions by

$$t_i = (Z_iu_iC_i)/\Sigma\ Z_ju_jC_j$$

For an ion dilute enough so that $C_i \ll C_j$, $t_i \simeq 0$,

and

$$J_i \simeq J_{d,i} = -D_i(dc_i/d_x)$$

Thus, Fick's first law (eq. 12) hold also for a diluted charged speices.

Variation of diffusion flux with time obeys to Fick's second law, i.e., for a plane, unidimensional process :

$$\frac{\delta C_i(x,t)}{\delta t} = D_i \frac{\delta^2 C_i(x,t)}{\delta x^2} \tag{14}$$

For a diffusion process with circular symmetry, Fick's second law is :

$$\frac{\delta C(r,t)}{\delta t} = D \frac{\delta^2 C(r,t)}{\delta r^2} + \frac{a}{r} \frac{\delta C(r,t)}{\delta r} \tag{15}$$

with $r = 0$ at the symmetry center
$a = 1$ for cylindrical symetry
$a = r$ for spherical symmetry

If r_o is the radius of the electrode, the second term in equation (15) is negligible when $t \ll r_o^2/\pi D$

Mass transport with convection

The diffusion layer

When the phase containing the electroactive species is moving relatively to the reaction interface, convection (or hydrodynamic transport) occurs. This phenomenon is less frequently encountered with the solid electrolytes than with the liquid ones, and takes place wether the electrode metal is in the liquid state or the phase containing the reactant is a fluid one (generally gaseous). Convection may be either natural or forced. Natural convection is caused by density variations resulting from composition modification or temperature gradient. Forced convection leads to reproducible results. It is obtained either by mechanical stirring or by gas circulation.

Results obtained with forced convection are frequently interpreted using the concept of diffusion layer, introduced by Nernst. A stagnant layer of a definite thickness δ, is assumed to existe at any solid-fluid interface. The species reacting on the interfacial reaction site is brought to this layer by stirring, and if stirring is sufficient, the reactant concentration on the layer surface is constant and

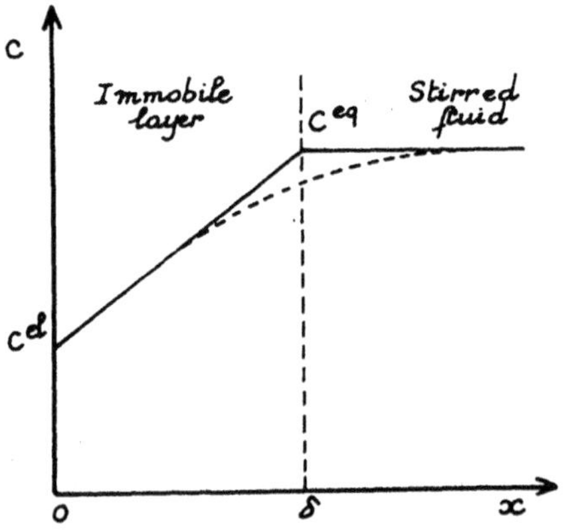

Fig. 6. The diffusion layer.

equal to the concentration in the bulk of the fluid. Transfer through the layer occurs by diffusion. Nernst supposed that the reactant concentration varies linearly with the distance from the interface. When stirring is stable, δ is constant and the concentration gradient at the interface is

$$\frac{\delta C}{\delta x}_{x=0} = \frac{C^{eq} - C^{el}}{\delta} \quad (16)$$

Combining equations 10, 13 and 16 gives

$$j = \pm \frac{nFD}{\delta} (C^{e} - C^{s}) \quad (17)$$

The current has a time-independant, finite value (steady-state current). This is caused by the permanent supply of material at the surface of the diffusion layer. For $C^{el} = 0$, electrolytic current reaches a limiting value, proportional to the equilibrium concentration :

$$j_1 = \pm (nFD/ D) C^{eq} \quad (18)$$

Applications :

1) Stationnary diffusion with liquid metallic electrode

ETSELL and FLENGAS |7| studied oxygen solubility in molten copper and silver. An impervious ZrO_2 - CaO tube, closed at one end is immersed in liquid metal. The inner surface of the tube is platinized and a platinum wire serves as a conducting lead. Air or oxygen are used as reference. Another metallic wire establishes the contact with liquid metal. The tube is vibrated vertically.

Oxygen is dissolved in the metal. When a potential more negative than the equilibrium potential is applied to the electronic conductor, dissolved oxygen is reduced at the boundary electrolyte-metal, following :

$$\underset{\text{in M}}{O} + \underset{\text{in M}}{2e} \rightarrow \underset{\text{e.s.}}{O^{2-}}$$

At temperatures greater than 1000°C the charge transfer reaction was supposed reversible, i.e., $I \ll 2 I_{01} I_{02}$ in expression (9). As for the derivation of equation (3), expression (9) becomes then :

$$\exp|(\alpha_1+\alpha_2)f\eta| - (C_o^{el}/C_o^{eq}) \exp| -(2-\alpha_1-\alpha_2)\ f\eta| \simeq 0$$

or

$$\exp 2\eta = C_0^{el}/C_0^{eq} \qquad (19)$$

Combining equations (17), (18) and (19) yields :

$$\eta = (2.3RT/2F) \log (1 - I/I_e) \qquad (20)$$

A typical current voltage curve, obtained by ETSELL and FLENGAS is reported on fig.7. Plots of the overvoltage vs log $(1-I/I_e)$ were found linear. At 1000°C and 1200°C, the experimental slopes were 0.130 and 0.153V, corresponding to the 2.3RT/2F values equals to 0.126 and 0.146V, resp. The variation of the limiting current with the oxygen concentration was linear also. All these results show that the overall process is controlled by a steady-state mass transfer.

2) Stationnary diffusion with a gaseous phase.

When the reactants are presents in a gaseous phase, good conditions for reproducible results are more difficult to obtain. The electrode Pt, O_2, Stabilized ZrO_2 has been the most studied among the system involving a gaseous phase. Three main cell devices have been used :

- the foil electrode (BROOK, PELZMANN and KROGER|8|): a platinum foil is pressed against the electrolyte surface, under load, and heated - before use- at 1350°C,
- the point electrode (KLEITZ|9|) : the tip of a platinum wire is maintained against the electrolyte. Gaseous flow is coming from an alumina tube, besides the platinum wire.

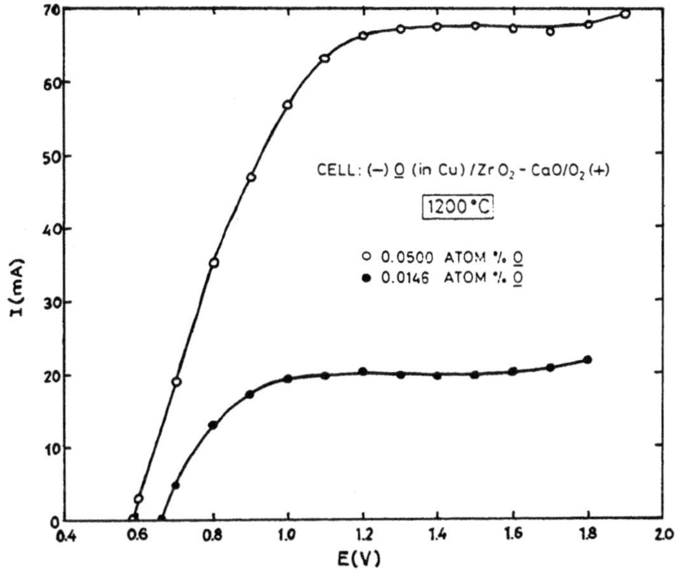

Fig. 7. Reduction of oxygen dissolved in liquid metal, under steady state conditions (ref.7)

- the porous electrode, obtained by sputtering or by decomposing platinum paint, or chloroplatinic acid by heating, (cf. reference 3 or 10 for a review of the various kinds of electrode systems).

Dense platinum electrodes look like liquid electrodes : convection in the gaseous phase brings oxygen at the outer electrode surface where it is dissociated following $O_2 \rightarrow 2O$. Then oxygen atoms diffuse through the platinum foil towards the metal electrolyte interface (presumably in the grain boundaries). Limiting current were observed, the value of which varied with the square root of the oxygen pressure (supporting the hypothesis of the dissociation of the oxygen molecule), and reciprocally with the foil thickness. On the other hand, the equation of the I. η curve obeyed to relation (20), establisehd for steady-state currents. Thus, the hypothesis of a stationnary diffusion state was proposed by the authors.

KLEITZ obtained current-voltage curves using a linearly changing potential. The obtained curves have the characteristics of a steady-state, diffusion controlled process, as frequently quoted from this work [3.8.10]. However, the conditions for a stationnary diffusion are not well defined in the case of a point electrode, and Kleitz proposed in fact a mechanism involving a limitation by the kinetics of the oxygen molecule dissocation .

The case of porous electrode is much less simple. Owing to the existency of pores, interdiffusion of oxygen and inert gas like Argon in the gaseous phase may take place, as pointed out by Kroger [11]. Then proportionnality between limiting current and oxygen pressure may be observed [12] depending on the size parameter of the pores.

Stationnary diffusion states are difficult to obtain in the case of a gaseous phase, and it is perhaps more convenient to use non stationnary state techniques, which provide similar indications and may give informations on the kinetics of the charge transfer process, and on the interference of coupled chemical reaction.

Mass transport without convection

This transport mechanism has its origin in the concentration gradient caused by electrolysis. The reactant flux is derived by integrating the two Fick's law, with the convenient initial and boundaries conditions, imposed by the excitation mode of the cell.

a. Chronoamperometry

1. General equations

This method involves the application of a potential step to the working electrode and the measurement of the resulting current as a function of time. The conditions for the derivation of the current variation are the following :

- Initial conditions : the distribution of diffusing species O and R is homogeneous at t=0. Thus, initial conditions are :

$$C_0(x,0) = C_0^{eq} \qquad \text{and } C_R(x,0) = C_R^{eq}$$

- Boundaries conditions : there is no change in the concentration of reactants far from the electrode (semi-infinite diffusion) :

$$C_0(\infty,t) = C_0^{eq} \qquad \text{and } C_0(\infty,t) = C_0^{eq}$$

At the electrode surface, the concentration gradient is related to the current :

$$- nFD_O(\partial C_O/\partial x)_{x=0} = I = nFD_R(\partial C_R/\partial x)_{x=0}$$

The I value being itself related to the applied overvoltage by the Butler Volmer relation (equation 2).
The current variation has been derived using Laplace transforms method |13|. General solution is :

$$I(t) = I_i \exp(\lambda^2 t).\mathrm{erfc}(\lambda\sqrt{t}) \qquad (21)$$

where

$$I_i = I_I|\exp(\alpha nf\eta) - \exp(-(1-\alpha)f\eta|$$

$$\lambda = \frac{I_0}{nF}\left(\frac{\exp(\alpha nf\eta)}{C_R^{eq}\sqrt{D_R}} + \frac{\exp(-(1-\alpha)f\eta)}{C_O^{eq}\sqrt{D_O}}\right)$$

$$\mathrm{erfc}\, u = 1 - \frac{2}{\sqrt{\pi}}\int_0^u e^{-z^2}\,dz$$

Equation (21) has two limiting values :

- for short times : $\lambda\sqrt{t} << 1$

$$I(t) = I_i(1 - \frac{2\lambda}{\sqrt{\pi}} t) \qquad (22)$$

  When diffusion process is not established, charge transfer kinetics limits the overall reaction rate. The limit of I(t) when t approaches zero value allows the study of charge transfer kinetics (I_i is the current given by the Butler Volmes relation, without diffusion influence).

- for long times $\lambda\sqrt{t} >> 1$

$$I(t) = \frac{\Lambda}{\sqrt{\pi}} \cdot \frac{1}{\sqrt{t}} \qquad (23)$$

$$\Lambda = nF\frac{1-\exp(-nf\eta)\quad C_O^{eq}\, C_R^{eq}\sqrt{D_O\, D_R}}{C_O^{eq}\sqrt{D_O} + \exp(-nf\eta)\; C_R^{eq}\sqrt{D_R}}$$

Parameters of electrochemical kinetics (α and I_O) have been vanished. The process is only diffusion controlled.

2) Application : iodine reduction at the interface Pt/ Ag_4I_5 [14]

Like $RbAg_4I_5$ potassium silver iodide is a silver ion conductor. The working electrode was of the point type. An iodine atmosphere was maintained over the electrolyte. It was obtained by flushing an argon flow over thermostated solid iodine. Iodine pressure varied from 10^{-6} to $10^{-3.5}$ atm. After some time, iodine apparently dissolves in the solid, presumably at the grain boundaries.

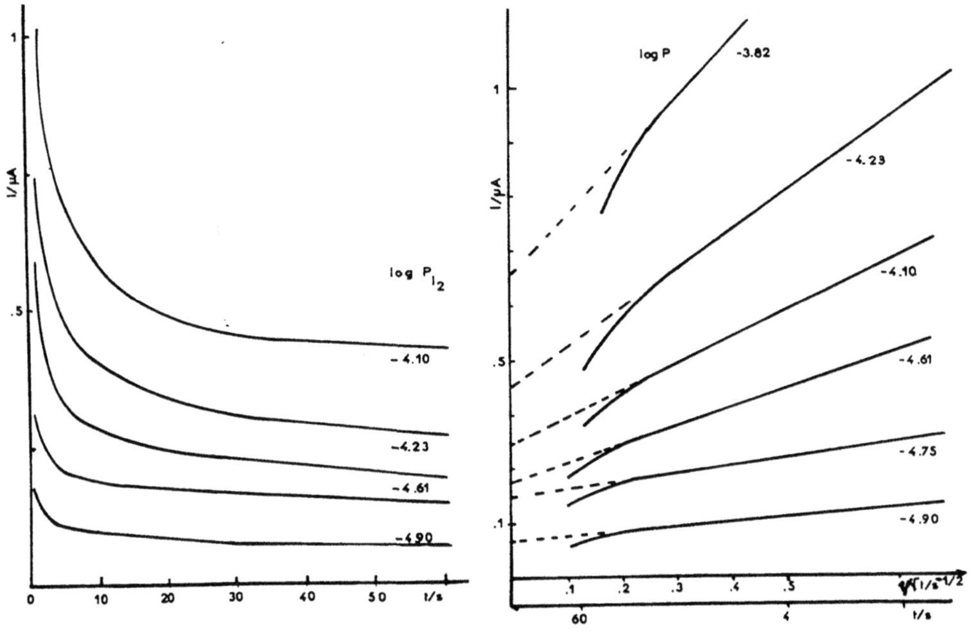

Fig. 8. Chronoamperometric studies at the interface $Pt, I_2/KAg_4I_5$ (ref. 14)

At time t = 0, an overvoltage of -0.2V is applied to the working electrode. Typical current variations are reported on fig. 8 .

Overall reduction reaction is

$$I_2 + 2e + 2Ag^+ \rightarrow 2AgI(s)$$

for which two mechanisms are possible :

$$\left.\begin{array}{l} I_2(g) \rightleftarrows I_2\ (ads) \\ I_2(ads)+2Ag^+ + 2e \rightarrow 2AgI(s) \end{array}\right\} \quad I$$

and

$$\left.\begin{array}{l} I_2(g) \rightleftarrows 2I\ (ads) \\ I(ads)+Ag^+ +e \rightarrow AgI(s) \end{array}\right\} \quad II$$

In both cases, there is only one diffusing species, I_2 or I. Equation (23) simplifies into :

$$I(t) = L/\sqrt{t} \qquad \text{with } L = nFC_0^{eq}\sqrt{\frac{D_0}{\pi}}\ (\exp(nf\eta)-1)$$

for $\eta = -0.2V$, $\exp(nf\eta) \ll 1$ and

$$I(t) = -nF\ C_0^e\ \frac{D_0}{\sqrt{\pi}}\ \frac{1}{\sqrt{t}} \qquad \text{(Cottrel equation)} \quad (23)$$

Plots of I(t) vs $1/\sqrt{t}$ are given in fig. (8b).

The fact that the observed straight lines does not intersect the origin of the axis was interpreted by the influence of a spherical diffusion, the contact between electrode and electrolyte being realized only on some points. Spherical diffusion adds one corrective term to equation (23), independant of time and equal to :

$$i = -nFAD_0C_0^{eq}\ /ro$$

where ro is the radius of the spherical electrode.

The slope of the straight lines increases with increasing iodine pressure. It is proportional to the concentration of oxidizing species C_0^{eq}, that is to $P(I_2)$ or $\sqrt{P(I_2)}$ if mechanism I or mechanism II is observed. Plot of the slope vs $P(I_2)$ and $\sqrt{P(I_2)}$ shows that mechanism II is the most probable, and that iodine is probably adsorbed on the form of iodine atoms.

3) Influence of double layer capacitance.

The variation of the electrode potential necessitates the modification of the charge of the double layer.

A capacitive current I_{dl} is added to the faradaic current :

$$I_{dl} = - (\Delta V/R)\exp(-t/RC)$$

Usually, I_{dl} becomes negligible at very short time.

b. Chronocoulometry

Chronocoulometric method looks like chrono amperometric method. But instead of the current, the charge passed through the circuit is recorded.

Faradaic current being proportional to $1/\sqrt{t}$, Faradaic charge is proportional to $\sqrt{t}$:

$$Q = 2nFC_0^{eq}\sqrt{\frac{D_0}{\pi}} \cdot \sqrt{t}$$

Double layer is charged at very short times and the corresponding charge Qdl is kept in memory on the graph, because the charge accumulation is recorded.

Further more, if the reactant is in equilibrium between adsorbed form at the interface electrode-electrolyte (that is, if there is an accumulation of O at the interface) adsorbed O, which does not require diffusion step prior to reduction, and is present in a limited amount, reacts sharply at the very short times, like the charging of the D.L. Thus the total charge is

$$Q = Q_F + Q_{D.L.} + Q_{ads.}$$

Zero time intercept of the variation Q = f(Vt) gives

$$Q_{DL} + Q_{ads}$$

Capacitance charge is obtained by a blank experiment. However, adsorbed species may change the double layer capacitance. The best is to reverse the potential step at some time τ. Theoretical calculations have been done by Christie and al |15|. For times greater than τ, the variation of $Q_\tau - Q$ must be plotted vs $\sqrt{t-\tau} + \sqrt{\tau} - \sqrt{t}$, giving a straigth line.

- If R formed diffuses and is reoxidized, the intercept is now :

$$a_o Q_{ads} + Q_{dl}$$

a_o being a small term, accounting for the R species formed by reduction of adsorbed O. a_o value has been calculated (-0.07)

- If R is not reoxidized, $Q = Cte = Q_{dl}$.

Chronocoulometric method allows determination of both d.l. capacitance and reactant interface excess.

c. Linear Sweep voltammetry

1. General equations.

In linear sweep voltammetry, the working electrode potential is driven linearly with time. Initially, the rate of the electrode reaction is very low and practically no current flows through the cell. When the potential decreases, the reaction starts and the current increases. There is a depletion of the electroactive species near the electrode surface and mass transfert becomes the rate determining step. The current potential curve has the shape shown on fig. 9 . The caracteristics parameters of such a voltammogram are the peak current I_p and the peak potential E_p. These quantitites have been related to the diffusion coefficient, and to the imposed parameter, that is the scan rate v. For a reversible reaction O + ne $\rightarrow$ R, involving two diffusing species, the value of the peak current is given by the equation |16| :

$$I_p = 0.452 \left(\frac{n^3 F^3 D}{\pi RT} \right)^{1/2} S\, C_O^{eq} .\, v^{1/2} \tag{24}$$

and the peak potential by

$$E_p = E° + (RT/2nF) \ln(D_O/D_R) - 1.1(RT/nF) \tag{25}$$

If only species O is allowed to diffuse, the equations are

$$I_p = 0.541 (4n^3F^3D/\pi RT)^{1/2} S\, C_O^{eq}\, v^{1/2} \tag{26}$$

$$E_p = E° + (RT/nF) \ln C_O^{eq} - 0.85(RT/nF) \tag{27}$$

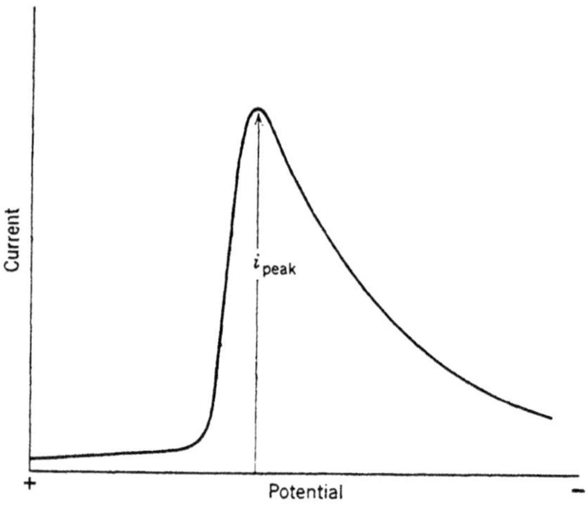

Fig. 9. Reversible current-potential curve for linear sweep voltammetry.

Thus, the peak current is proportional to the square root of the scan rate and to the concentration of electroactive species.

2. Application.

Iodine reduction at the interface Pt/KAg_4I_5

Experimental conditions were the same as for chronoamperometry. Obtained curves are shown on fig.10 for a scan rate value equal to 0.5 s^{-1}. Peak current were observed, the values of which were studied in terms of scan rate v and iodine pressure values. The variation of the peak current I_p, as a function of $v^{1/2}$ are straight lines, intersecting the origin of axis. The slope of these lines, theoretically proportional to the equilibrium concentration of reactant, plotted versus the square root of $P(I_2)$ gave straight lines, confirming that iodine atoms are diffusing species.

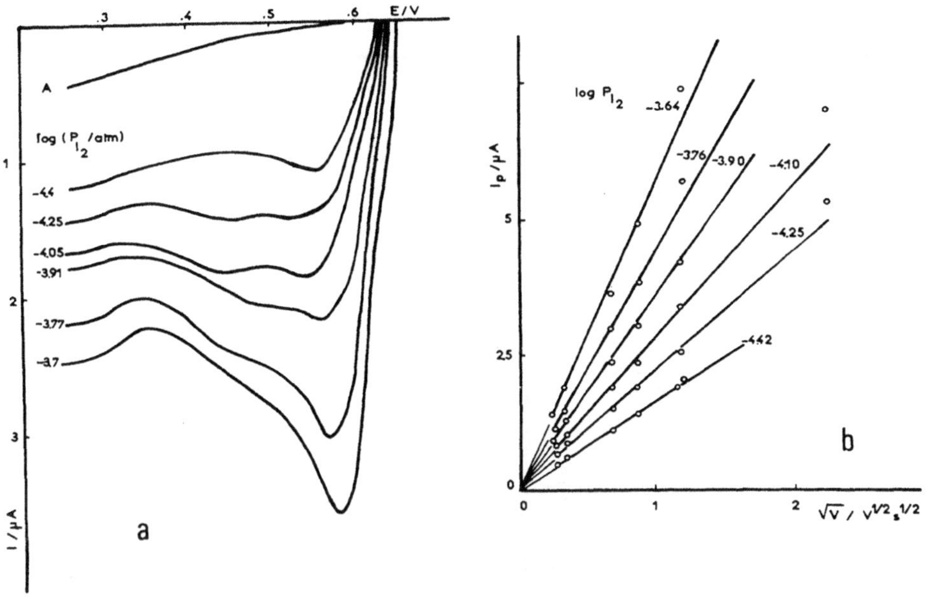

Fig. 10. Linear sweep voltammetry studies at the interface Pt, I_2/KAg_4I_5 (ref. 14)

3. Influence of double layer capacitance.

On figure 11 a are reported the current variations when no iodine pressure is maintained in the cell. The curves show current plateaus for potential values ranging from 0.3 to 0 V. These currents are caused by double layer charging. If a constant double layer capacitance is assumed, the d.l. current is

$$I_{dl} = v\, C_{dl}(1 - \exp(-t/RC))$$

When the ratio t/RC is high, I_{dl} is constant and proportionnal to the scan rate, allowing an estimation of $C_{dl} \simeq 0.8\ \mu F\ mm^{-2}$ (fig. 11b). In the presence of iodine, d.l. capacitance may change and this fact is perhaps the origine of the irregularities observed on experimental curves.

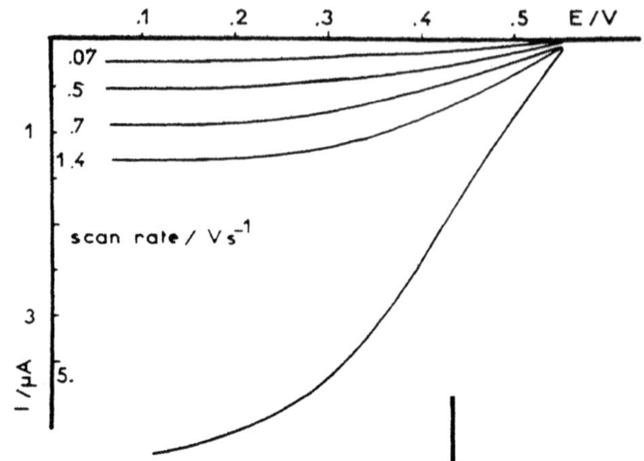

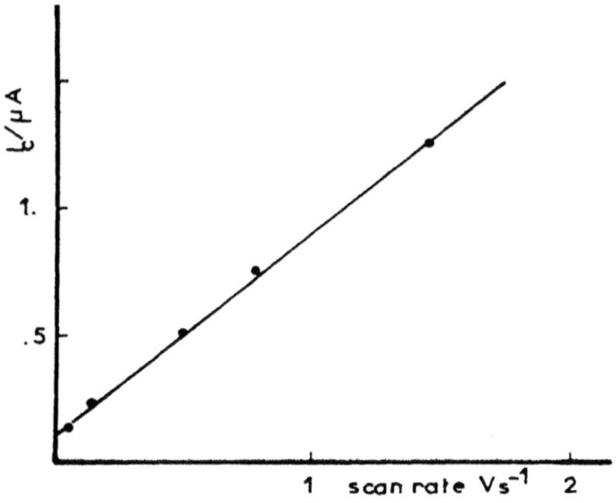

Influence of double layer capacitance (ref. 14).

d. Chronopotentiometry.

1. General equations.

A current step, of intensity I_o, is applied to the working electrode at t = 0, and the electrode potential is recorded. An abrupt varation of the potential is observed at the time τ, called transition time.

During the reduction reaction $O + ne \rightarrow R$, the flux of O at the electrode surface, proportional to I_O, is constant. The corresponding boundary condition is

$$I_O/nFD_O = \partial C_O(0,t)/\partial x \qquad (28)$$

Another boundary condition expresses that the sum of the fluxes of substances O and R is equal to zero at the electrode (if R is allowed to diffuse) :

$$D_O \times \partial C_O(0,t)/\partial x = D_R \,\partial C_R(0,t)/\partial x \qquad (29)$$

The limit conditions are the same as for chronoamperometry. Solution of the linear diffusion equations leads to the value of the transition time (Sand equation) :

$$\tau^{1/2} = (D_O\pi)^{1/2} nFSC_O^{eq}/2I_O \qquad (30)$$

For reversible charge transfer reaction, the values of the surface concentrations at any time (given by the solution of the diffusion equations) may be substituded into the Nernst equation (eq.3'), yielding the E - t relation :

$$E = E° + (RT/2nF)\ln(D_R/D_O) + (RT/nF)\ln\{(\tau^{1/2}-t^{1/2})/t^{1/2}\}$$

$$\left(\text{or}\quad E = E° + \frac{RT}{2nF}\ln\frac{D_R}{D_O} + \frac{RT}{nF}\ln\frac{\tau^{1/2}-t^{1/2}}{t^{1/2}}\right)$$

If only O diffuses, one obtains :

$$E=E°+(RT/nF)\ln C_O^e + (RT/nF)\ln(1-t^{1/2}/\tau^{1/2}) \qquad (31)$$

For $t > \tau$, another reduction reaction takes place, (i.e., the reduction of the electrolyte itself) and the potential has a limited variation.

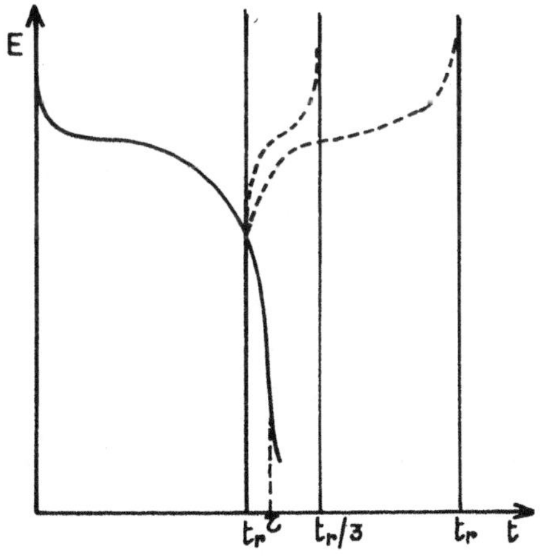

Fig. 12. Chronopotentiometric curve with reverse current step.

Reverse current step.

At a time t_r, less than τ the cathodic reduction current is replaced with an anodic oxidation current, causing the reoxidation of the R species obtained during reduction. A reverse potential time curve is observed, with a reverse transition time τ_R. For the case in which $i_c = -i_a$, one obtains the relation

$$\tau_R = t_r/3 \qquad (32)$$

independent of the absolute value of the applied current intensity and on the diffusion coefficient values.

If R is an insoluble species, it does not diffuse during the cathodic step and τ_b is equal to t_r. If R is more or less adsorbed on the electrode surface, τ_b has an intermediate value between t_r and $1/3\ t_r$.

2) Applications

Hladik and coworkers |17| studied the reduction of cadmium chloride and nickel chloride in solid LiCl-KCl eutectic. The WE was metallic wire, sealed in a glass tube, and arased, for only the cross section of the wire be into contact with the electrolyte. Reverse transition times are equals to direct electrolysis times. Thus nickel metal is deposited at the electrode surface during reduction. Reverse transition time is about equal one third of the direct electrolysis current. In order to explain this fact, Hladik suggested the formation of cadmium I-chloride diffusing from the electrode to the electrolyte.

The case of electroactive species dissolved in metallic electrode was considered by Rickert and El. Miligy |18|, who determined the diffusion coefficient of oxygen in liquid copper. The experimental cell was about the same as that of Etsell and Flengas, except that the metal was inside the electrolyte tube, and the tube was not vibrated. The authors plotted the potential variations vs

$$\ln \frac{\tau^{1/2} - t^{1/2}}{1/2}$$

and obtained a good linear relation ship. Knowing the

current value, the electrode surface and the oxygen concentration, they determined the diffusion coefficient of oxygen in copper.

Effet of double layer charging process.

The main disadvantage of chronopotentiometric measurements is the double layer charging process (another source is the reaction of surface oxides or extraneous adsorbed impurities).

The double layer charging current is related to the rate of potential change by the expression

$$i_{d.l.} = C_{d.l.} \frac{dE}{dt} \qquad (33)$$

where $C_{d.l.}$ may be some function of E. Thus, only a part of the applied current is used for the electrode reaction. The part of the charging process increases with decreasing transition time(dE/dt increases), and with low reactant concentration (C_{dl} remaining app. the same). An illustration of effect of double layer charging is given on fig.13.

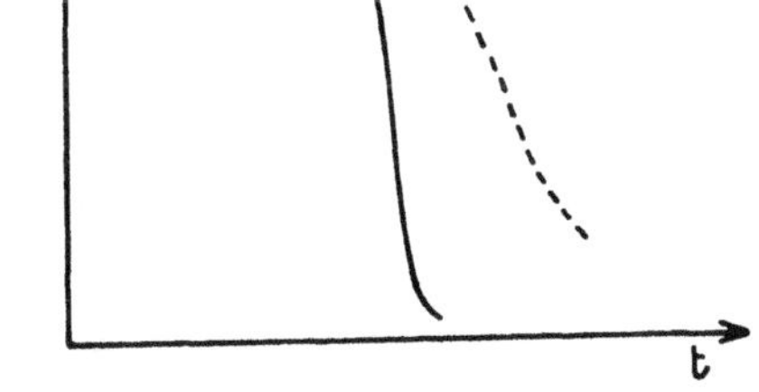

Fig. 13. Influence of double layer capacitance on chronopotentiometric curve.

Different approaches have been proposed to correct the transition time from the doube layer effect :

- Use of a blank cell, matched with the working cell, giving a cell current equal to I_{dl} (if electrodes are exactly similar), which is added to the working cell current [19].
- Electronic devices, rejecting in the working electrode circuit a current proportionnal to dE/dt (a constant value of C_{dl} is then assumed.)
- Graphical techniques, which are empirically established, to compensate the I_{dl} perturbation [20,22]

All these techniques have some uncertainties and chronopotentiometric method can be used accurately when the double layer charging process is negligible, that is for long transition time and high concentration levels.

COUPLED CHEMICAL REACTION

Introduction

Influence of coupled chemical reactions corresponds to steps 2 and 6 in the general scheme of electrochemical reaction. Several mechanisms may occur :

- Chemical reaction preceding charge transfer (CE Mechanism)

$$Y \rightleftarrows O \qquad O + ne \rightleftarrows R$$

- Chemical reaction following charge transfer (EC Mechanism)

$$O + ne \rightleftarrows R \qquad R \rightleftarrows Z$$

- Catalytic reaction :

$$O + ne \rightleftarrows R + Z \qquad Z \rightarrow O$$

- Chemical reaction occuring between two charge transfer reactions (ECE Mechanism) :

$$O + ne \rightleftarrows R \qquad R \rightarrow X \qquad x + n'e \rightleftarrows Y$$

If the rate of the chemical reaction is too low, comparatively to mass transport processes, the flux of electroactive reactant is a function of the rate of chemical reactions. The influence of this phenomenon on the response to electroanalytical excitations has been studied for reactions in liquid electrolytes, and several reviews have been published [23,24]. However, only a few studies of such influences are available for solid electrolytes, and two examples will be given, as illustration of the phenomenon.

Application :

Oxido-reduction of silver at the Ag | Ag_3SI interface.

TAKAHASHI and YAMAMOTO [25], and later, KARPACHEV, OVSHINNIKOV and YUSHINA [26] studied oxidation and reduction of silver metal electrodes at the interface Ag/Ag_3SI. Obtained current-voltage curves are reported on fig.14, curve 1. Curve 2 represents the current-voltage curve for a charge-transfer controlled reaction. The overall electrochemical raction is

$$Ag \rightleftarrows Ag^+ + e$$

TAKAHASHI and YAMAMOTO assumed the reaction takes place at interstitial sites of the lattice of the solid, and the initial charge transfer reaction is :

$$Ag \rightleftarrows Ag_i^+ + e$$

This reaction is followed by an insertion of Ag_i^+ in a normal site, the insertion reaction being the rate determining step :

$$Ag_i^+ + V_{Ag}{}^+ \underset{k_b}{\overset{k_f}{\rightleftarrows}} Ag_{Ag}^+$$

The diffusion equation is then :

$$\frac{\partial C}{\partial t} = D \frac{\partial^2 C}{\partial x^2} + v$$

where C is the Ag_i^+ concentration and v the rate of chemical formation of Ag_i^+ :

$$v = k_b \ C(Ag_{Ag}^+) - k_f \ C(V_{Ag}{}^+)C$$

The concentration of normal silver ions and silver vacancies were assumed to be constant and the rate v is related to the equilibrium value $\overline{C}$ of C by:

$$v = v_o(1-C/\overline{C}) = v_o(1-u)$$

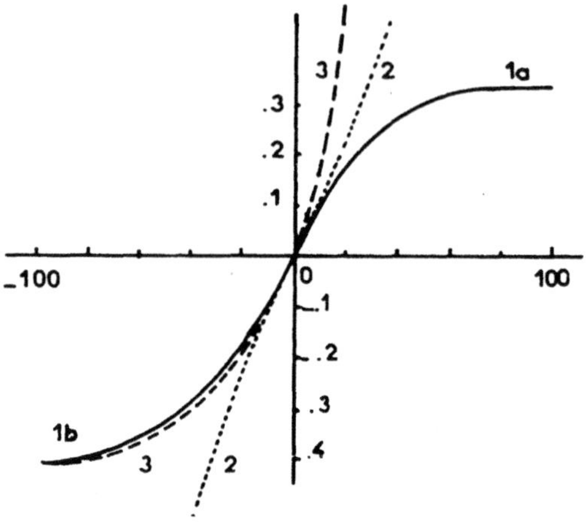

Fig.14. Oxido-reduction of silver at the Ag/Ag_3SI interface (curve 1A, ref. 25, curve 1B ref.26)

for stationnary current, diffusion equation is simply :

$$\frac{d^2u}{dx^2} = - \frac{v_o}{D\overline{C}} (1 - u)$$

the integration of which gives the value of $\frac{du}{dx}$:

$$\left(\frac{du}{dx}\right)^2 = - \frac{2v_o u}{D\overline{C}} \left(1 - \frac{u}{2}\right) + K$$

K is calculated from the boundary condition $u(\infty) = 1$, $\frac{\partial u(\infty)}{\partial x} = 0$ and

$$\frac{du}{dx} = \left(\frac{vo}{D\overline{C}}\right)^{1/2} (u - 1)$$

the current density is then :

$$j = - D \left(\frac{dC}{dx}\right)_{x=0} = -F \sqrt{v_o D\overline{C}} \left(\frac{C(0)}{\overline{C}} - 1\right)$$

where C(0) is the Ag_i^+ concentration at the electrode surface using expression (3) for a reversible reaction:

$$\eta = \frac{1}{f} \ln \frac{C_{(0)}}{\overline{C}}$$

results in current-overvoltage relation :

$$\eta = \frac{1}{f} \ln \left(1 - \frac{j}{F\sqrt{v_0 D\overline{C}}}\right) \tag{34}$$

Theoretical curve, corresponding to equation (34) is compared with the experimental ones (fig. curve 3). There is a good agreement in the cathodic part of the curve, but no limiting current is theoretically observed at high positive overvoltages (anodic part). To explain this discrepancy, Takahashi and Yamamoto supposed there is a great increase of interstitial silver ions in the vicinity of the anode. These ions react with a negative ion to precipitate a substance of very low ionic conductivity, like AgI, which blocks the electrode.

Equation (34) looks like a stationary-state diffusion controlled current-overvoltage curve. It is a general case for first order coupled chemical reaction. For reaction of higher order, relations are different.

REFERENCES.

|1| A.A. VLCEK, Progress in Polarography, Vol. I., Interscience, New York, 1962, p. 269.

|2| T.N. ANDERSEN and H. EYRING, Physical Chemistry, Vol.IXA/ Electrochemistry, Academic Press, New York 1970, H. EYRING Ed..

|3| S. PIZZINI, M. BIANCHI, A. CORRADI and C. MARI, J. Appl. Electrochem., 4 (1974), 7.

|4| R.D. ARMSTRONG, T. DICKINSON, H.R. THIRSK and R. WHITFIELD, J. Electroanal. Chem., 29(1971), 301.

|5| K.J. VETTER, Electrochemical kinetics, Academic Press, London, New York, 1967.

|6| A.T. FILAYEV, S.V. KARPACHEV and S.F. PAL'GUEV, Electrochemistry of Molten and Solid Electrolytes, Vol. 4, p. 161, Consultants Bureau, New York, 1967.

|7| T.H. ETSELL and S.N. FLENGAS, J. Electrochem. Soc., 119 (1972), 198.

|8| R.J. BROOK, W.L. PELZMANN and F.A. KROGER, J. Electrochem. Soc., 118(1971), 1890.

|9| M. KLEITZ, Thesis, Grenoble University, 1968.

|10| R.E.W. CASSELTON, J. Appl. Electrochem., 4, (1974), 25.

|11| F.A. KROGER, J. Electrochem. Soc., 120(1973), 75.

|12| T.H. ETSELL and S.N. FLENGAS, Ibid., 118, (1971), 1890.

|13| H. GERISCHER and H. VIELSTICH, Z. Phys. Chem., N.F., 3 (1958), 16.

|14| P. ROLLAND, Thesis, ParisVI University, 1974.

|15| J.H. CHRISTIE, R.A. OSTERYOUNG and F.C. ANSON, J. Electroanal. Chem., 13(1967), 236.

|16| R.S. NICHOLSON and I. SHAIN, Anal. Chem., 36(1964), 706.

|17| J. HLADIK, Y. POINTUD and G. MORAND, J. Chim. Phys., 66, (1969), 113.

|18| H. RICKERT and A.A. EL MILIGY, Z. Metallkunde, 59(1968),635.

|19| W.D. SHULTS and al. Anal. Chem., 37(1965), 1415

|20| P. DELAHAY and T. BERZINS, J. Am. Chem. Soc., 75(1953),2486.

|21| W. REINMUTH, Anal. Chem., 33(1961), 485.

|22| R.W. LAITY and J.D.E. MCINTYRE, J. Am. Chem. Soc., 87(1965), 3806.

|23| E.R. BROWN and R.F. LARGE, "Cyclic Voltammetry, AC. Polarography and related techniques", in "Techniques of Chemistry", Vol. I, Part IIA, "Electrochemical Methods", A. Weissberger, Ed. p. 423.

|24| R.W. MURRAY, "Chronoamperometry, Chronocoulometry and chronopotentiometry", Ibid., p. 591.

|25| T. TAKAHASHI and O. YAMAMOTO, Electrochim. Acta, **11**, (1966) 911.

|26| S.V. KARPACHEV, Yu. M. OVCHINNIKOV and L.D. YUSHINA, Electrokhim, 7, (1971), 382.

DISCUSSION

GERISCHER : Your reaction mechanisms assume that neutral species like O or J are present at the metal-electrolyte interface. In what way do these species reach the interface ? Are they present in the bulk of the two phases? Has their diffusion a great influence on the kinetics ?

VEDEL : Neutral species are adsorbed either on the electrode surface or at the grain boundaries of the solid. They are brought to the interface by diffusion. Diffusion may be the rate determining step for the reaction kinetics.

WAGNER : An alternative possibility for the current at infinite time shown in your plot of log i vs $1/\sqrt{t}$ is that the AgI exhibits electronic leakage. We have carried out somewhat similar experiments on the diffusion of copper into single crystalline gold and can relate the leakage current (log i vs $1/\sqrt{t}$ at $t \to \infty$) to hole conduction in CuCl.

VEDEL : If the equilibrium potential E_{eq} of the cell

$$Ag/KAg_4I_5/Pt(I_2)_g$$

is plotted versus the I_2 partial pressure, straight lines are obtained, the slopes of which are always greater than the theoretical Nernst slope (at various temp.). In the case of electronic conductivity, the slopes would be less than the theoretical ones and this hypothesis was not considered.

IMPEDANCE DIAGRAMS FOR SOLID ELECTROLYTE CELLS

R.D. Armstrong

Electrochemistry Research Laboratories, School of Chemistry, University of Newcastle-upon-Tyne, Newcastle upon Tyne, NE 1 7RU, (England)

It is the purpose of this paper to suggest a simple model for the metal-super ionic conductor interphase which can explain some of the experimental observations which have been made on these systems. The model is essentially a refinement of one discussed earlier (1). Super ionic conductors are taken to be (typically) sodium β-Al_2O_3 and Ag_4RbI_5, i.e. solids which have very high ionic conductivities (up to 10^{-1} Ω^{-1} cm^{-1}), negligible electronic conductivities and very high concentrations of mobile species ($\sim$ 10 mol l^{-1}). In what follows it will be assumed initially that the solid consists of immobile anions and cations which are all equivalent and all mobile.

CASE 1. BLOCKING ELECTRODES WITH NO SPECIFIC ADSORPTION

In the first case the metal will be considered blocking to both anions and cations e.g. the C-Ag_4RbI_5 interphase at potentials between 50 mV and 600 mV anodic to Ag/Ag^+. The model proposed is shown in figure 1. The metal surface is adjacent to a layer of cations and anions (layer 1) which is in turn adjacent to another layer of anions and cations (layer 2). In both layers the number of anions is the same (equivalent to a charge of q_-C cm^{-2}) and equal to that which would be found in the bulk of the solid. In the second layer electroneutrality holds so that the cation charge (q_{2+}) is equal to q_-. However in the first layer there is a deviation from electroneutrality i.e. $q_{1+}+q_- = q_s$ so that there is a nett charge q_s on the electrolyte side of the interphase. This charge q_s is balanced by a charge $q_m = -q_s$ on the metal surface. In layer 1 it is assumed that the cations are present in normal lattice sites and that energetically these

M. Kleitz and J. Dupuy (eds.), Electrode Processes In Solid State Ionics, 261–275.

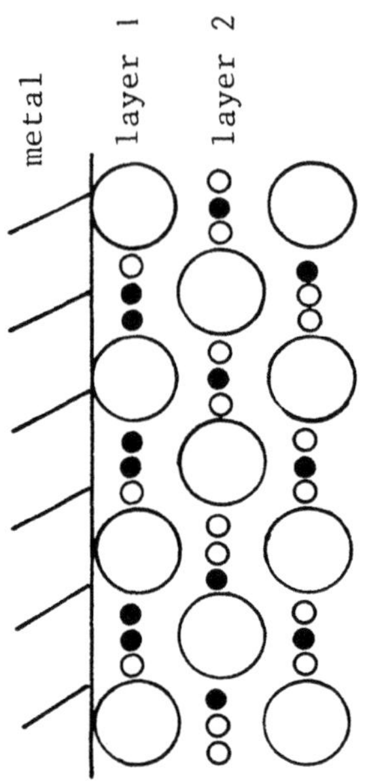

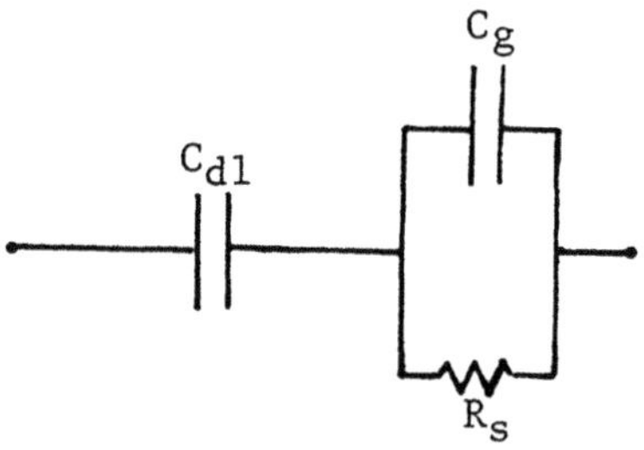

Fig.1 : Model for metal-solid electrolyte interphase blocking situation. no specific adsorption (•) cations (◯) anions (o) vanct cation sites

Fig.2 : Equivalent circuit for model of fig.1

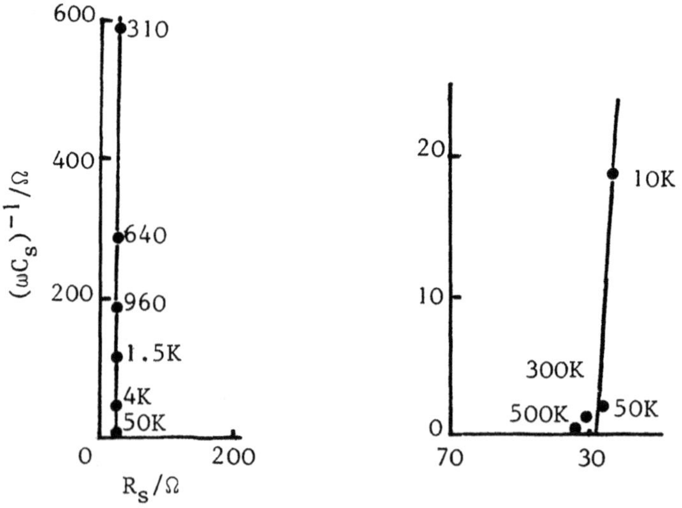

Fig.3 : Impedance of single crystal β-alumina, length 3.0 cm with vacuum deposited gold electrodes area 3.0 sq cm.

sites are only slightly disturbed by the presence of the metal surface. The assumption that the space charge extends only one atomic distance can be justified from the fact that in these solids the Debye length is always less than the size of an atom. Of course, there is also the requirement that $|q_m| < 20\ \mu C\ cm^{-2}$ so that the deviation from stoichiometry in the first layer is never sufficient to change significantly the chemical potential of the cation, i.e. we require $|q_m|/|q_-| < 0.1$, where q_- might be 200 $\mu C\ cm^{-2}$.

The model of figure 1 leads directly to the equivalent circuit of figure 2 where C_g is a geometric capacity (2) arising from the finite dielectric constant (ε_r) of the solid electrolyte and R_s is its bulk resistance. (The bulk dielectric constant and conductivity can be assumed to be frequency independent over a limited frequency range. It should be noted however that ε_r reflects the behaviour of the ions during acceleration as well as the fixed parts of the crystal lattice.) The impedance is supposed to be measured between the metal electrode and a metal wire reference electrode placed at a distance l from the metal electrode. (l is typically > 1 mm). (This reference electrode could be a silver wire in the case of C-Ag_4RbI_5.) We require that

$$C_g = \varepsilon_o \ \varepsilon_r A/(l-r_1) \simeq \varepsilon_o \ \varepsilon_r A/l$$

$$R_s = l/A\sigma$$

and

$$C_{dl} = \varepsilon_o \ \varepsilon_r' A/r_1$$

where σ is the conductivity of the material, ε_r' is the effective dielectric constant of the electrode/first layer region and r_1 is the distance between the centres of charge of the cations in the first layer and the electrode surface. C_g is not placed across the total cell since the displacement current between the metal electrode and layer 1 would then be counted twice over. Thus C_g can be thought of as arising from the summation of the double-layer-type capacities between layer 1 and layer 2, layer 2 and layer 3, etc. At low frequencies (e.g. for C-Ag_4RbI_5 at 25°C < 1 MHz) where C_g has no influence, the equivalent circuit would be simply C_{dl} in series with R_s and this is what is found experimentally (1). (see figure 3). The value of C_{dl} predicted on this model can never exceed $\sim$ 50 $\mu F\ cm^{-2}$ and should be independent of the metal electrolyte potential difference. It is however obvious that the equilibrium positions of metal cations in the first layer will be to some extent a function of q_m so that a more realistic expectation might be that C_{dl} will slowly vary with applied potential (figure 4). It is worth noting that such a possibility means that measurements on cells of the form C/Ag_4RbI_5/C lead to

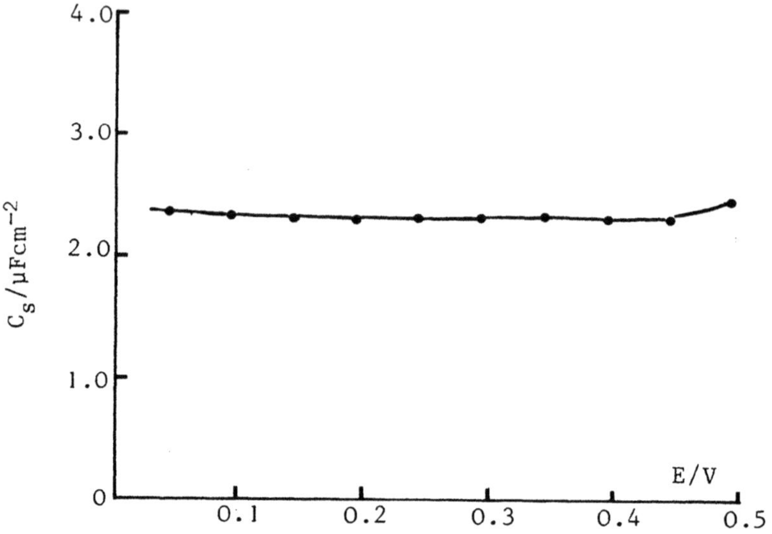

Fig.4 : C_{dl}-E, for C/AgI at 175°C, 6 x 10^5 kgm^{-2}, 3000Hz.

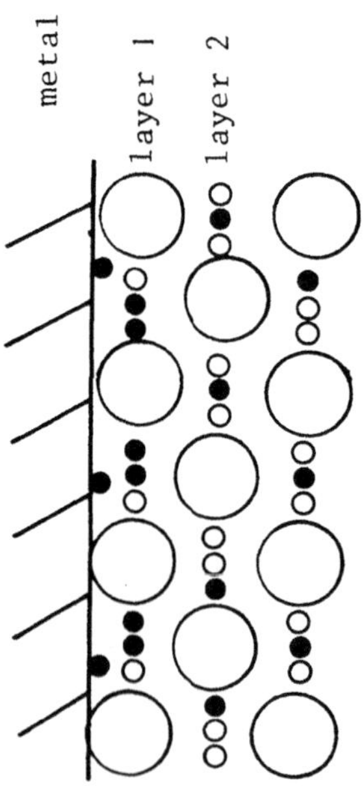

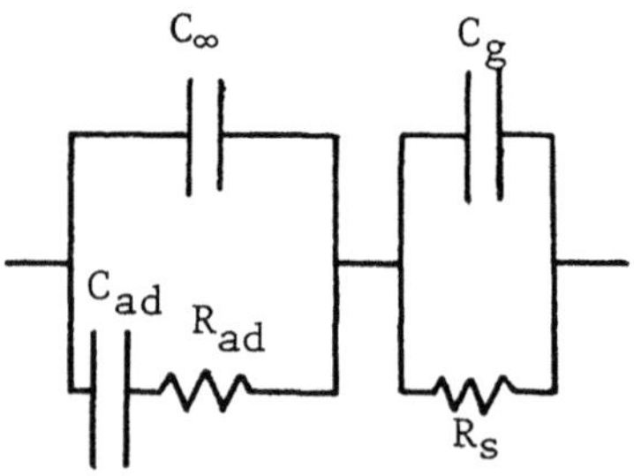

Fig.5 : Metal-Solid electrolyte interphase with specific adsorption.

Fig.6 : Equivalent circuit for model of fig.5 blocking case.

some ambiguity since the value of C_{dl} derived from such a cell (without a reference electrode) is appropriate to an unknown metal-electrolyte potential difference.

CASE 2. BLOCKING ELECTRODES WITH SPECIFIC ADSORPTION

A second possibility which must now be considered is that in layer 1 (of figure 1), because of the proximity of the metal surface some abnormal positions become possible for the metal cations (as in figure 5). They may be positions which are simply closer to the electrode than the normal ones but in addition a fraction (3) (λ) of the cation charge (z_+e) may be neutralized by partial electron transfer from the metal. This is still the blocking case with no equilibrium between metal cations in the bulk of the metal and the different metal cations in the bulk of the electrolyte. Thus the interphase might be that between Pt and Ag_4RbI_5 at potentials between 0 mV and 600 mV anodic to Ag/Ag^+. If the number of cations in abnormal sites is Γ_A (mol cm^{-2}) then the charge transferred to the cations from the metal will be $-\lambda\ \Gamma_A z_+ F$ (C cm^{-2}). It is convenient to define a quantity Q such that $Q = q_m - \lambda\ _A z_+ F$ where q_m is the charge remaining on the metal surface. The equivalent circuit for this situation is that of figure 6. At limiting low frequencies a double layer capacity will be found as

$$(C_{dl})_{\omega \to 0} = \frac{dQ}{dE} = \left(\frac{\partial Q}{\partial \Gamma_A}\right) \frac{d\Gamma_A}{dE} + \left(\frac{\partial Q}{\partial E}\right)_{\Gamma_A} \qquad (1)$$

It is probable that $(\partial Q/\partial E)_A = C_\omega$ will still be potential independent and 50 $\mu F\ cm^{-2}$. However, the term

$$\left(\frac{\partial Q}{\partial \Gamma_A}\right) \frac{d\Gamma_A}{dE} = C_{ad}$$

is not limited by simple parallel plate capacitance considerations and could quite easily lead to C_{dl} values as high as 1,000 $\mu F\ cm^{-2}$. (This figure comes from the analogous case 4 of Pt/1 M $HClO_4$aq. where ther is specific adsorption of H^+ and a degree of partial charge transfer). All that can be said with certainty is that there is a limit on the magnitude of $\int_{E_o}^{E} C_{dl} dE$ (E_o is the potential of zero charge) since only one monolayer of specifically adsorbed cations is possible.

In the equivalent circuit (figure 6) there are two capacitances constituting C_{dl} corresponding to the two parts of equation (1). This recognises the possibility that specific adsorption may be an activated process and require a characteristic time $\tau = C_{ad}R_{ad}$ for completion.

Some considerations similar to those above have been given by Mitoff and Charles (5). However, as has been pointed out by MacDonald (6) they have erroneously assumed that a Warburg impedance would be present. This cannot be the case since the interfacial capacitance is linked in a direct manner with the concentration of metal cations in the first layer i.e. the potential difference cannot be varied independently of this concentration and this is required for the normal Warburg impedance.

Raleigh (7) has suggested that for the Pt-AgBr interphase some specific adsorption and partial charge transfer to Ag^+ is involved, whilst our own results (1) suggest that this is also true to some extent for Pt-Ag_4RbI_5 (figure 7).

CASE 3. NON BLOCKING ELECTRODES

We shall now consider the situation where the possibility that metal cation exchange can occur between the metal and the electrolyte exists e.g. Ag-Ag_4RbI_5. The double layer structure which would be anticipated depends on the magnitude of the exchange current for cation transfer between the metal and the cations in the first layer adjacent to the surface. For a substance which has a conductivity of 10^{-1} Ω^{-1} cm^{-1} the charge transfer resistance for cation transfer between two adjacent cation layers is $\sim 10^{-6}$ Ω cm^2, since the sum of these charge transfer resistances must lead to the macroscopic conductivity. Now provided that the charge transfer resistance for cation transfer between the metal and layer 1 is very much greater that this value, which is very probable, the equilibrium structure of the double layer described in Section 1 (figure 1) will still hold, with the charge transfer between metal and electrolyte simply introducing a slight perturbation of the system. Since the metal-electrolyte potential difference can only be changed by changing the concentration of metal cations in the first layer, the normal Warburg impedance as found in Ag/1 M $HClO_4$ + 10^{-3} M $AgClO_4$ cannot exist so that the equivalent circuit for this situation is that of figure 8. Experimental measurements (8) on the system Cu/CuBr + NR_4Br suggest an approximate conformity with this circuit (figure 9).

For the case where the charge transfer resistance for cation movement between the metal and the electrolyte is less than that between adjacent layers in the bulk of the electrolyte the interfacial impedance becomes negligible, and the equivalent circuit is simply R_s in parallel with C_g. It is possible that this is the case for the cell Na/β-Al_2O_3/Na when the ceramic surfaces are smooth (at 150°C) (9).

It is interesting to note that for charge transfer across the metal-solid electrolyte interphase, where the charge being

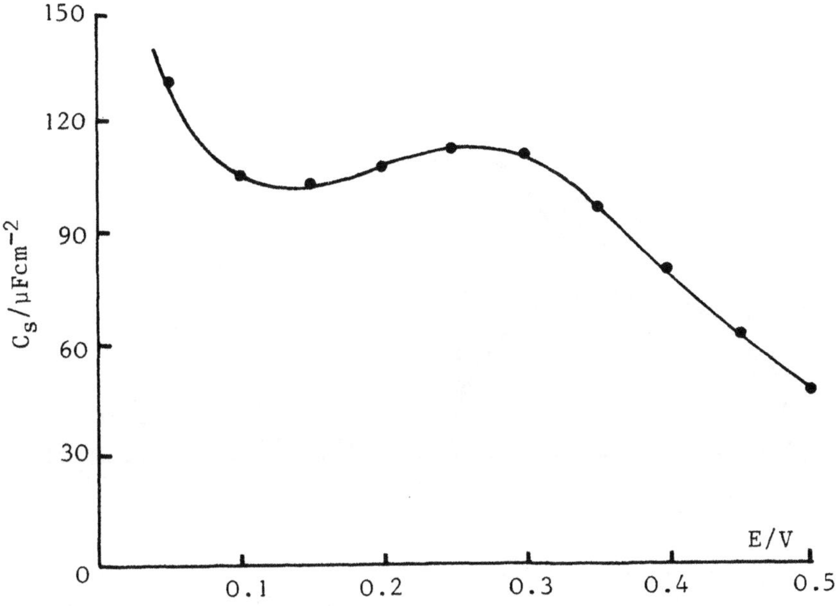

Fig.7 : C_{dl}-E, for Pt/AgI at 200°C, 4 x 10^5 kgm^{-2}, 3000 Hz.

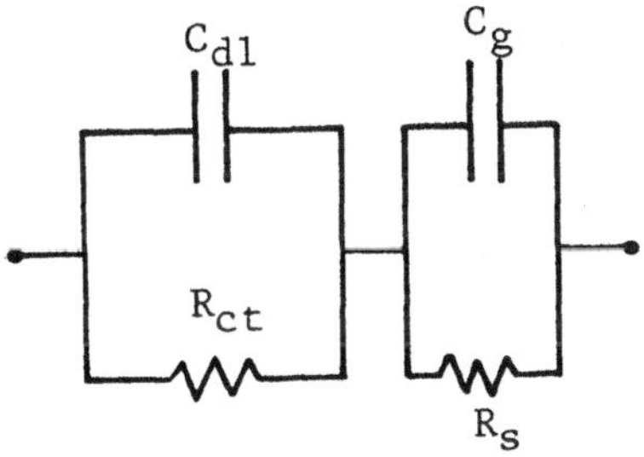

Fig.8 : Equivalent circuit for non-blocking case e.g. Ag-Ag_4RbI_5 R_{ct} = charge transfer resistance for Ag^+ cations between the metal and the electrolyte, assumed to be a single step reaction.

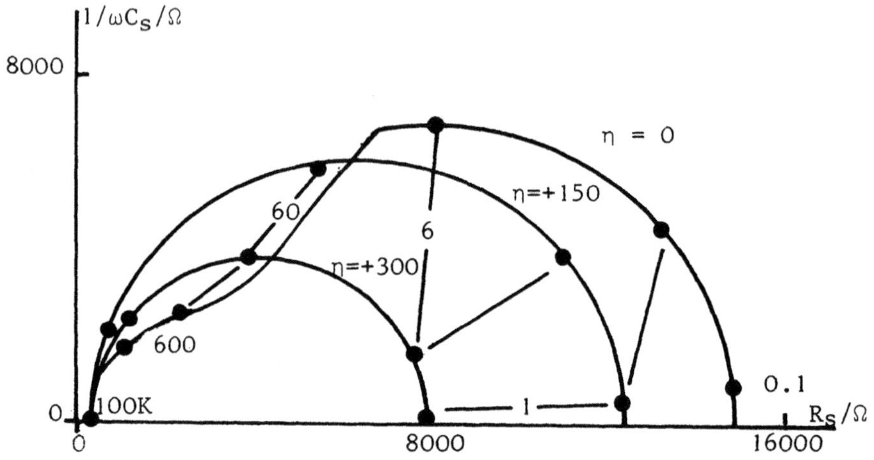

Fig.9 : The impedance of the Cu/Cu^{+} reaction at the interphase $Cu/Cu_7C_7H_{15}N_4Br_8$ at various overpotentials η (w.r.t. Cu/Cu^{+}) ; electrode area 0.2 cm^2. Bulk resistance (to reference electrode-) 140 Ω.

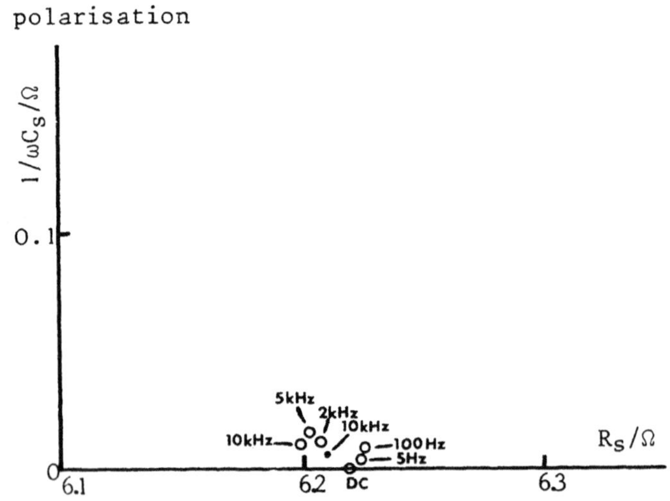

Fig.10 : Complex impedance of polished disc at 150°. (●) Wheatstone bridge (O) P.S.D. Electrode area 4.5 cm^2.

transferred is the mobile cation of the electrolyte one would not expect a Tafel relationship between the current and the overpotential. This is because the concentration of Ag^+ in layer 1 must be varied if the potential is to be varied. Instead the relationship would be, for metal deposition (using the model of figure 1)

$$i = \Gamma_{Ag^+} \exp(\alpha nF\eta/RT) \qquad (2)$$

where Ag^+ is the concentration of Ag^+ in layer 1. This can be rearranged to give

$$i = i_o \left[1 + \frac{C_{dl}\eta}{|q_-| - \Delta E C_{dl}}\right] \exp\left(\frac{\alpha nF\eta}{RT}\right) \qquad (3)$$

where ΔE is the potential difference between the point of zero charge and the reversible Ag/Ag^+ potential. (C_{dl} has been taken to be potential independent in this case). Under many circumstances the term in square brackets will approximate to unity and a normal Tafel relation will be found. In fact experiments on the Ag-Ag_4RbI_5 interphase (10) suggest that a Tafel relation does indeed hold. Equation (3) can be regarded as the analogue of the Frumkin $\emptyset_2$ correction in aqueous electrochemistry.

It should be noted that the assumption that the transfer of say a silver ion between silver metal and the electrolyte in a single step reaction is likely in most cases to be a gross simplication. In fact it could well be a two or three step reaction with intermediate adatoms. In such a case (11) the interfacial impedance will be much more complex than that corresponding to a single resistance (R_{ct}) in parallel with a double layer capacity. Another possibility is that two or three dimensional nucleation is involved.

CASE 4. SITUATIONS WHERE A NORMAL WARBURG IMPEDANCE CAN ARISE

In cases 1-3 a conventional Warburg impedance cannot arise, as discussed earlier. However, there are certain situations where the equivalent circuit of figure 11 (except for the effects of finite cell thickness) would be appropriate. These include :

(i) A solid electrolyte with immobile anions and two different sorts of cations, one of which is present at a low concentration, both ions having similar mobilities (figure 12). Such an electrolyte might be Na-β-Al_2O_3 with a small fraction of the sodium ions substituted by Ag^+ ions. For the interphase Ag/Ag-Na-β-Al_2O_3 the circuit of figure 11 is to be expected since Na^+ cations can act as a support electrolyte for Ag^+ (For the interphase Pt/Ag-Na-β-Al_2O_3 at blocking potentials the equivalent circuit

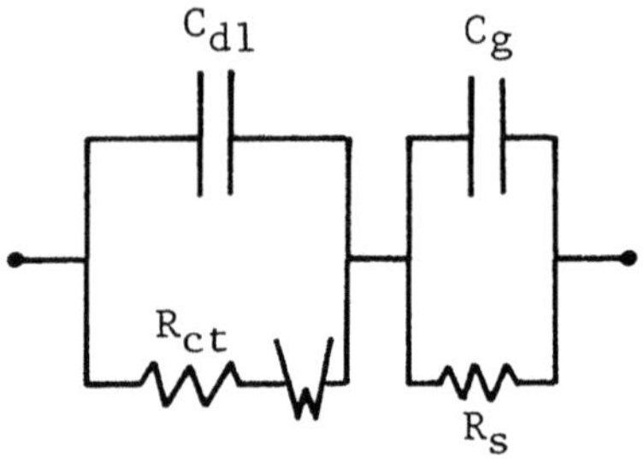

Fig.11 : Conventional electrochemical equivalent circuit, with the addition of C_g effect of finite cell size is ignored

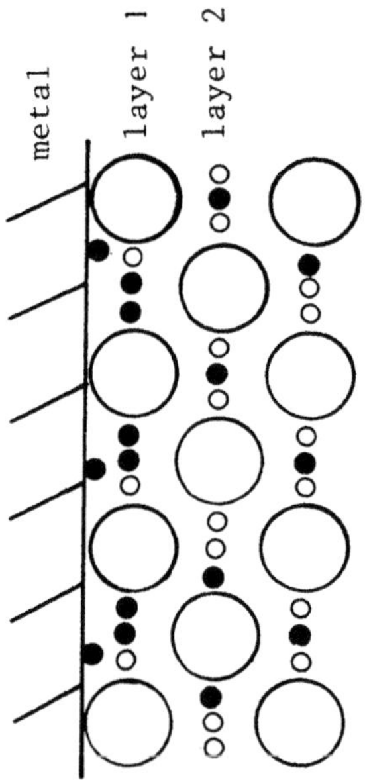

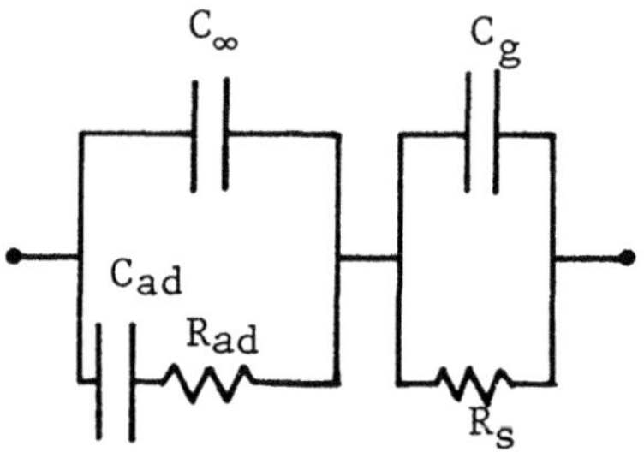

Fig.12 : Metal-solid electrolyte interphase with specific adsorption

Fig.13 : Equivalent circuit for model of fig.12 blocking case.

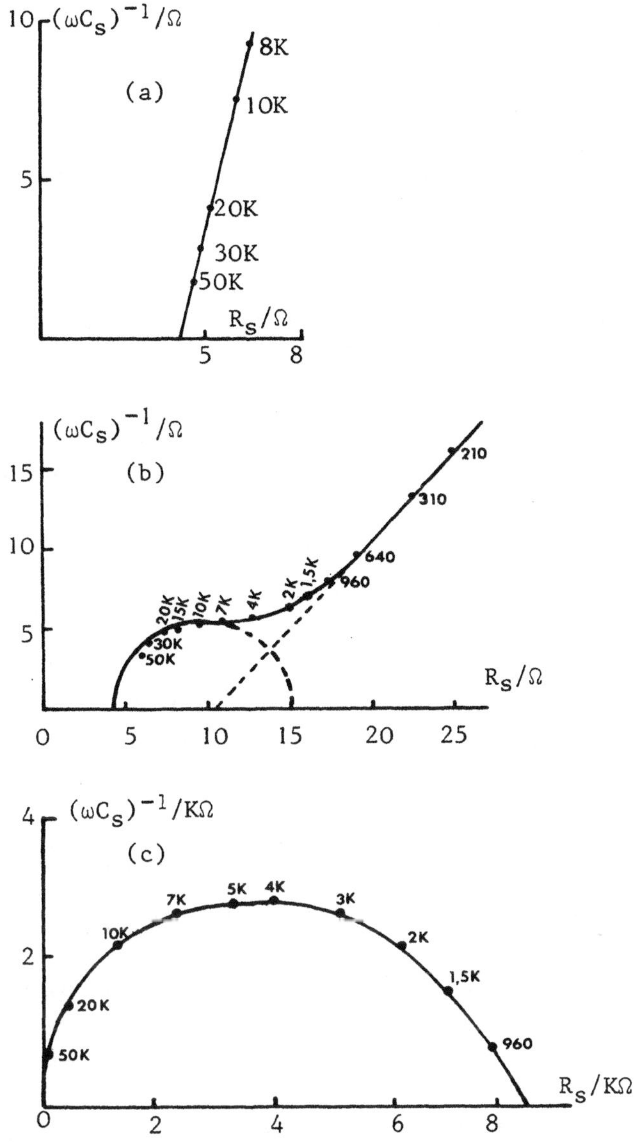

Fig.14 : Complex impedance plane displays for the Pt/Ag_4RbI_5 interphase at potentials vs Ag.Ag^+ of (a) 300 mV (b) 650 mV and (c) 685 mV. Electrode area 0.2 cm^2

would be that of figure 13, i.e. that of Lorenz and Mockel (12) for the adsorption of organic molecules).

(ii) A solid electrolyte where the products of the electrode reaction can diffuse without being influenced by electrical fields. This includes Pt-Ag_4RbI_5 between 640 and 680 mV on Ag/Ag^+ when the diffusion of electrochemically generated iodine gives rise to a Warburg impedance (13) (figure 14).

SUMMARY

A simple model for the interphase between metals and super ionic conductors such as Ag_4RbI_5 is proposed.

REFERENCES

(1) R.D. Armstrong and R. Mason, J.Electroanal.Chem., 41 231 (1973)

(2) J.R. Macdonald, Phys.Rev., 92 4 (1953)

(3) J.W. Schultze and K.J. Vetter, J.Electroanal. Chem., 44 63 (1973)

(4) A.N. Frumkin, Advan.Electrochem., 3 287 (1963)

(5) S.P. Mitoff and R.J. Charles, J.Appl.Phys., 43 927 (1972)

(6) J.R. Macdonald, J.Appl.Phys., 44 3455 (1973)

(7) D.O. Raleigh, Electroanal.Chem. 6 87 (1973)

(8) R.D. Armstrong, T. Dickinson and R. Whitfield, J.Electroanal. Chem., 39 257 (1972) ; P.Willis, unpublished work.

(9) R.D. Armstrong, T. Dickinson and J. Turner, J.Electroanal. Chem., 44 157 (1973)

(10) R.D. Armstrong, T. Dickinson, H.R. Thirsk and R. Whitfield, J.Electroanal. Chem., 29 301 (1971)

(11) R.D. Armstrong and R.E. Firman, J.Electroanal. Chem., 45 257 (1973)

(12) W. Lorenz and F. Mockel, Z. Elektrochem., 60 507, 939 (1956)

(13) R.D. Armstrong, T.Dickinson and P.Willis, J.Electroanal. Chem., 48 47 (1973)

DISCUSSION

VOÏNOV : You said the Debye length you calculate in the case of solid electrolyte is smaller than atomic dimension and consequently diffuse layer effects are irrelevant. Presumably, you use for this calculation the Debye length formula developed for the case of aqueous electrolytes. As you have stressed, in many solid electrolytes, only one species can migrate and in that case the solution of Poisson's equation is not the same as in the case of liquid electrolytes. As long as you do not have this solution and have shown it can be approximated by an exponential, and that this exponential is the same as in the case of liquid electrolytes, it seems to me difficult to calculate a Debye length in solid electrolytes.

ARMSTRONG : The concept of a Debye length holds in β-alumina just as in aqueous electrolytes. However, there are certain differences, such as the possibility of finding a Mott-Schottky exhaustion layer when there is only one mobile carrier (as in β-Al_2O_3).

MACDONALD : According to your model for C_{dlm} the thickness of the equivalent plane-parallel capacitor should be about an ionic radius. Since you evidently do not observe any appreciable dielectric saturation, the effective dielectric constant must be of the order of unity. Using a unity dielectric constant, do you actually calculate from your capacitance data a thickness of the order of an atomic radius ? If not, some aspects of your model must be questioned.

ARMSTRONG : The evidence for the parallel plate-model is principally that C_{dl} varies only slight with E. Magnitudes are difficult to compare since we don't generally know the true contact area.

RALEIGH : Some of your double layer capacitance data, such as for silver halides on glassy carbon, seem too low (tenths of $\mu F/cm^2$) for a 1 st-ions-layer capacitance. I would suggest that one way one can get a voltage-independent parallel-plate-type capacitance is by an impurity layer, say of oxygen or nitrogen molecules between the electrode and electrolyte and that this is a possibility that is always difficult to rule out.

ARMSTRONG : This is a possibility which it is difficult to entirely exclude.

FARRINGTON : Both structural and NMR results indicate that sodium ions occupy non-equivalent sites within the beta alumina conduction-plane. It is unclear exactly how this non-equivalency is manifest on the microscopic level whether at unit cells or in larger domains. Similar non-equivalency does not seem to be the case in β''-alumina. I submit that detailed models of interfacial behaviour in β-alumina must take into consideration these structural data presently available despite their ambiguities. Complete site equivalency and ion mobility should not be assumed for β-alumina.

DELL : 1. Is the Na/β-Al_2O_3 interface free of polarization and capacitive double layer effects at all current densities ?

2. Could the contrast between the observed behaviour of the Na/β-Al_2O_3 interface and the Cu/CuBr organic interface stem from the fact that in the first case one has a liquid metal and in the second case a solid metal ?

ARMSTRONG : 1. So far as we can see, yes under clean conditions.

2. Yes. I think that the contact is likely to be more intimate in the case of a liquid metal.

WAGNER : You made a careful distinction between the apparent Warburg effect for an electrolyte such as sodium, β-alumina or CuBr and the Warburg effect for Rb Ag_4I_5. Would you comment further on these differences in terms of mechanism(s) ?

ARMSTRONG : For metal/metal ion exchange reactions such as Na/Na-β-Al_2O_3 or Ag/Ag_4RbI_5 a Warburg impedance is not possible because there is only one mobile species. However, when an uncharged species such as I_2 is involved in the electrode reaction ($I_2 + 2e > 2I^-$) a new possibility arises and one finds a Warburg impedance.

GREEN : Have you tried evaporated silver contacts in order to determine in conjunction with your other work, the influence of contact pressure, contact contamination (e.g. an intermediate layer of oxide), contact crystallography on the polarization data for silver deposition at the Ag/$RbAg_4I_5$ interface ?

ARMSTRONG : Yes, and with evaporated Pt on Ag_4RbI_5 the reaction is so fast that only Ohm's law can be found.

KROGER : A few years ago Vander Meulen and I, found that

upon deposition of Ag on C or Pt from AgCl, the Ag formed, had an activity > 1, from which we estimated the dimensions of the deposited particles. It would be of interest to carry out the same measurements for the case studied by you when you know the particle size. Have you done that and does the excess activity correspond to the particle dimensions ?

ARMSTRONG : For freshly deposited Ag on Pt (from Ag_4RbI_5) we have done this and the activity difference generally corresponded to a potential difference of 5-10 mV. This would be consistent with a particle size of around 1 µm.

BONANOS : You have found that at certain frequencies the relative dielectric constant of your superionic conductor becomes negative. We know that this vould correspond to absorption of energy. Are you suggesting that ions are actually vibrating in unison with the applied field at this frequency ?

ARMSTRONG : The mobile ions in superionic conductors vibrate at frequencies between 20 cm^{-1} and 100 cm^{-1} (spectrochemica Acta 30 A (1974) 1213-1218).

ION TRANSFER AT THE INTERFACE BETWEEN AN ELECTRONIC AND IONIC CONDUCTOR

H. Gerischer

Fritz-Haber-Institut d. Max-Planck-Gesellschaft
1 Berlin 33 (Dahlem), Faradayweg 4-6

I. Theory

At the interface between a metal or semiconductor and an electrolyte (liquid or solid), the initial and final states of an ion transfer reaction are extremely different with respect to their chemical interaction with the surroundings. In the electronic conductor the binding is predominantly metallic or covalent while in the electrolyte the electrostatic forces are prevailing. This results in an energy barrier of often considerable height since an intermediate state has to be passed in which neither this nor the other type of interaction is fully developed.

The kinetics of ion transfer at such an interface can be treated as an adiabatic reaction with an energy barrier to be passed by the ion, the height of which gives the activation energy. However, only in case of a contact between two liquid phases, one single pathway over an energy barrier is sufficient to describe the reaction. In case of a solid-liquid or solid-solid interface a number of different reaction paths are possible due to the presence of different structural positions for the reacting particles. This implies serious complications for the kinetics. For a discussion of the difference between a metal and a semiconductor, it is not necessary to consider all structural details. I want to make only a few general remarks about this point.

The kinetics of the simplest case, the liquid-liquid interface are well known and follow the Volmer-Butler equation which can be written, with emphasis on the

M. Kleitz and J. Dupuy (eds.), Electrode Processes in Solid State Ionics, 277–308.

voltage dependence and neglecting concentration polarization,

$$(1) \quad j = j_o \left[\exp\left(\alpha \frac{ze_o}{kT}\eta\right) - \exp\left(-\beta \frac{ze_o}{kT}\eta\right)\right]$$

$$\eta = \Delta\phi_\infty - \Delta\phi_o \text{ (overvoltage)}; \quad \alpha+\beta = 1$$

$$j_o = k_+ \cdot N_+ \cdot ze_o \exp\left(-\frac{{}^oE_A^+}{kT}\right) = k_- \cdot N_- \cdot ze_o \exp\left(-\frac{{}^oE_A^-}{kT}\right)$$

j_o = exchange current density; ${}^oE_A$ = activation energy at equilibrium

At solid electrodes in liquid electrolytes, although one finds frequently current voltage curves of the same exponential type as equ. (1), the physical meaning is not as simple as in the liquid-liquid case. We all know the important function of the kink sites (half crystal position) for the growth and dissolution of a crystal. Other surface sites can act as intermediates in the ion transfer reaction, like the various adsorption positions on crystal faces of different orientation. Nucleation phenomena complicate the situation at higher rates considerably in the crystal growth case. In the cathodic dissolution case, dislocations and other lattice imperfections play a decisive role. The result is that a potential variation affects various parameters in different reaction steps to a different degree and every particular material exhibits its individual features. This is similar for solid semiconductors but they have in addition some special features in common which are not found at metals.

The most difficult case is the solid-solid interface. The generation and diffusion of vacancies is here in most systems the decisive factor. But very little is known of the kinetics in most of such reactions. The energy for the formation of vacancies is relatively high in semiconductors compared to metals. They are only generated at high temperatures or by especially strong forces as applied in mechanical treatments. If the electrolyte needs also the formation of vacancies for ion transfer, the situation becomes so complicated that reactions are to be expected only at very high temperatures. I shall restrict my lecture therefore to the solid-liquid interface which has been studied with semiconductors in much detail at normal temperatures [1,2].

There are two most pronounced differences in the

electrochemical behaviour between semiconductors and metals.

1. The much lower concentration of mobile charge carriers in a semiconductor causes a much more extended distribution of electric excess charge underneath the surface (diffuse space charge layer). Therefore, the variation of electric field strength in the surface which changes height of the energy barrier for ion transfer is much smaller, often negligible.

2. The coordination number is smaller as in metals although lattice energies are similar. The average bond strength of a single bond is therefore much higher. Bonds between neighbouring atoms are strongly directed and have large covalent character. The energy barriers for a displacement of atoms from normal binding positions are high.

These two characteristics explain a great deal of the differences between semiconductor and metal electrodes.

The space charge layer in semiconductor electrodes

Fig. 1 compares the potential distribution at a semiconductor/electrolyte and a metal/electrolyte interface. The extension of the space charge layer in the semiconductor is a function of the mobile charge concentration in the bulk and can be approximated by a "Debye-length" L:

$$(2) \quad L = \left(\frac{\varepsilon\varepsilon_o \, kT}{2 \, e_o^2 \, N_{mob}} \right)^{1/2} = \frac{85 \cdot \sqrt{\varepsilon}}{\sqrt{N_{mob}}} \, cm^{-3}$$

N_{mob} = number of electron (or holes) per ccm

ε_o = permittivity of vacuum ($8.854 \cdot 10^{-14} \frac{A \cdot s}{V \cdot cm}$)

ε = relative dielectric constant

The exact distribution of charge and potential can be derived from electrostatic theory and the laws of electron distribution statistics, solving the Poisson-equation with the appropriate boundary conditions. For a one-dimensional charge extension

(3) $$\frac{d^2\phi}{dx^2} = -\frac{\rho(x)}{\varepsilon\varepsilon_o}$$

(4) with $\rho(x) = e_o\left(p(x)-n(x) + \sum_i N_{D_i^+}(x) - \sum_i N_{A_i^-}(x)\right)$

ϕ = electric potential

ρ = electric charge per ccm

p = hole concentration ; n = electron concentration

$N_{D_i^+}$ = concentration of ionized donors D_i with energy $E_{D,i}$

$N_{A_i^-}$ = concentration of ionized acceptors A_i with energy $E_{A,i}$

boundary conditions:

(5) $x\to\infty$: $\rho = 0$; $\frac{d\phi}{dx} = 0$

electron distribution

(6) $$n(x) = n_o \cdot \exp\left(-\frac{e_o\Delta\phi_x}{kT}\right)$$

$$p(x) = p_o \cdot \exp\left(\frac{e_o\Delta\phi x}{kT}\right)$$

as long as Boltzmann statistics can be applied ($n \ll N_C$ and $p \ll N_V$)

$\Delta\phi_x = \phi_\infty - \phi(x)$; $n_o (p)_o$ = electron (hole) concentration in the bulk

(7) $$n(x)\cdot p(x) = n_i^2 = N_C \cdot N_v \cdot \exp\left(-\frac{\Delta E_{gap}}{kT}\right)$$

n_i = electron concentration in intrinsic semiconductor

N_C, N_V = effective density of states in conduction or valence band.

(8) $$N_{D_i^+} = N_{D_{i,o}} \cdot \left[1 + \exp\left(\frac{E_F - E_{D,i}(x)}{kT}\right)\right]$$

(9) $$N_{A_i^-} = N_{A_i} \cdot \left[1 + \exp\left(-\frac{E_F - E_{A_i}(x)}{kT}\right)\right]^{-1}$$

E_F = Fermi energy

(10) $E_{D_i}(x) = E_{D,i} - e_o \cdot \Delta\phi_x$

$E_{A,i}(x) = E_{A,i} - e_o\ \Delta\phi_x$

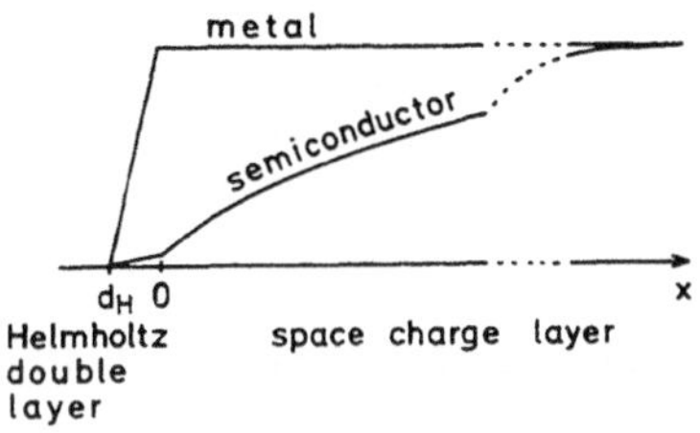

Fig. 1 - potential distribution at interface

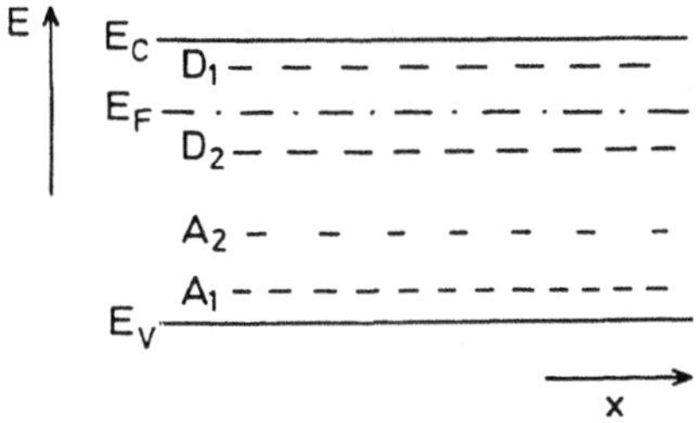

Fig. 2 - energy terms in semiconductor

Fig. 2 gives a schematic presentation of the position of the characteristic energy levels in the energy band scheme in the absence of electric fields.

The solution of the Poisson-equation can be obtained analytically in some special cases, numerically in most other systems, as found in a large literature (see e.g. [3,4,5]). One general relation, derived from the first integration of the Poisson-equation is very helpful since it connects the field strength in the surface with the overall amount of excess charge in the space charge region.

$$(11)\quad \left(\frac{d\phi}{dx}\right)_s = \frac{1}{\varepsilon\varepsilon_o}\int_o^\infty \rho(x)\,dx = \left[\frac{2}{\varepsilon\varepsilon_o}\int_o^{\Delta\phi_s} \rho(\Delta\phi)\,d\Delta\phi\right]^{1/2}$$

$$= \frac{q_{sc}}{\varepsilon\varepsilon_o}$$

q_{sc} = excess electric charge in semiconductor index s means surface

$\Delta\phi_s = \phi_\infty - \phi(x=o)$ = voltage drop between bulk and surface

Space charge layer capacity

Instead of solving these equations for the full charge and potential distribution it is often easier to derive the differential capacity between the space charge layer and the electrolyte C_{sc} in dependence on $\Delta\phi_s$

$$(12) \quad C_{sc} = \frac{d\, q_{sc}}{d\,\Delta\phi_s} = f\ (\Delta\phi_s) \quad (\mathrm{Farad/cm^2})$$

C_{sc} can easily be measured. With this relation the unknown $\Delta\phi_s$ can be obtained which is most important for kinetics since it controls the surface concentration of electrons and holes (compare equ. (6)) as long as equilibrium is maintained.

Three typical situations shall be considered which are shown in Fig. 3.

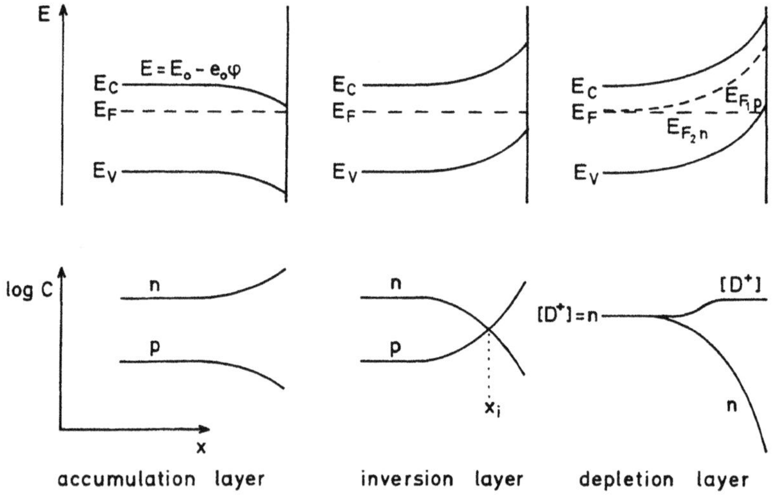

Fig. 3 - characteristic situations for space charge layer; above: position of band edges
below: concentration of charge carriers

The space charge capacity of an intrinsic semiconductor behaves like the diffuse double layer (Gouy) capacity in dilute electrolytes, namely:

$$(13) \quad C_{sc} = \left(\frac{2\varepsilon\varepsilon_o e_o^2 n_i}{kT}\right)^{1/2} \cosh\left(\frac{e_o\Delta\phi_s}{2\ kT}\right)$$

Analytical solutions can also be derived for doped semiconductors with small band gap which maintain equilibrium between electrons and holes. Fig. 4 gives as an example the result for an intrinsic and an n-type Ge-electrode.

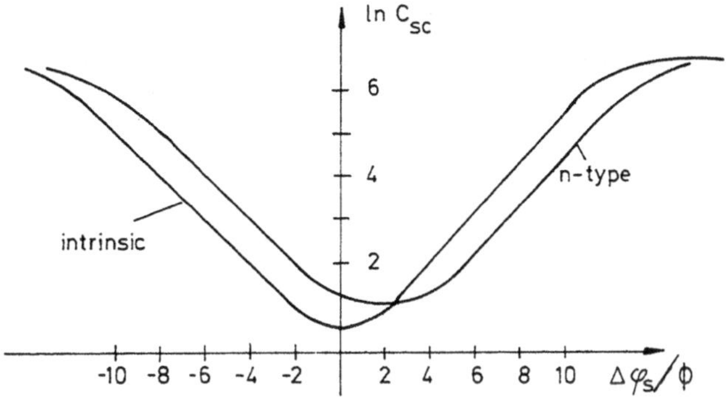

Fig. 4 - space charge capacity of 2 Ge-electrodes in dependence on voltage drop in space charge (theoretical)

For comparison, we see in Fig. 5 the result of experiments with intrinsic Germanium [6].

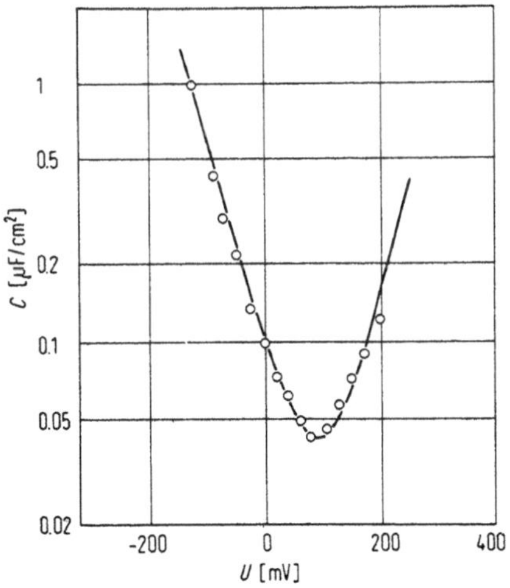

Fig. 5 - theoretical curve and experimental points of space charge capacity for intrinsic Ge-electrode

From such a measurement the potential drop $\Delta\phi_s$ in the space charge layer can be derived for every applied electrode potential U measured in a galvanic cell against a reference electrode.

Another simple case is the depletion layer in semiconductors with a wide band gap where no equilibrium is maintained between electrons and holes due to a too slow rate of electron-hole pair generation. In this case, we have a parabolic relation between space charge capacity and $\Delta\phi_s$, the so-called Mott-Schottky equation [7,8].

$$(14) \quad \left(\frac{1}{C_{sc}}\right)^2 = \frac{2\,kT}{\varepsilon\varepsilon_o\, N_D \cdot e_o^2}\left(\frac{e_o\Delta\phi_s}{kT} - 1\right)$$

The capacity plot of $\frac{1}{C^2}$ versus the applied voltage allows an extrapolation to $\Delta\phi_s = \frac{kT}{e_o}$ or $\Delta\phi_s = 0$, indicating in this way the energy position of the semiconductor Fermi level in the potential scale of a galvanic cell.

In Fig. 6 such a plot is seen for two differently doped ZnO-electrodes (n-type) [9]. From the slope of such a curve one can calculate the density of donors in such a semiconductor (compare equ. (14)).

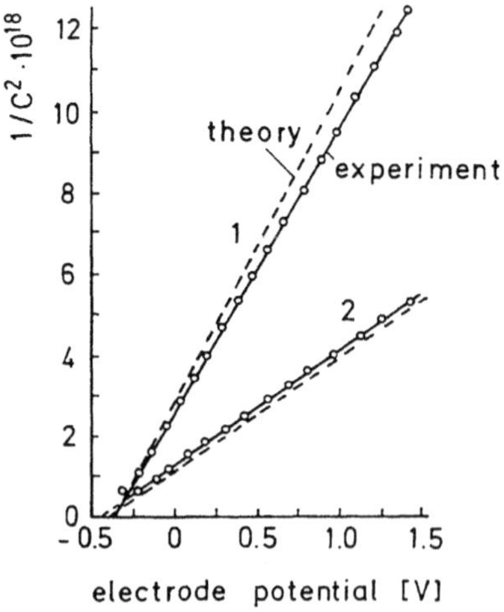

Fig. 6 - capacity of ZnO-electrodes in Mott-Schottky plot; (1) $0.59\Omega^{-1}cm^{-1}$; (2) $1.79\Omega^{-1}cm^{-1}$.

Absolute and electrochemical potential scale

The potential U in the electrochemical scale for which $\Delta\phi_s = 0$ is called the "flat band potential" because no electric potential drop is then found in the bulk of the semiconductor ($U_{f.b.}$). It is the zero point of charge for the semiconductor electrode which, besides by capacity measurements, can also be determined from the surface photovoltage effect [10,11].

If the flat band potential $U_{f.b.}$ is known, one can relate the energy terms E of the semiconductor to the energy scale U of an electrochemical cell versus a suitable reference electrode. In solid state physics, the reference state of electron energies is usually the vacuum level, if one wants to compare different solids with each other. The electrochemical potential scale versus a reference electrode is fully in parallel with this absolute scale having only the opposite sign because of the negative charge of the electron.

$$E_{abs} = \text{Const} - U_{electrochem.} \tag{15}$$

In aqueous solution with a normal hydrogen electrode as a reference system, the constant is about -4.5 eV [12].

The relation between the energy position of the band edges and the flat band potential is in the electrochemical scale,

$$U_c = U_{f.b.} - \frac{kT}{e_o} \ln \frac{N_c}{n_o} = \text{Const} - E_c \tag{16}$$

$$U_v = U_{f.b.} + \frac{kT}{e_o} \ln \frac{N_v}{p_o} = \text{Const} - E_v \tag{17}$$

$$E_c - E_v = \Delta E_{gap} \tag{18}$$

Influence of space charge on energy barrier in surface

What are the consequences of the existence of a space charge layer for electrode reaction kinetics? First of all, the electrical field strength in the surface is small and is very little changed by a variation in the potential applied. The height of the energy barrier for ion transfer is therefore little affected.

The principal reason for the voltage dependence of electrolytic processes at metal electrodes is obviously missing.

Experience has, however, shown that many charge transfer reactions at semiconductors are very much voltage dependent. The reason for this is directly connected with our second main difference between metals and semiconductors, the high strength of a single bond in the semiconductors.

Chemical bond of surface atoms and kinetics

To give an impression of the differences between metals and semiconductors, let us look at the table of Fig. 7. I have compiled there data for various materials and calculated average single bond energies. We see that the bond strength of semiconductors is usually 2-4 times as high as in most metals.

material	atomisation energy	coordination number	average single bond energy	band gap
	eV		eV	eV
Ag	2.95	12	0.24	—
Cu	3.52	12	0.29	—
Au	3.82	12	0.32	—
Pt	4.21	12	0.35	—
Fe	5.88	8(+6)	ca.0.53	—
Ta	8.10	8(+6)	ca.0.73	—
Ge	3.91	4	0.98	0.67
Si	4.70	4	1.18	1.11
C	7.40	4	1.85	5.4
GaAs	6.36	4+4	0.79	1.35
GaP	7.40	4+4	0.93	2.24
CdS	5.58	4+4	0.70	2.41
ZnO	7.10	4+4	0.89	3.20
TiO_2	19.4	6+3+3	1.61	3.04

binding energies

Fig. 7 - table of characteristic energies

This has the consequence that the energy barrier for ion transfer through the interface on an adiabatic path in the electronic ground state is extremely high. The situation is only altered if the electronic system reaches an excited state in which the chemical bond is weakened. We must be aware now, that the presence of holes and electrons in a semiconductor already means the presence of excited electronic states; A hole in the bond-system of a solid means that an electron has been excited from a binding quantum state to a state with higher energy in the conduction band. This latter state has in most cases antibonding character. The presence of holes in the valence band and of electrons in the conduction band corresponds therefore to an excited electronic state. In other words, electrons and holes in semiconductors are defects in the band system of the crystal.

Although these electron states are delocalized in the lattice of a semiconductor, one can assume for kinetic considerations that a statistical probability exists of their presence in a single chemical bond, as well in the bulk as in the surface. The concentration terms n and p or n_s and p_s have just this statistical meaning. Treating these electron states like individual species, as is common in semiconductor statistics, we have a measure for the concentration of weakened bonds in the concentration of electrons and holes. Such defective bonds of the lattice are much more reactive than the intact bonds and that is the way in which the reaction rate can here be influenced by the applied voltage. The surface concentration of electrons and holes, that is, the number of reactive sites on the surface, varies with the voltage. Instead of changing the height of the energy barrier as at metal electrodes, the applied voltage changes the number of preferential surface sites for the formation of ions. We shall see that the voltage variation can in this way be even more effective than by varying the energy barrier.

Anodic reactions

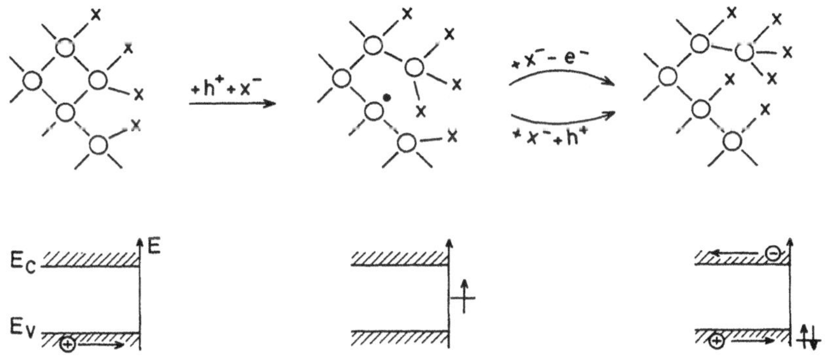

Fig. 8 - mechanism of anodic attack of a single surface bond

Let us discuss this at first for an anodic process which can be studied most easily experimentally. Fig. 8 gives a model for the kinetics of anodic attack on a chemical bond in the surface of a semiconductor, showing the two necessary steps for oxidation. While the first step definitely needs the interaction with a hole, the second can also occur chemically and is then connected with an injection of an electron. This will be analyzed in the following pictures for elemental and compound semiconductors [13,14].

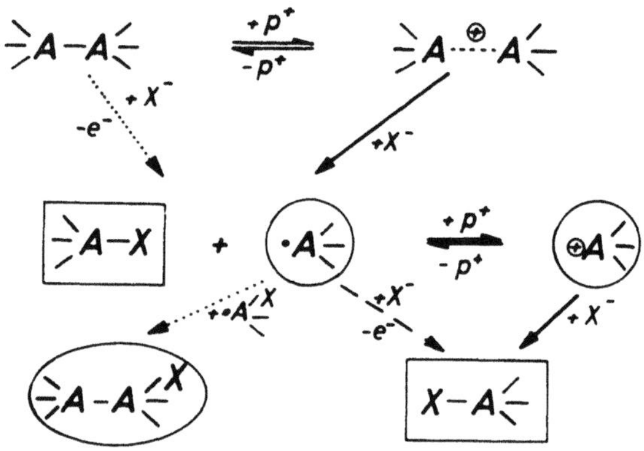

Fig. 9 - mechanism of anodic attack on single bond in elemental semiconductor

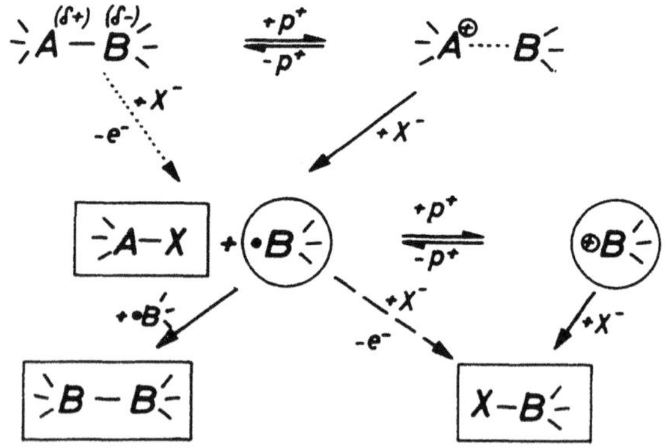

Fig. 10 - mechanism of anodic attack on single bond in compound semiconductor

Fig. 9 and 10 give reaction schemes for the anodic oxidation of semiconductors. The energetics which control the reaction rates are schematically described in Fig.11.

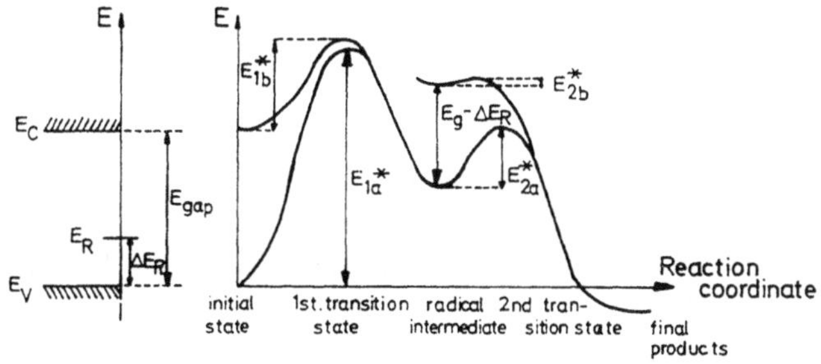

Fig. 11 - energy profile for anodic ion transfer

The first step is always rate determining since it has by far the higher activation barrier. There are two possible reaction paths which we have to compare.

(19) $-\overset{|}{\underset{|}{A}}-\overset{|}{\underset{|}{A}}- + x^- \longrightarrow -\overset{|}{\underset{|}{A}}-x + -\overset{|}{\underset{|}{A}}\cdot + e^-$

(20) $\nu \sim N_c \cdot \exp\left(-\frac{E_{1a}^*}{kT}\right)$

(21) $-\overset{|}{\underset{|}{A}}-\overset{|}{\underset{|}{A}}- + h^+ + x^- \longrightarrow -\overset{|}{\underset{|}{A}}-x + -\overset{|}{\underset{|}{A}}\cdot$

(22) $\nu \sim p_s \cdot \exp\left(-\frac{E_{1b}^*}{kT}\right)$

The second path (21) has the much smaller activation energy and will dominate if the surface concentration of holes is not extremely small. p_s is directly related to $\Delta\phi_s$ (compare equ. (6)) as long as equilibrium for holes is maintained between bulk and surface. This is certainly so for p-type specimen. In such cases, the current will increase exponentially with the applied voltage according to:

(23) $j \sim p_o\left(\exp \frac{e_o \Delta \phi_s}{kT}\right) \sim \exp\left(\frac{e_o(U-U_{f.b.})}{kT}\right)$

In comparison with a metal electrode, a Tafel type current voltage curve is formulated in equ. (23) with a charge transfer coefficient of $\alpha = 1$.

We have neglected in this derivation that some parts of the applied voltage may be lost in other parts of the double layer and not appear in the space charge layer of the semiconductor. This happens indeed by various reasons in reality. Equ. (23) is, however, the simplest case and some systems come very close to this ideal situation.

Cathodic reactions

The cathodic case can be treated in a very similar way [13,14]. The reaction scheme for the cathodic decomposition of a compound semiconductor is shown in Fig. 12, the energy profile in Fig. 13.

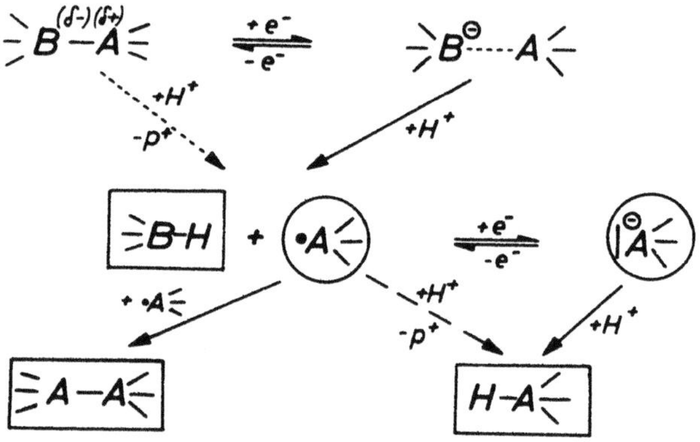

Fig. 12 - mechanism of cathodic attack on single bond in compound semiconductor

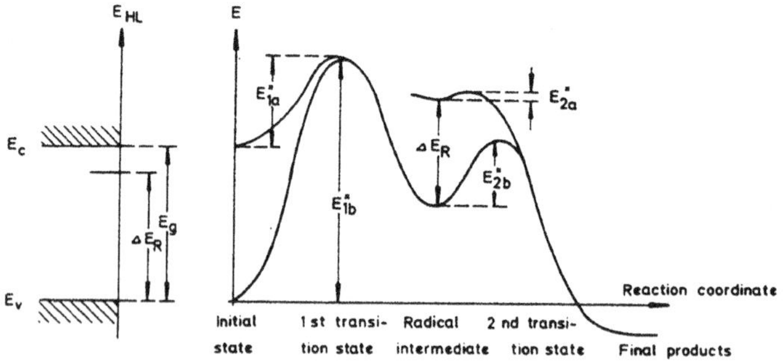

Fig. 13 - energy profile for cathodic ion transfer

The competitive reaction paths and their rates are,

(24) $\quad -\overset{|}{\underset{|}{B}}-\overset{|}{\underset{|}{A}}- + H^{+} \longrightarrow -\overset{|}{\underset{|}{B}}-H + -\overset{|}{\underset{|}{A}}\cdot + h^{+}$

(25) $\quad \nu \sim N_{v} \exp\left(-\frac{E^{*}_{2a}}{kT}\right)$

(26) $-\overset{|}{\underset{|}{B}}-\overset{|}{\underset{|}{A}}- + e^- + H^+ \longrightarrow -\overset{|}{\underset{|}{B}}-H + -\overset{|}{\underset{|}{A}}\cdot$

(27) $$\nu \sim n_s \cdot \exp\left(-\frac{E^*_{2b}}{kT}\right)$$

Again the second step (26) has the lower activation barrier and will predominate for not too small n_s. In n-type specimen, the kinetics will give an exponential voltage dependence with a formal cathodic charge transfer coefficient of $\beta = 1$.

(28) $$-j \sim n_o \cdot \exp\left(-\frac{e_o \Delta\phi_s}{kT}\right) \sim \exp\left(-\frac{e_o(U-U_{f.b.})}{kT}\right)$$

We have discussed in equ. (24-27) the case of cathodic decomposition of a semiconductor and not the cathodic deposition. The latter process is very difficult to achieve because the formation of a very much localized chemical bond from ions which have strongly bonded ligands in solution is a very unlikely reaction. The influence of electrons accumulated in the surface will be small on this step if not a preceding reduction in homogeneous solution is possible, what would change the kinetics drastically.

N- and p-type electrodes

We have learned from this discussion that anodic processes on semiconductors are controlled by holes, cathodic ones by electrons. They can occur only on such specimen where they are available in the surface. This causes a distinct difference in the electrode behaviour of p- and n-type material, which is schematically shown in Fig. 14.

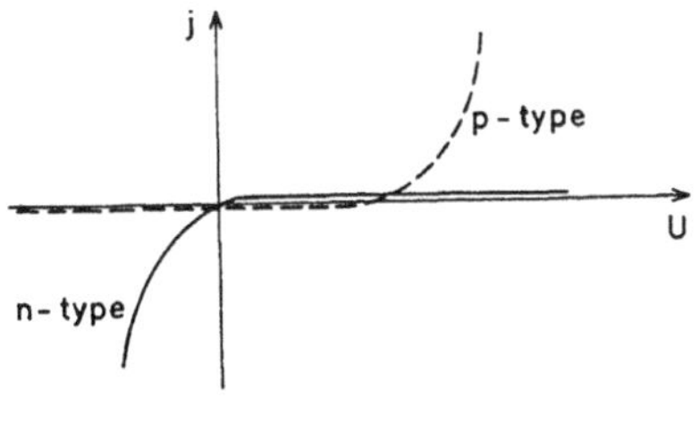

Fig. 14 - typical current voltage curves for n- and p-type semiconductor electrodes

I have not discussed a number of complications which are important in reality, like the modification of the double layer structure by adsorption of ions, the change in double layer structure by surface states or by the degeneracy of electrons or holes in accumulation layers. This is not necessary for the understanding of the principles, but some of this will show up in the following discussion of some experimental examples.

II. Experimental Techniques and Experiences

Current-voltage curves

We have seen from the theoretical discussion that the rates of electrode reactions at semiconductors should be controlled by the presence of electrons or holes in the surface. This has indeed been confirmed by many experiments for which I shall give now a few examples. The classical system was the study of the Germanium electrode by Brattain and Garrett who observed at first the pronounced difference in the electrode behaviour of p-type and n-type specimen [15]. An example from our own measurements [16] is given in Fig. 15.

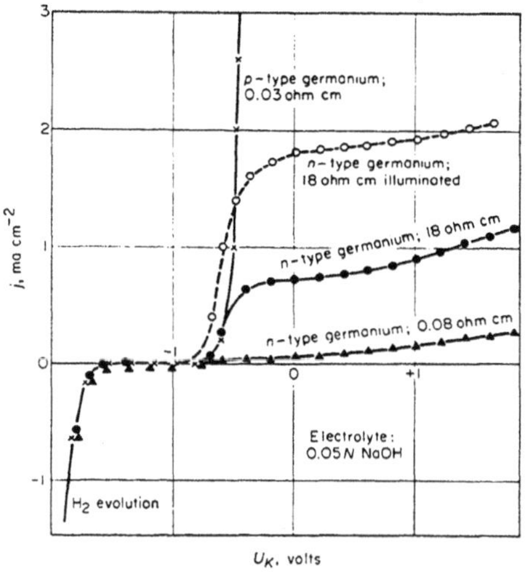

Fig. 15 - Current-voltage curves of various Ge-electrodes [16].

p-type specimen showed the expected exponential increase of the current with the applied anodic voltage but the slope of the logarithmic current voltage curves was not 60 mV per decade of the current but higher. This indicates that not all of the applied potential change is left in the space charge layer but some is found in the Helmholtz double layer. Such a behaviour is often found and can be attributed to accumulation of charge in surface states. In spite of this deviation from the simple theory it could be confirmed by Boddy [17] that the rate of anodic dissolution of Germanium is strictly controlled by the concentration of holes in the surface. Fig. 16 shows this result and indicates how the concentration of holes in the surface was derived from simultaneous capacity measurements. The capacity at high enough frequencies is controlled by the space charge layer and it corresponds for Germanium electrodes with the theoretical expectations. We have seen in the previous lecture, that then the voltage drop in the space charge layer and with it the surface concentration of electrons and holes can be derived from the measured capacity values.

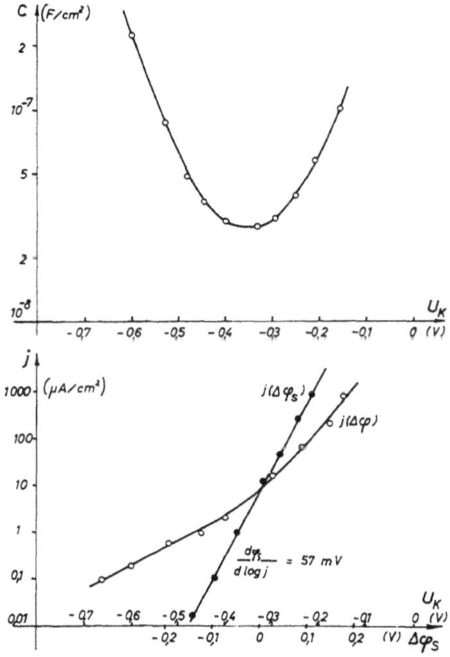

Fig. 16 - Current-voltage curve of intrinsic Ge-electrode plotted versus potential drop in space charge

Silicon behaves very similar in contact with aqueous solution [18]. Oxide-layers are formed during anodic dissolution which prevent a straight forward analysis of the current-voltage curves. The situation is much simpler in hydro-fluoric acid solutions [19] as shown in Fig. 17. The clear difference between n-type and p-type specimen in their anodic and cathodic behaviour is even shown in systems where an oxide layer is found where the rate of oxide formation on an n-type specimen is limited by the generation of holes in the surface [18]

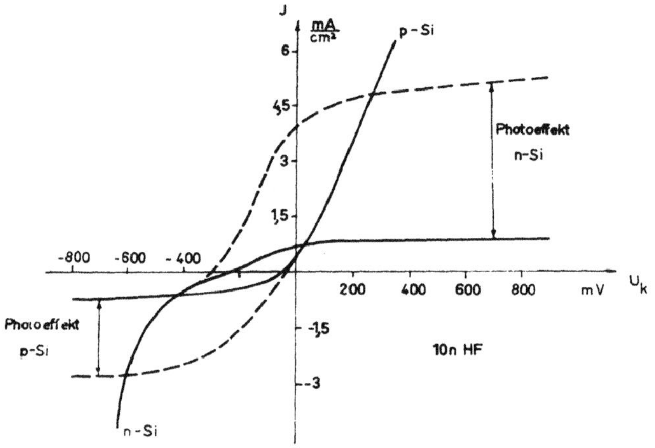

Fig. 17 - Current-voltage curves of Si-electrodes in HF

Anodic currents at n-type specimen and cathodic currents at p-type specimen are controlled by diffusion and thermal generation of minority carriers. The height of these so-called "saturation currents" is given by

$$j_{sat} \approx e_o \left(\frac{D_m}{l_m} \cdot N_m + g_v \, \delta_{sc} + g_s \right) \tag{29}$$

$l_m = \sqrt{D_m \tau_m}$ = mean free path of minorities

D_m, τ_m = diffusion coefficient and lifetime

g_v = volume generation rate of minorities

g_s = surface generation rate of minorities

δ_{sc} = extension of space charge layer

Generation of minority carriers

The generation of minorities and their transport to the surface can be increased by various means. This is most important for semiconductors with a wide band gap where practically no minorities are available or are generated at normal conditions and j_{sat} is zero. The most important way of generating minority carriers is light absorption. The light energy must exceed the band gap to generate electron-hole pairs by an electronic transition from the valence to the conduction band. If this excitation process occurs within the space charge layer the electrical field present there separates the generated charge carriers driving one to the surface, the other to the bulk.

A great effect on the rate of ion transfer reactions can only be expected if the minorities are moved to the surface. An additional photo-generation of majorities has little influence on their concentration if this is already relatively high. Large photocurrents are therefore found when a depletion layer is formed, namely for n-type specimen at anodic reactions and p-type specimen at cathodic ones. Some typical photo-current spectra for anodic bias of n-type specimen and cathodic bias of p-type specimen are shown in Fig. 18. One sees that the photocurrent spectrum reveals the light absorption properties of the semiconductor and indicates clearly the energy of the band gap by a steep increase of the photocurrent.

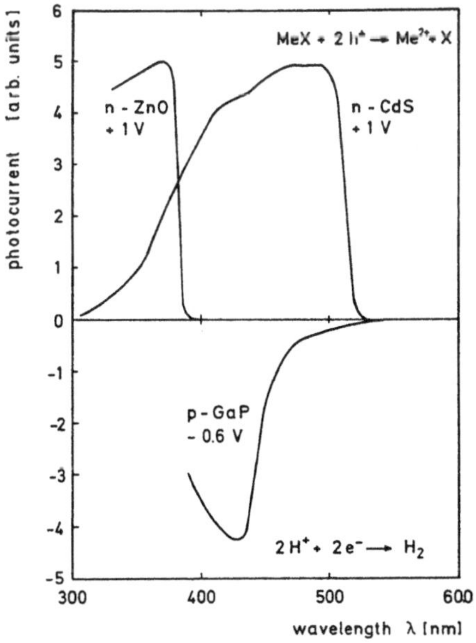

Fig. 18 - photocurrent spectra of n- and p-type semiconductor electrodes

The real quantum yield depends, however, not only on the light absorption properties of the specimen but also on the depth of the space charge layer and on other solid state properties [20]. This is shown in Fig. 19 where the fate of an electron-hole pair generated within the space charge layer is demonstrated. Light which is absorbed in greater depth outside of the space charge layer will have little influence. The minority carriers which are generated in the field free bulk will rapidly disappear by recombination since their lifetime in an excess of majority carriers is very short.

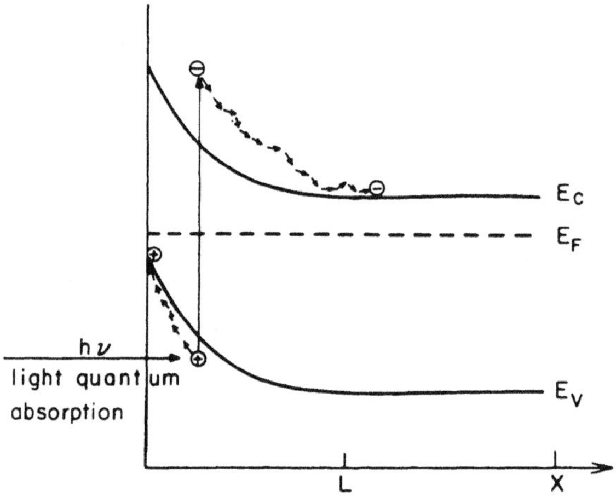

Fig. 19 - Generation and separation of electron-hole pairs in depletion layer

The photocurrent therefore depends on the extension of the space charge layer in relation to the penetration depth of the light. A maximum quantum yield for the photocurrent is therefore reached with increasing voltage when the depth of the space charge layer exceeds the penetration depth of the light. Fig. 20 shows photocurrent-voltage curves at ZnO-electrodes for light of different wavelengths. Since light with lower quantum energy than the band gap is much weaker absorbed and penetrates therefore deeper into the crystal, we see in Fig. 20 that it needs higher voltages to reach the saturation current for light of wavelength above 380 nm which corresponds to the band edge in this semiconductor.

Another means of minority carriers generation is charge injection into the surface by suitable redox reactions [21]. Redox systems can either exchange electrons with the conduction band or with the valence band depending on the position of their electron exchange energy levels in relation to the band edges of the semiconductor [2,22].

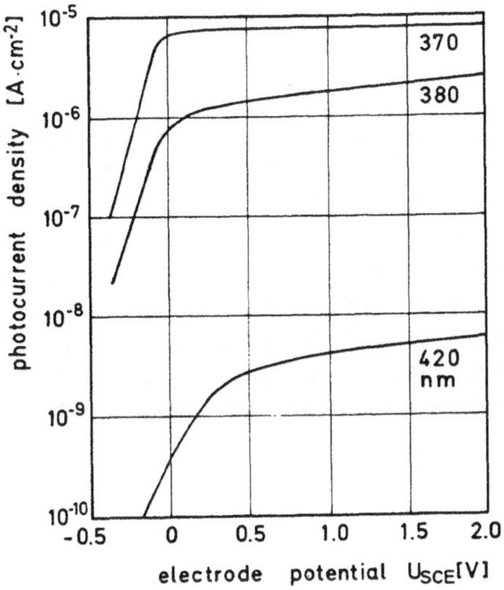

Fig. 20 - Photocurrents versus voltage for ZnO-electrode at illumination with light of different wave length

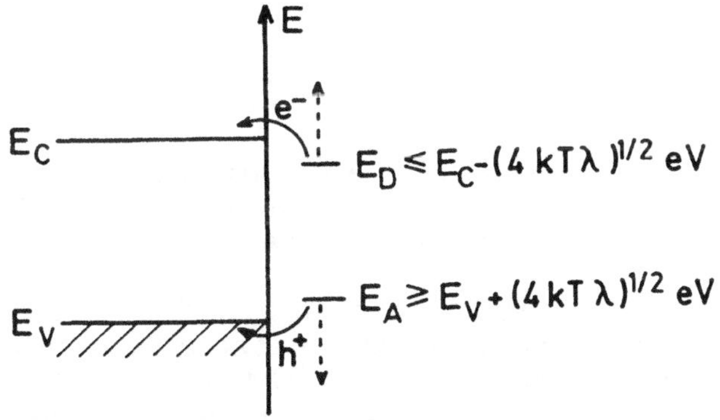

Fig. 21 - Energy term correlations for electron and hole injection by electron donors or acceptors

Fig. 21 gives an energy scheme which explains the conditions for these two different reactions. For our discussion we are only interested in the injection of electrons or holes. We see from this picture that an electron donor must have an electron energy level not too far below the conduction band edge for being able to inject electrons. On the other hand, an electron acceptor can only inject holes if his vacant electron level is located in the range or not too far above the valence band edge.

If such electron donors or electron acceptors are present in the neighbouring electrolyte, electrons or holes are injected into the surface, independent on whether we have n-type or p-type specimen. Drastic effects are found if minority carriers are injected. For instance, n-type specimen are rapidly oxidized in contact with hole injecting electron acceptors like p-type specimen by an applied anodic voltage. On the other hand, p-type specimen in contact with electron injecting redox species react like n-type specimen at cathodic bias.

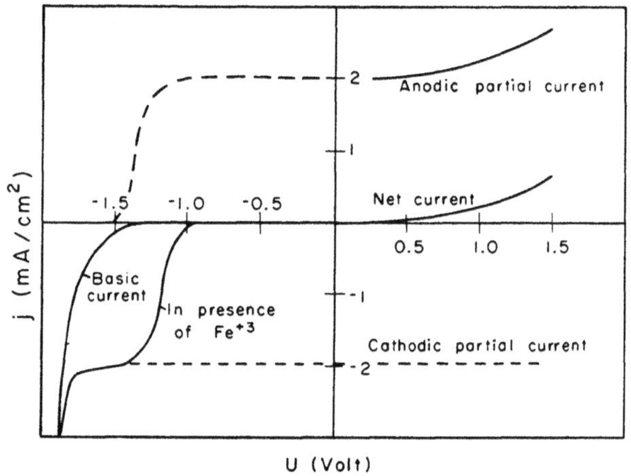

Fig. 22 - Overall and partial current-voltage curves of GaAs-electrode in contact with hole injecting electrolyte

Fig. 22 gives an example for the dissolution of n-type Gallium-Arsenide in the presence of a hole injecting electron acceptor [23]. While the anodic attack of these n-type specimen is very slow under anodic bias, the dissolution rate in such an oxidizing electrolyte is controlled by the rate of hole injection and in a wide range fully independent on the applied voltage. The analogous cathodic reactions have much less been studied since they are connected with a drastic change of the surface properties of the semiconductor when for instance a metal is deposited on the surface.

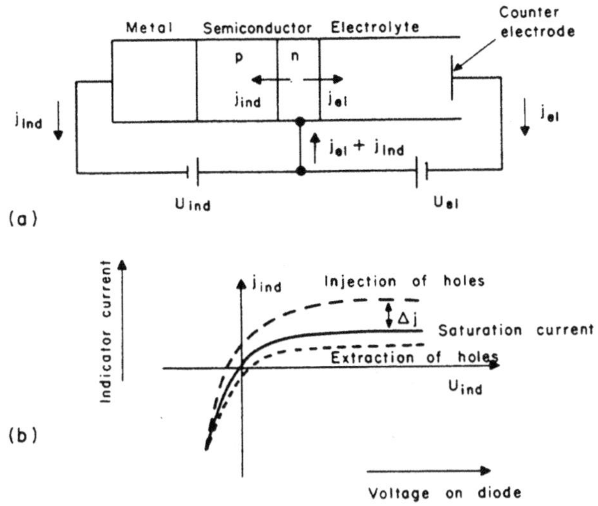

Fig. 23 - Schematic diagram of transistor type experiment for studying the influence of minorities on electrode reaction

A very ingenuous, unfortunately however rather seldom applicable method for the injection of minority carriers is the so-called "transistor method", developed by Brattain and Garrett [15]. Fig. 23 shows the principle. A thin slice of the semiconductor serves as the electrode, one side of which is contacted with another material to form a p-n-junction. If the distance between the surface of the slice used as electrode and the p-n-junction on the other side is in the order of the mean free path of the minority carriers, then injection or extraction of minority carriers through the p-n-junction diode will increase or reduce the supply of minority carriers for the electrode surface. This can clearly be seen in the influence of the diode voltage on the electrolytic current as shown in Fig. 23 for the case of a Germanium-electrode. Such a tool is very useful for quantitative measurements because the number of minority carriers reaching the electrode surface can be calculated and compared with the change of the electrolysis current.

A further quite general possibility of minority carrier generation offers the field-effect. If high electric fields can be reached in the surface tunnel transitions from electronic states at the edge of the valence band or from donors above the valence band to energy states in the conduction band become possible [24,25].

The necessary field strength is only reached in depletion layers. Since the highest field strength will always be obtained at the semiconductor surface, this kind of minority carrier generation starts in the surface. Fig. 24 gives the energy conditions in terms of the band model for an n-type specimen at anodic bias. It is assumed here that tunnel transitions occur from the valence band to the conduction band.

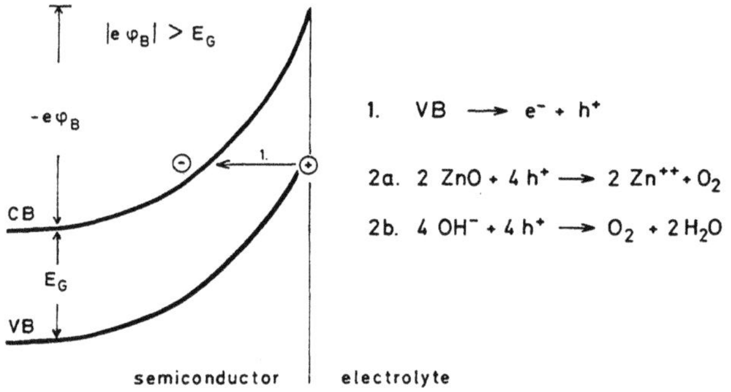

Fig. 24 - Hole generation by tunnelling mechanism in high electric field with consecutive electrode reactions

In many systems, tunnel transitions from surface states above the band edge of the valence band will begin earlier. Impurity states in the bulk above the valence band have the same effect. Since the field strength at the surface at a given voltage increases with the donor concentration which forms the excess charge in the depletion layer this kind of minority carrier generation begins at so much lower voltages so higher the donor concentration. Fig. 25 gives an example for this where steep current increase marks the beginning of hole generation via tunnelling.

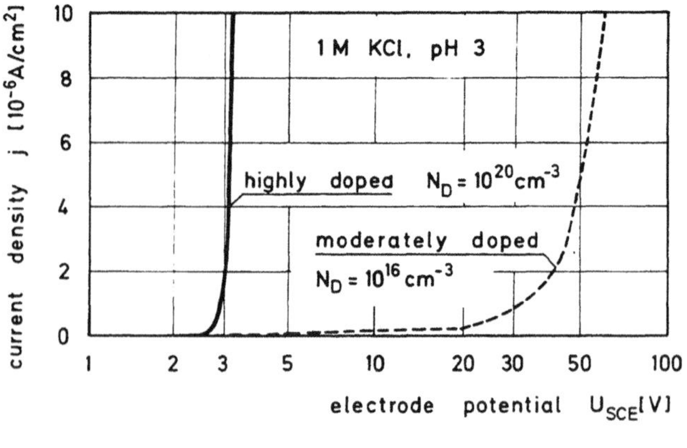

Fig. 25 - Current-voltage curves for ZnO-electrodes with high and low donor concentration

Mechanisms and structural effects

We have seen that the rate determining step in anodic decomposition of semiconductors consumes holes. The reaction itself contains a number of consecutive steps, some of which may be connected with the injection of electrons (compare Figs. 9,10). It is difficult to analyze the ratio between hole consumption and electron injection but in some cases this could be done. The table of Fig. 26 summarizes the results [13]. One sees that the anodic decomposition of all semiconductors with a wide band gap proceeds completely via consumption of holes, what means that also the intermediate stages are stable enough to stick to the surface until they can capture a hole. This hole-capturing, however, occurs obviously with great preference compared to the formation of a new intermediate.

	ΔE [eV]	Oxidative decomposition products	Contribution of holes [%]	Reductive decomposition products
Ge	0,66	Ge^{4+}	~ 58 %	GeH_4 (0,3 ‰)
Si	1,09	Si^{2+}; (Si^{4+})	~ 100 %	?
GaAs	1,35	Ga^{3+}, As^{3+}	~ 99,8 %	< 0,1 ‰ AsH_3
CdTe	1,5	Cd^{2+}, Te^{4+} (?)	?	Cd, $H_2Te \xrightarrow{\frac{1}{2}O_2} H_2O + Te$
CdSe	1,74	Cd^{2+}, Se	~ 96 %	Cd, H_2Se (?)
CuO	1,95	Cu^{2+}, $\frac{1}{2}O_2$	?	Cu, H_2O
CdO	2,2	Cd^{2+}, $\frac{1}{2}O_2$	?	Cd, H_2O
GaP	2,35	Ga^{3+}, P^0 oder P^{3+}	~ 100 %	?
CdS	2,4	Cd^{2+}, S	~ 100 %	Cd, H_2S
ZnO	3,3	Zn^{2+}, $\frac{1}{2}O_2$	~ 100 %	Zn, H_2O

Fig. 26 - Some data on decomposition reactions

This preference of hole-capturing by the intermediate stages can be understood if we look somewhat deeper into the structural details of the decomposition of a semiconductor crystal. Fig. 27 gives such a picture for the anodic decomposition of Cadmium-Sulfide. It is indicated in this picture that the attack of the crystal will only occur in kink sites which consist at this compound-crystal of a CdS-molecule. If, for instance, a Cd-atom in the surface attaches a nucleophilic ligand from the solution in a reaction induced by the presence of a hole its bond to the two neighbouring sulfide-atoms will be weakened and some kind of electronic surface state located within the band gap will be formed by the remaining unpaired electron. Such a surface state will be a very good scavenger for holes and the bond breaking is completed with assistance of another nucleophilic reactant. The remaining sulfur atoms will now interact with each other forming new chemical bonds which on the other hand weaken the remaining bonds to neighbouring Cd-atoms. This seems to be the way in which the decomposition propagates and islands of sulfur are formed on the surface. That such a decomposition step, however, happens on a smooth surface plane is much less likely. We see here how important structural defects will be for the ion formation on the surface of semiconductors.

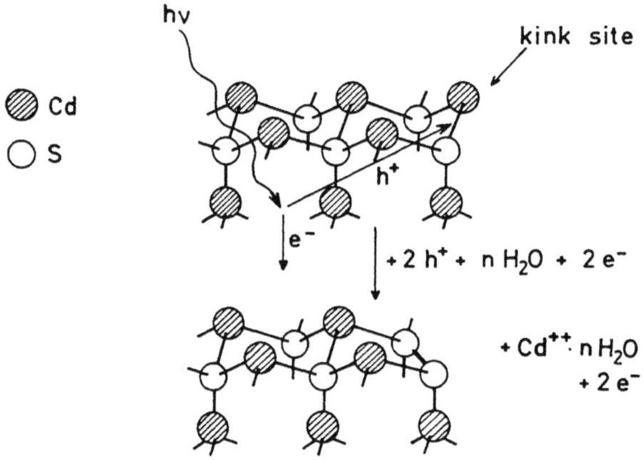

Fig. 27 - Crystal model for anodic decomposition of CdS

Another example for this large influence of structural factors has been found in the case of cathodic decomposition reactions, even with n-type specimen. Some n-type semiconductors like ZnO and CdS decompose at high cathodic polarization according to the mechanism explained previously (Fig. 12). The accumulation of electrons weakens the surface bonds and leads to decomposition into the metal and OH^-- or respectively SH^--ions. Fig. 28 gives an example for the p-type semiconductor Cu_2O where the decomposition is largely enhanced by illumination. It is interesting to note that mechanically damaged surfaces of Cu_2O can easily be reduced cathodically at dark in contrast to smooth chemically etched ones [26]. This can be explained by the assumption that the generation of mechanical defects produces centers for thermal generation of electrons and reduces simultaneously the energy barrier for the formation of a metal nucleus on the surface.

$$Cu_2O + 2h\nu + 2H^+ \longrightarrow 2Cu + H_2O$$

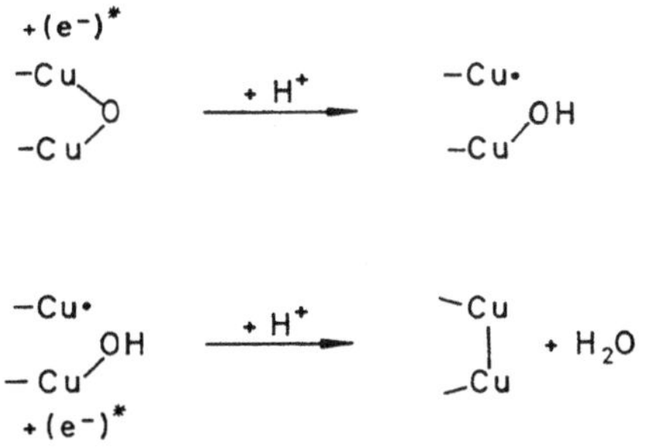

Fig. 28 - Reaction schema for cathodic decomposition of Cu_2O

It is our general experience that the cathodic decomposition of semiconductor is very much favoured by the creation of dislocations, for instance by mechanical polishing. Only when a high cathodic voltage is applied, the decomposition reaction begins also at smooth faces of semiconductors like ZnO and CdS.

Role of surface states

Another important observation is that parallel electrolytic reactions can help to prevent semiconductor decomposition. Such parallel reactions are favoured by the presence of surface states with energies below the conduction band or above the valence band edge. Because of their energy position such surface states will prevent the accumulation of electrons or holes at the surface but cause an increasing variation of the voltage drop in the Helmholtz double layer by the accumulation of charge directly on the surface of the semiconductor [14,27].

Electrons in surface states are reactive enough to interact with hydrogen ions from the electrolyte or other electrophilic reactants and vacant surface states easily react with nucleophilic reactants. How surface states catalyze electron transfer from the valence or the electron acceptors or from electron donors to the semiconductor [28] conduction band is shown in Fig. 29. The stability of TiO_2-electrodes against anodic photodecomposition may be related to the presence of surface states which allow to oxydize the electrolyte instead of decomposing the crystal. Another reason is certainly the very high bond strength in TiO_2-crystals which was shown in the table of Fig. 8.

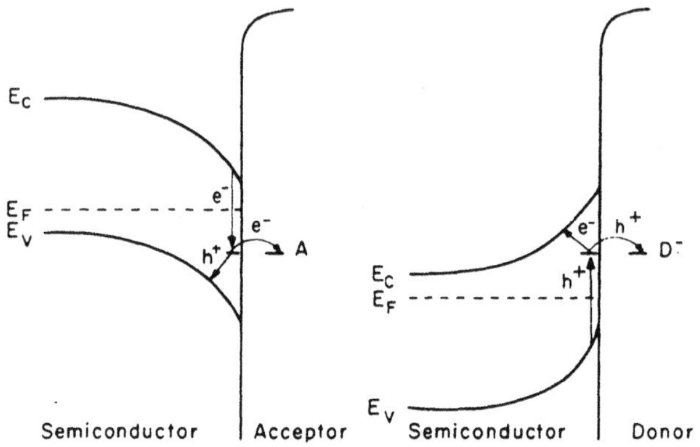

Fig. 29 - Catalysis of electron or hole transfer by surface states

Conclusion

I have concentrated my lecture on the reaction steps where ions are formed in the surface of semiconductors. We have seen that the electronic situation plays a very important role for this type of process. There are many further aspects, structural and chemical ones which are, however, less general. Very little is known, as to my knowledge, of the contact between a semiconductor and a solid electrolyte. But I am sure, any study of such a system will have to consider the same effects which one has found in the experiments with semiconductor electrodes in liquid electrolytes.

REFERENCES

(1) V.A. Myamlin and Y.V. Pleskov "Electrochemistry of Semiconductors" Plenum Press, New York 1967

(2) H. Gerischer in "Physical Chemistry" Vol. IX A (edited by H. Eyring, D. Henderson, W. Jost) Academic Press, New York 1970

(3) A. Many, Y. Goldstein, and N.B. Grover "Semiconductor Surfaces" North Holland Publ. Co., Amsterdam 1965

(4) R.H. Kingston and S.F. Neustadter J. Appl. Phys. 26, 718 (1955)

(5) M.J. Spaarney
"The Electrical Double Layer"
The International Encyclopedia of Physical Chemistry and Chemical Physics, Vol. 4,
Pergamon Press, Oxford 1972

(6) H. Gerischer, M. Hofmann-Perez and W. Mindt
Ber. Bunsenges. Phys. Chem. 69, 130 (1965)

(7) N.F. Mott, Proc. Royal Soc. (London)
A 171, 27 (1939)

(8) W. Schottky, Z. Phys. 113, 367 (1939)

(9) J.F. Dewald, Bell Syst. Techn. J. 39; 615 (1960)

(10) C.G.B. Garrett and W.H. Brattain
Phys. Rev. 99, 376 (1955)

(11) P.J. Boddy and W.H. Brattain
Ann. N.Y. Academy Sci. 101, 683 (1963)

(12) F. Lohmann, Z. Naturforsch. 22a, 813 (1967)

(13) H. Gerischer and W. Mindt
Electrochim. Acta 13, 1329 (1968)

(14) H. Gerischer, Surface Sci. 13; 265 (1969)

(15) W.H. Brattain and C.G.B. Garrett
Bell Syst. Techn. J. 34, 129 (1955)

(16) F. Beck and H. Gerischer
Z. Elektrochem. 63, 500 (1959)

(17) P.J. Boddy, J. Electrochem. Soc. 111, 1136 (1964)

(18) P.F. Schmidt and W. Michel
J. Electrochem. Soc. 104, 230 (1957)

(19) R. Memming and G. Schwandt
Surface Sci. 4, 109 (1966)

(20) J. Tauc, "Photo- and Thermoelectric Effects in Semiconductors", Pergamon Press, Oxford 1962

(21) F. Beck and H. Gerischer
Z. Elektrochem. 63, 943 (1959)

(22) H. Gerischer
Z. phys. Chem. (Frankfurt) 27, 48 (1961)

(23) H. Gerischer and I. Mattes
Z. phys. Chem. (Frankfurt), 49, 112 (1966)

(24) H. Kiess and R. Williams
J. Chem. Phys. 32, 1505 (1960)

(25) B. Pettinger, H.R. Schöppel, T. Yokoyama, and H. Gerischer
Ber. Bunsenges. Phys. Chem. 78, 1024 (1974)

(26) H.R. Schöppel and H. Gerischer
Ber. Bunsenges. Phys. Chem. 75, 1237 (1971)

(27) H. Gerischer
Advances in Electrochemistry and Electrochem. Engin. 1, 139 (1961)

(28) H. Gerischer, B. Pettinger, and M. Lübke
"Proceedings of the Symposium on Electrocatalysis", The Electrochemical Soc., 1974, p. 162

DISCUSSION

HENISCH : The calculated carrier concentrations at the surface are equilibrium concentrations. They can prevail only in the absence of current, or else in the presence of current provided the carrier lifetime is zero. Is it known to what extent departures from electronic equilibrium influence the results ? Have corresponding tests been made, e.g. by observations of electrode behaviour under illumination ? The degree of surface disorder should influence the situation, in as much as it controls the carrier generation rate.

GERISCHER : If the majority carriers are consumed (or generated) at the interface, the deviation of the space charge distribution from the exact equilibrium situation is small in many cases. This can be (and has been) checked by capacity measurements with super-imposed high frequency a.c. or by the absence of considerable photoeffects. The situation is fully different if minority carriers react in the surface. In such cases the complicated transport equations have to be solved if one wants to know the surface concentrations. But this is often impossible due to incomplete knowledge of rate constants, etc. However, even in such cases, mechanisms for reactions can be derived from reaction order, with respect to minority carrier generation for instance, by generating or injecting minorities from outside.

Surface disorder has a great influence on minority generation rate. This can be seen in a variation of saturation currents for reactions which consume minorities.

WAGNER : Would you please comment further on what the role of surface damage is on the mechanism of reduction of, e.g. Cu_2O ?

GERISCHER : It is not fully clear how this occurs. However, a reasonable assumption seems to be that the formation of a metal nucleus is enhanced by the presence of dislocations in the surface. The Cu-OH bond may also be weaker at such deformed areas and OH^- might more easily dissociate from the surface if an electron can be captured from the bulk. Whenever this reaction occurs the remaining surface Cu-atom will react with a H^+ ion on H_2O molecule to form a Cu-H bond with capturing another electron. At an intact surface, finally H_2 will be developed by $Cu\text{-}H+H^++e^- \rightarrow H_2$ while in a dislocation, neighbouring Cu-atoms will interact and aggregate.

PIZZINI : It seems that in the case of TiO_2 the oxygen evolution takes place under illumination without any titanium dissolution. Apparently water molecules adsorbed at the surface could be involved in the primary reaction step, and we, in fact, observed basification of the titanium oxide comportment. Could not well be this the case of ZnO, where, due to the basification (local) of the solution, dissolution of the crystal ZnO takes place instead of the anodic dissolution of Zn ?

GERISCHER : Even at ZnO electrodes, analysis has shown that lattice decomposition and oxidation of H_2O by holes are competitive reactions, the yield of the latter increasing with the role of hole generation's. This seems to indicate an increasing probability of OH^- interaction with holes when more oxygen vacancies or more lattice defects are generated in the surface. A chemical dissolution mechanism can be excluded since this has been observed in higly alkaline solutions where the protons formed by electrolysis react rapidly with OH^-. The stability of TiO_2 against anodic decomposition is in my opinion caused by the strong bonding of the Ti-atoms (ions) which keeps the Ti-atoms in their lattice position even in presence of a large number of oxygen defects (vacancies) in the surface.

VOÏNOV : Solid electrolytes of interest are electronic insulators as well as ionic conductors and I wonder how and when the semi-conductor or insulator band model breaks down if ions start moving in such an electronic insulator ?

GERISCHER : This depends on whether the interface remains blocking for the mobilized ions or not. In the first case, the increase of mobile charge carriers modifies only the charge distribution in the space charge layer but the electronic energy levels in the surface (and in the bulk) are little affected. The same holds for the electronic processes. In the second case the chemical composition will change and the situation becomes very complicated. No general predictions can be made without knowing relative rates of electronic and ionic processes in the surface.

CONTACTS ON IONIC SOLIDS

Heinz K. Henisch
Materials Research Laboratory and Department of Physics, The Pennsylvania State University, University Park, Pennsylvania 16802, USA.

INTRODUCTION

There are, of course, many aspects of contact theory which are common to ionic and homopolar solids. Of these, only matters which are frequently overlooked or else in need of further clarification will be discussed in this paper. In addition, there are matters which arise from the ionic character of the contacted solid itself, and these can have different forms of origin: (a) the special nature of surface states on ionic solids, (b) the generally large ($\gg kT$) activation energy of donor and acceptor states in such materials, (c) the fact that such materials are often near-insulators, or in more modern terminology, "relaxation semiconductors", and (d) the importance of trapping effects. All these factors have specific consequences for the manner in which charge is transported across a metal-semiconductor boundary, and for the propagation of contact-related effects into the interior of the material. These effects, whether large or small in specific cases, always involve non-equilibrium carrier concentrations. In some practical contexts, the task is to maximize the non-equilibrium, in others to minimize it. In both it is necessary to understand what is happening and why. These problems are discussed below.

THE FREE SURFACE

Since (metal) contacts are necessarily made with (semiconductor) surfaces, the nature of surfaces as such is highly relevant. In principle, the considerations involved are (i) surface

M. Kleitz and J. Dupuy (eds.), Electrode Processes in Solid State Ionics, 309–329.

geometry, (ii) surface contamination by foreign matter, (iii) deformation of the band structure, and (iv) the effect of surface states.

(i) Surface Geometry (including surface homogeneity). These factors govern the relationship between apparent and real contact area, and govern electrical behavior inasmuch as they are responsible for the existence of geometrical spreading resistances.[1] In principle (but not always in practice and never completely), the effects can be minimized by working on cleaved surfaces, and by the use of evaporated or sputtered electrodes which adhere uniformly to the semiconducting substrate.

(ii) Surface Contamination. The choice between measurements on 'ideal' surfaces (e.g. surfaces freshly cleaved in ultrahigh vacuum), and measurements on 'real' surfaces (e.g. cleaned by routine laboratory methods) represents a classic dilemma, and the decision must always depend on context and purpose. Inasmuch as the effect of monolayer contaminants may be considered to be a modification of the surface state structure, it can be discussed under that heading.

(iii) Deformation of the Band Structure. Solids are known to have a surface tension, which shows itself as a small change of lattice spacing. Correspondingly, there must be band structure effects of great complexity (band gap, nonisotropy), but very little is known about them nor, indeed, about the corresponding strain patterns. In addition, it is necessary to envisage modifications which arise from contact pressure. Even small contact thrusts (e.g. of a few grams) generate enormous contact pressures when the contact area is small, pressures which frequently correspond to the yield point of the contacting metal! Band structure information derived from contact measurements may thus legitimately differ from bulk parameters derived by other methods. For these and many other reasons, pressure contacts are best avoided.

(iv) Effect of Surface States. The relationship between density and position of Tamm states and the ionicity of the solid is discussed elsewhere in this volume. For present purposes, it is sufficient to recall that surface states (of whatever origin) provide additional accommodation for charge carriers with energies which correspond to forbidden states in

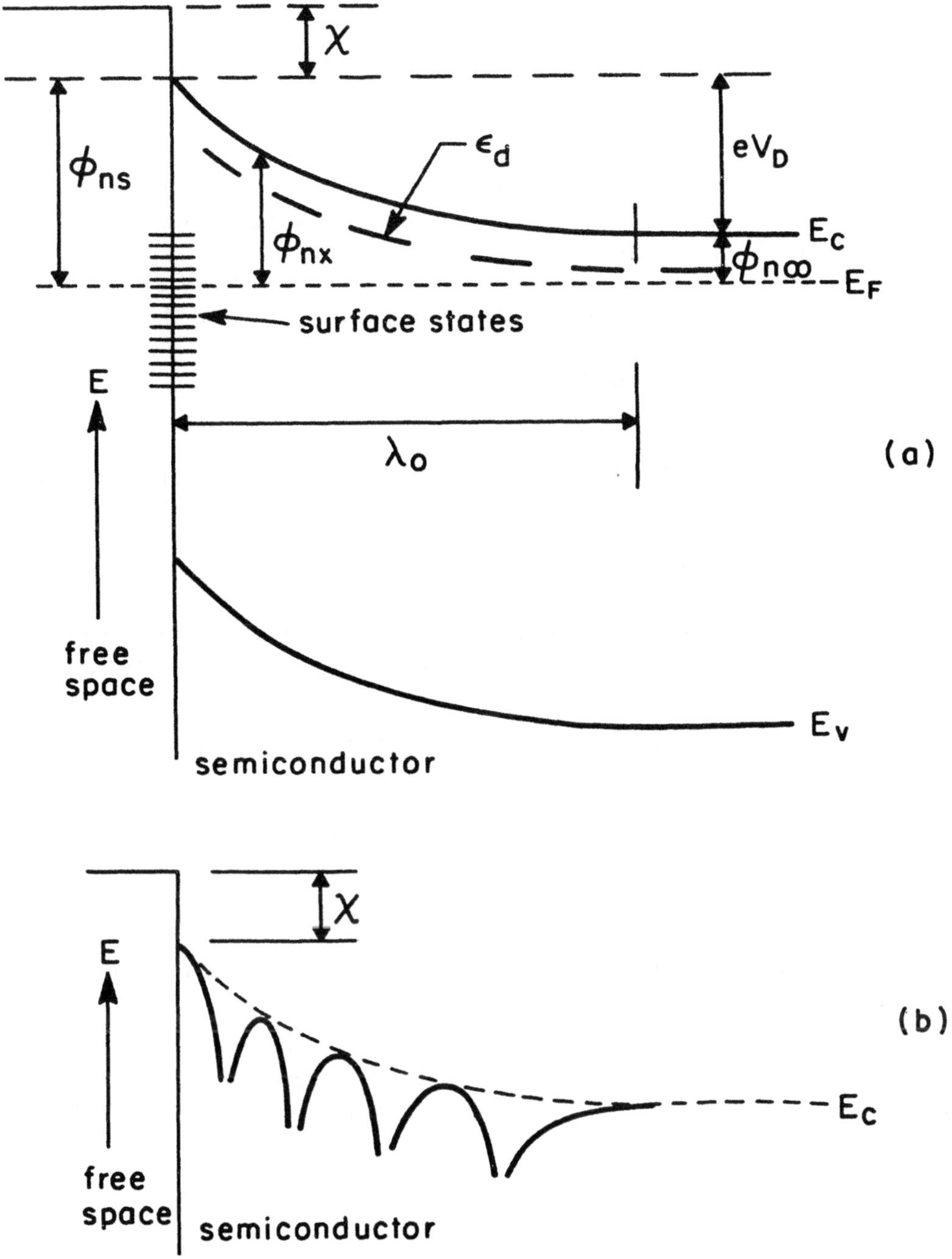

FIG. 1

Energy profile of barrier on a free surface: (a) "traditional" form of the Schottky barrier; (b) form in the presence of discontinuous space charge (schematic). χ = electron affinity. E_F denotes the Fermi level. n-type material.

the bulk. As a result, the existence of surface states give rise to a double layer. Relative to the bulk, the surface charge may be positive or negative in principle, but it is here assumed that we are dealing with n-type material and that the surface charge is negative. As the surface acquires charge, its electron energy increases, and so does the number of surface states which are necessarily unoccupied. On these conflicting tendencies depends the final equilibrium (Fig. 1a). The well known Schottky relationship connects barrier height (V_D) and barrier thickness (λ_o) in a simple manner (which is based, however, on a number of delicate assumptions):

$$eN_d\lambda_o^2/2 = V_D \tag{1}$$

where N_d is the concentration of donors.[2] It is derived by the integration of Poisson's equation in familiar form. For every value of ϕ_{ns} on Fig. 1a, there is a corresponding diffusion potential V_D, and a corresponding surface charge, which is customarily considered fixed. It need not be so, if the local electronic equilibrium is displaced by the passage of current, a fact which has received very little attention.

The barrier thickness λ_o evidently has the character of a screening length. How, then, does it differ from the conventional Debye length? The two calculations are based on somewhat different models. That for the Debye length sets no limit to the local carrier concentration; the barrier calculation which leads to eqn. (1) does set such a limit, namely the charge density accommodated in donors, and this assumed to be constant throughout the barrier. Accordingly, the Debye length defines an exponentially decaying potential contour, whereas the present barrier system yields a parabolic system.

Equation (1), the corner-stone of Schottky's contact rectification theory, provides useful general insight, but is based on too many assumptions to be universally acceptable. The following complications can and often do occur:

Complication A: Equation (1) is based on the assumption that the charge density $\rho = eN_d$ for $x < \lambda_o$ and $\rho = 0$ for $x > \lambda_o$. In fact, the charge density can never become zero in an abrupt manner at any particular value of x. The contribution of the free carriers must also be taken into account, all the more for barriers which are low and thick. Accordingly, we must have

$$\rho = e[N_d - n(x)], \tag{2}$$

with the free carrier concentration

$$\left.\begin{aligned} n(x) &= n_{\infty}\exp\left[-\frac{(\phi_{nx}-\phi_{n\infty})}{kT}\right] \\ \text{and} \qquad -\phi_{nx}/e &= V_x \end{aligned}\right] \tag{3}$$

When these terms are entered into Poisson's equation, it becomes clear that the barrier no longer has a well defined thickness. As expected, it does not end at $x = \lambda_0$, but disappears asymptotically as $x \to \infty$.

Complication B: The above arguments are based on the assumption that $\rho = eN_d$, which implies that the ionization of donor centers is complete in the bulk as well as the barrier. In ionic solids, this is not usually correct. Accordingly[3], the charge density becomes somewhat more complicated than eqn. (2), and has to be written

$$\rho/e = N_d - \frac{N_d}{1 + \frac{1}{2}\exp(\Delta E_d/kT)} - n_{(x)} \tag{4}$$

where $\Delta E_d = \phi_{nx} - E_d$, and by now the solution of Poisson's equation is possible only by numerical methods.

Complication C: Image force considerations lead to a well-known correction (lowering) of the barrier height which, in the conventional form, is valid only for rare electron transits. When current densities are high, the image force correction is no longer associated with the familiar field-pattern of interacting spheres. In effect, the importance of the image force correction diminishes as the current density increases. The fact that the medium is not a pure dielectric but is somewhat conductive has the same consequences. It should also be borne in mind that the dielectric constant which enters into image force corrections must be the high frequency value, and not the static value, because of rapid electron transit.

Complication D: The charge density which governs the barrier height and thickness must in principle include a correction for the contribution of minority carriers. The effect of this contribution is to make the barrier thinner, particularly for very small values of x. Accordingly, the barrier also becomes more transparent in its upper regions, which can be regarded as a change of effective barrier height.

Complication E: The above analysis assumes that the average

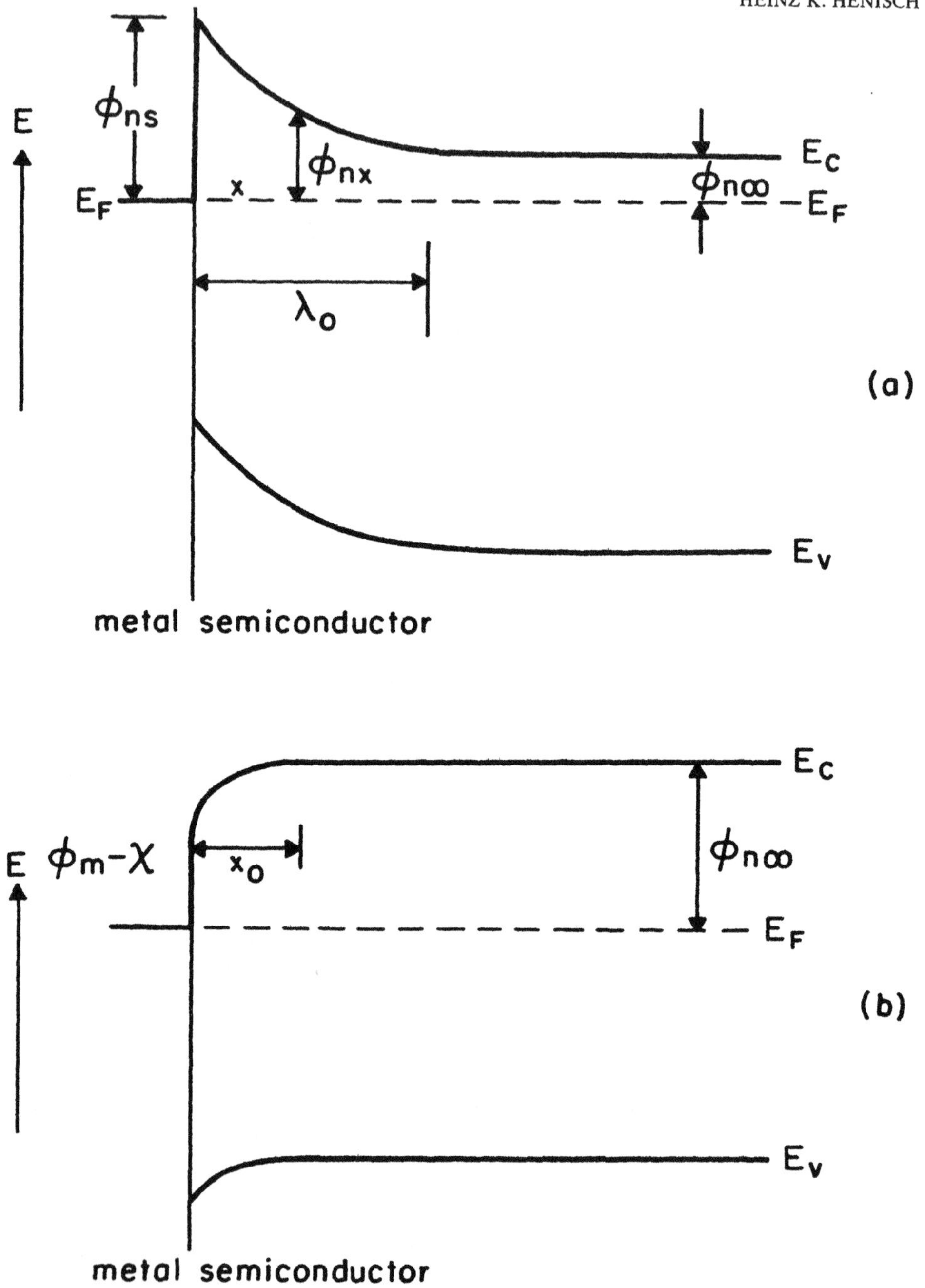

FIG. 2

Energy contours of metal-semiconductor contacts. (a) Electron loss by semiconductor, in the presence of surface states; (b) electron gain by semiconductor, in the absence of surface states. n-type material. ϕ_m = thermionic work function of the material.

distance between neighboring donors (namely $N_d^{-1/3}$) is much smaller than λ_o. In practice, this assumption is often not fulfilled, which means that the charge density cannot be handled as a continuum. As a result, the barrier profile is much more complicated than suggested by Fig. 1a, probably nearer to that shown in Fig. 1b. The exact profile would vary from z-section to section and would reflect the random distribution of the ionized centers. Though Schottky(2) pointed this out as long ago as 1942, the matter has been entirely ignored ever since. It awaits refined analysis now. One of the intuitively obvious results is that the current density through contacts cannot be uniform over a macroscopic contact area in such a situation.

CONTACTS WITH METALS

Because initially coincident Fermi-levels are highly unlikely, practically all contacting processes involve a charge exchange between the contacting members, and hence the establishment of a contact barrier. If there were no surface states, the barrier height would be given simply by the difference of electron affinities. Accordingly, different metals would be associated with contact barriers of different height on a given semiconductor. For the familiar homopolar materials Ge and Si, no such correlation has ever been found, and it was indeed this fact which first led Bardeen to re-examine the theory of surface states.(4) When there is no correlation, it is necessary to conclude that the semiconductor is screened from outside electrical influences by a two-dimensional sheet of charge on the surface, the charge being accommodated in surface states. When a charge exchange takes place between metal and semiconductor for the establishment of electronic equilibrium, it involves in the first instance only carriers in and from surface states, and as long as it does so, the internal field pattern remains undisturbed. If the charge exchange demanded by the equilibrium conditions were larger than the charge which can be accommodated in the available surface states, then the screening would be imperfect. Some correlation between contact properties and the electron affinity (work function) of the contacting metal would then be expected, but not the complete correlation associated with surfaces which are free of surface states. Figure 2a shows the contact situation for electron loss by the semiconductor in the presence of surface states.

One of the most important discoveries of the last decade is the correlation between barrier height ϕ_{ns} and the ionicity of the

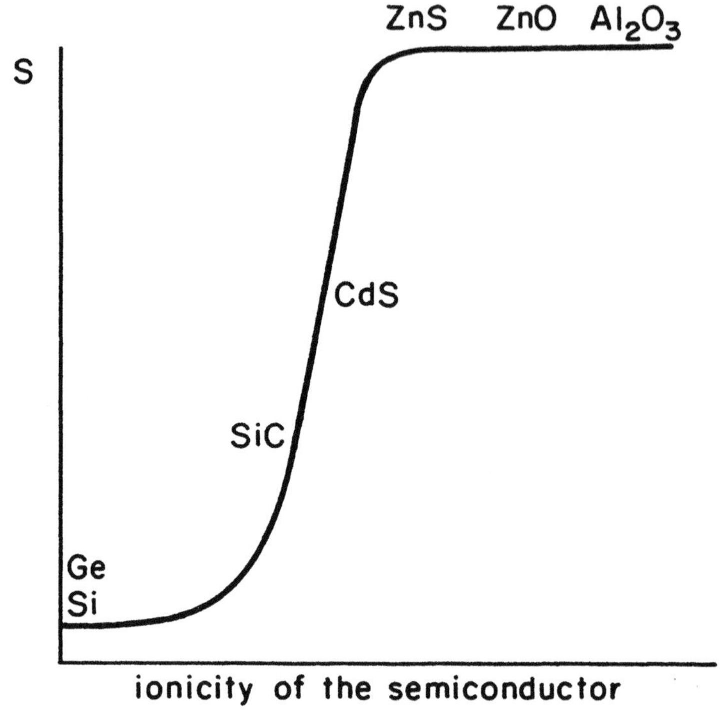

FIG. 3

Schematic relationship between S (defined as $d\phi_{ns}/d\chi_m$) and ionicity of the semiconductor, where χ_m is the electronegativity of the contacting metal. [After Kurtin, McGill and Mead(8).] Curve supported by observations on many other materials.

contacted solid, as schematically illustrated by Fig. 3. This is, of course, the outcome of a correlation between ionicity and surface states (density and energetic position). The overall result is that surface states are less important the greater the ionicity, and one of the implications is that long-range order is also less important. Moreover, one is tempted to conclude that surface contamination is less important. These results have been found to be closely related to optical surface properties. The pioneering work was done by C. A. Mead and co-workers(5-8) in America, and by Turner and Rhoderick(9) and co-workers in England. One of the potential applications stems from the prospect of modifying ionicity near the surface with a view to establishing specific and desirable contact properties.(10)

The case of electron gain is more often (but not always with justification) discussed in terms of a material free from surface states (Fig. 2b). There is again a screening length, and a band curvature which is, however, limited to a few multiples of kT. This case was first considered by Mott and Gurney.(11) Its character will be further considered below.

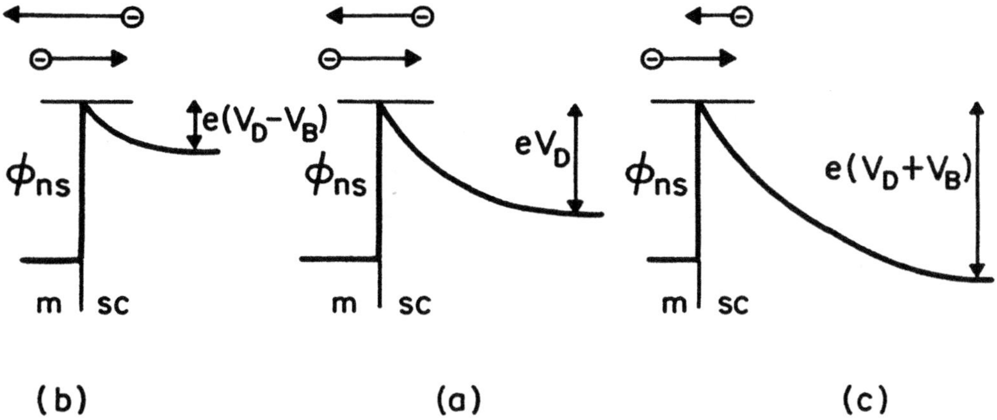

FIG. 4

Contact rectification through asymmetric distortion of a Schottky barrier: (a) equilibrium; (b) metal positive, large (forward) current; (c) metal negative, small (reverse) current. n-type material.

CURRENTS THROUGH BARRIERS

When external voltages are applied to a barrier of the kind shown on Fig. 2a, the contour is asymmetrically distorted depending on polarity; accordingly, we have current rectification (Fig. 4). In the traditional picture, space charges which arise from the moving carriers themselves are neglected; we tend to be satisfied with a calculation of λ which differs from λ_o only inasmuch as the total voltage on the barrier then differs from V_D. In general, such a procedure cannot be justified, and the full transport equations must be used for calculating currents and local fields:

$$\left.\begin{array}{l} J = J_n + J_p \\ J_n = J_{n\,\text{field}} + J_{n\,\text{diff.}} \\ J_p = J_{p\,\text{field}} + J_{p\,\text{diff.}} \\ d^2V/dx^2 = -\rho/\varepsilon \end{array}\right] \qquad (5)$$

and appropriate boundary conditions.

Indeed, whereas the transport conditions as such are noncontroversial (except in the presence of high-field effects[12] on the carrier mobilities), the boundary conditions present a number of important problems. Accordingly, it is useful to envisage certain limiting cases, which arise as described below. Even then, the only really satisfactory conclusions are those derived by com-

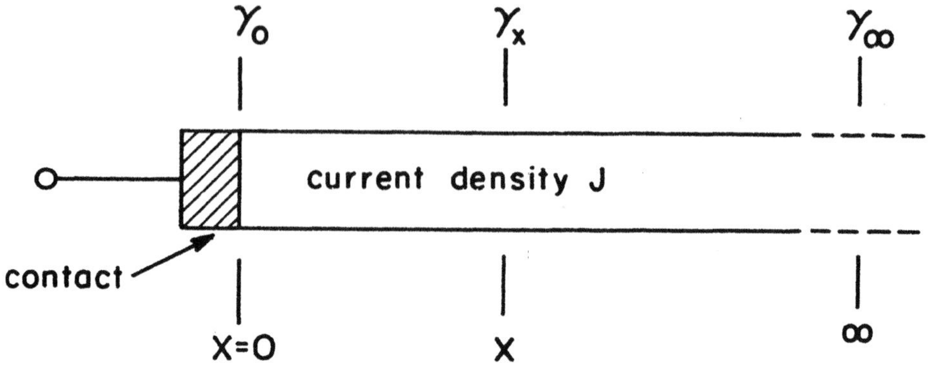

FIG. 5

Definitions of the current composition ratio $\gamma_{\infty} = J_{p\infty}/J$, and of the minority carrier injection ratio $\gamma_0 = J_{po}/J$, for n-type material.

puter techniques, applied to systems without "simplifying approximations". Even then, it must be remembered that the "classical" approach implied by these equations is only an approximation. Currents over low barriers (e.g. V_D comparable with kT/e) cannot be treated in this way but must be analyzed in wave-mechanical terms.[13]

For a system of the kind shown on Fig. 5, we define[13] the current composition ratio in bulk as $\gamma_{\infty} = J_{p\infty}/J$, with emphasis on the role of the minority carriers. Of course, J_{px}/J may have different values in the vicinity of contacts, and its particular value at x = 0, namely $\gamma_0 = J_{po}/J$, is called the minority carrier injection ratio. It is obvious that $\gamma_0 = \gamma_{\infty}$ is an exceedingly unlikely state of affairs, and even if it were to apply on one current density J, it would not apply under other conditions. As a result of $\gamma_0 \neq \gamma_{\infty}$, every form of specimen discontinuity (in any parameter relevant to electrical conduction) leads in the presence of current to departures from the equilibrium carrier concentrations, and these departures propagate over some distance into the semiconducting medium. Accordingly, every type of current-carrying contact involves carrier concentration gradients, and therefore diffusion currents which imply departures from ohmic conduction. There is therefore no such thing as an 'ohmic' contact; all one can hope to achieve is a contact of sufficiently low resistance for its non-linearity to be unimportant. In a trivial sense, of course, all contacts are 'ohmic', as long as the applied voltages remain within the range $V_B = \pm kT/e$, but this is saying no more than that all curves are straight lines when considered in sufficiently small segments. (The popular practice of equating 'ohmic' with 'low resistance' is based on deplorable confusion but is probably too ingrained to yield to reason.)

For forward and reverse voltages applied, the system offers the following possibilities, as long as we are dealing with a conventional ('lifetime') semiconductor, i. e. one in which the minority carrier lifetime is greater than the dielectric relaxation time.

	Minority Carriers		Majority Carriers	
		Forward Voltages		
$\gamma_0 > \gamma_\infty$	injection	$\Delta p > 0$	accumulation	$\Delta n > 0$
$\gamma_0 < \gamma_\infty$	exclusion*	$\Delta p < 0$	extraction	$\Delta n < 0$
		Reverse Voltages		
$\gamma_0 > \gamma_\infty$	extraction*	$\Delta p < 0$	exclusion	$\Delta n < 0$
$\gamma_0 < \gamma_\infty$	accumulation*	$\Delta p > 0$	injection	$\Delta n > 0$

of which the first is, of course, the basis of the conventional transistor. The remaining effects could also give rise to "transistor-like" phenomena in principle, though effects which imply negative values of Δp are a priori unfavorable, because they permit only minimum modification of the conductive properties. The selection of one or other of the above possibilities guides the choice of boundary conditions for subsequent calculations.

It must now be admitted that, as far as is known, no computer computation to the above degree of generality has ever been performed. To that extent, contact theory may still be expected to yield results which are entirely new, and in some contexts (see below) it has already done so.

MAJORITY CARRIER INJECTION

Figure 2b shows the energy contour at a contact in which the establishment of equilibrium demands an electron gain by the semiconductor. The customary analysis of voltage-current characteristics neglects minority carriers, on the assumption that the band gap is sufficiently large to permit this. We obtain

---

*Terms originally coined by P. C. Banbury, University of Reading.

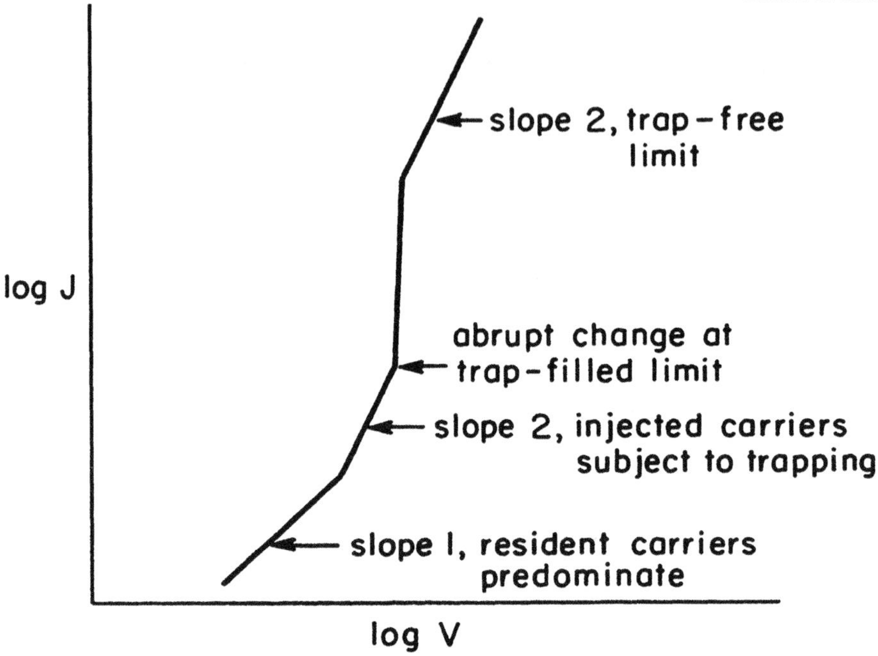

FIG. 6

Schematic voltage-current relationship for majority carrier injection into a material containing traps, in the absence of a Schottky barrier. [After Lampert and Mark(15).]

majority carrier injection into the semiconductor in this case when the semiconductor is made positive with respect to the metal contact, but it would be quite wrong to believe that majority carrier injection as such is limited to this model. It can also occur (as suggested above) when a Schottky barrier is present, a possibility which conventional forms of rectification theory neglect entirely. The appropriate calculations remain to be made.

In the absence of a barrier, the voltage-current relationship in the presence of majority carrier injection depends in a straight-forward way on whether the material contains traps or not. The trap-free case is, of course, the simplest.(15) If resident carriers could be neglected, the injected carriers themselves would yield the characteristic $J \alpha V^2$ relationship, called the trap-free limit. [The corresponding relationship for vacuum is the familiar Child's Law, namely $J \alpha V^{3/2}$. It differs from the present case because the mean free path of electrons in vacuum is comparable with the dimensions of the system, whereas, in a solid, it is of course much smaller.] In practice, the resident carriers are not always negligible, and their contribution is a linear region near the origin of the J-V characteristic. Non-linearity sets in when the concentration of injected carriers becomes comparable with that of the resident carriers. This case and the following have been elaborately treated by Lampert and Mark.[15]

In the presence of traps (Fig. 6), the current is bound to be smaller; how much smaller depends on the trap depth and concentration. It can be shown that, as a first approximation, a material containing only shallow traps behaves as if its majority carrier mobility were not μ but $\theta\mu$, where θ is the ratio of free carriers to carriers in traps.

The case of deep traps is more complicated. When carriers are first injected, they are trapped, but a stage can be reached when all the accommodation is full, and that stage is marked by an abrupt change of electrical properties. It has been called the trap-filled limit, and is marked by a sharp increase of current, to the value given by the pure square law relationship. So sharp is the expected change that it might easily be misinterpreted as dielectric breakdown. The deeper the traps, the sooner they are completely filled. The trap-filled limit is then observed at lower voltages. Because detailed features of the voltage-current characteristic are related to trap structure in a more-or-less direct way, measurements on such systems, followed by a Lampert-Mark type of analysis, can be highly informative.[15]

MINORITY CARRIER INJECTION

We are here concerned with currents which involve minority carrier movement into the material under the conditions $\gamma_0 > \gamma_\infty$. If the material concerned is a near-insulator to the extent which permits all resident carriers to be neglected, then the effects of minority carrier injection are in principle no different from those of majority carrier injection. In each case, the only carriers which move are "external", and the conduction process is controlled by the space-charge and trapping considerations outlined above. The case becomes more complicated and more interesting when resident and injected carriers interact, and this is the situation which will now be discussed.[16]

When two types of carriers of concentrations n and p are present, the relationship $pn = n_i^2$ constant represents the equilibrium conditions as far as recombination is concerned, but

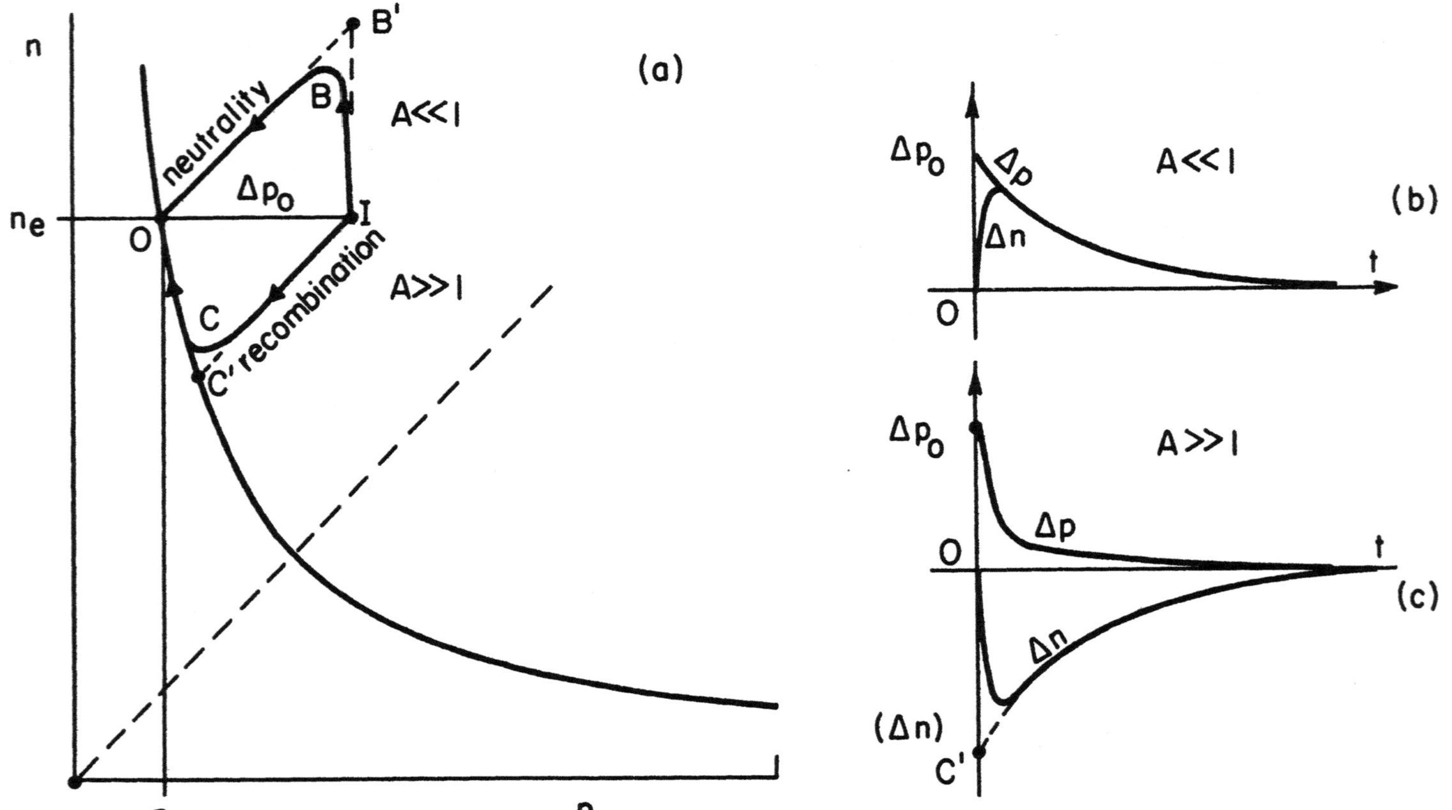

FIG. 7

Response of a trap-free system to a sudden minority carrier increment Δp_0. (a) Return to equilibrium (at point O) in the concentration plane. (b) Carrier concentrations versus time in the lifetime case. (c) Carrier concentration versus time in the concentration case. [After Popescu and Henisch.(16)].

does not necessarily represent equilibrium as far as charge is concerned.

Figure 7 envisages a spatially homogeneous case which shows how a trap-free system reacts to the sudden imposition of a small minority carrier increment Δp_0. At the point I, there are two possibilities, depending on the relative magnitude of the carrier lifetime τ_0 and the dielectric relaxation time of the material, or, more accurately, the dielectric relaxation time of the majority carriers, τ_{Dn} (= $\varepsilon/e\mu_n n_e$). It is convenient to define a quantity A, by the equation

$$A = \tau_{Dn}/\tau_0 \; (1 + p_e/n_e) \tag{5}$$

p_e and n_e being the equilibrium concentrations. Figures 7b and c then show how a trap-free system would relax for $A \ll 1$ and $A \gg 1$. The situation $A < 1$ is called the "lifetime" case; $A > 1$ is called the "relaxation" case. $A = 1$ is a very special case in which the influx of majority carriers (in response to a sudden increment Δp_0) is just balanced by recombination. This means that $\Delta n = 0$ at all times. $A = 1$ then represents the boundary between two (trap-free) operating regimes.

In a practical situation (current through a contact) the carrier distribution is not, of course, homogeneous, but some considerations corresponding to Fig. 7 will apply at every point. In the trap-free lifetime case, Δp and Δn assume various values in the space between the contact plane and the undisturbed bulk material, but are both positive. Accordingly, we experience the <u>increase</u> of effective local conductivity which is familiar from all the considerations of transistor physics. In no circumstances (within the framework of the 'lifetime' model) can $\Delta n > \Delta p$.

For the trap-free $A > 1$ case, Δn is always negative and can, numerically, be much larger than Δp. Accordingly, one might expect an <u>increase</u> of total resistance, in sharp contrast with the lifetime case. However, such an expectation is superficial, since carrier concentrations as such do not govern the total current; they govern only the field-current component. In addition, we have diffusion currents, and their effect is to keep the total resistance down. The overall result has been shown (by computer solution of the transport equations) to be a smaller-than-bulk resistance, qualitatively similar to that of the lifetime case, but with a different voltage-dependence. Thus, in the relaxation case, the total resistance remains linear over an applied voltage range greatly in excess of kT/e. Such an extended linearity could easily (but quite wrongly) be diagnosed as the absence

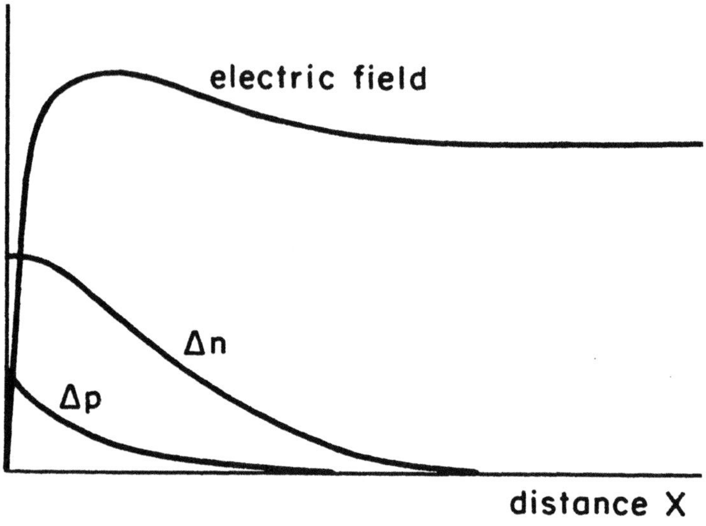

FIG. 8

Schematic relationship between local field, the free carrier concentration increments Δp and Δn and distance x from the contact, for sufficiently high levels of minority carrier injection into an n-type (relaxation or lifetime) semiconductor containing a sufficient number of traps. [After Popescu and Henisch.(17)]

of contact effects altogether. Indeed, such linearity constitutes a popular criterion for the absence of injection, which must now be discarded.

All this applies only to the trap-free case. The case with traps differs in two ways:

(a) Recombination in the presence of traps cannot be characterized by a constant lifetime τ_o, independent of injection level, and

(b) Trapping implies a space charge consisting of immobile carriers.

Of the two effects, the second has been shown to be enormously more important than the first. An analysis of complete generality is not yet available, but the case of a single trapping level, equally "accessible" to electrons and holes, has been calculated.(17) The traps capture injected minority carriers, and these constitute a positive space charge, which has the effect of attracting additional majority carriers into the region. Δp and Δn are therefore both positive in this case also, but Δn can be very much greater than the free hole concentration Δp. The overall effect is then governed by the fact that electrons experience a sharp concentration gradient, as shown in Fig. 8. This gradient would, by itself, produce a current in the direction opposite to that of the net-current under consideration, and this current must evidently be prevented from flowing. It _is_ so prevented by a high local field, i. e. a field higher than that prevailing in the undisturbed bulk for the total current under consideration. This, in turn, implies a total resistance _greater_ than that expected from bulk considerations alone. This effect was originally predicted by van Roosbroeck, but for very different reasons (exhaustion of the majority carriers). The self-consistent computer solutions have demonstrated that the field which stops "back diffusion" is the actual cause. There are observations on GaAs which can be interpreted as experimental confirmation of these findings. A detailed voltage-current characteristic remains to be calculated for this case, the expectation being that it will likewise prove linear over an extended range of applied voltages. Of course, $A = 1$ is then no longer the boundary between positive and negative increments of Δn; a more appropriate criterion is $\approx A_0/(M_0 + 1)$ where A is the value of A for zero excitation and M_0 the trap concentration. Indeed, the resistance increase due to "back diffusion" is expected to occur even more prominently in lifetime than in relaxation semiconductors. In practice, A_0 and M_0 are not, of course, independent quantities. The net effect of their interrelationship is to make the outcome less dependent on trap concentration than one would otherwise expect.

The computer solutions so far available are based on the assumption that the field E_0 at $x = 0$ is zero, and that $\gamma_0 = 1$. If, for any actual contact configuration $\frac{1}{2} < \sim \gamma_0 < 1$, then the same type of effects are expected, but to a smaller extent. The assumption $E_0 = 0$ is more serious, inasmuch as it implies that there is no Schottky-type barrier or equivalent. Present theory is thus best suited to deal with _junctions_ under forward bias, and this limits the use of the available models in the interpretation of experiments. These models are nevertheless thought to be of considerable importance, (i) for our understanding of the processes involved, and (ii) because of their practical bearing, e.g. in connection with the internal resistance of solar cells.

REFERENCES

1. H. K. HENISCH, _Rectifying Semiconductor Contacts_, Section 7.8. Oxford, Clarendon Press (1957).
2. W. SCHOTTKY, Z. f. Phys. _118_, 539 (1942).
3. E. A. GUGGENHEIM, Proc. Phys. Soc. A _66_, 121 (1953).
4. J. BARDEEN, Phys. Rev. _71_, 717 (1947).

5. C. A. MEAD, Solid State Electronics 9, 1023 (1966).

6. C. A. MEAD, Appl. Phys. Letters 6, 103 (1965).

7. S. KURTIN and C. A. MEAD, J. Phys. Chem. Solids 29, 1865 (1968).

8. S. KURTIN, T. C. McGILL and C. A. MEAD, Phys. Rev. Letters 22, 1433 (1969).

9. M. J. TURNER and E. H. RHODERICK, Solid State Electronics 11, 291 (1968).

10. V. L. RIDEAUT, Solid State Electronics 18, 541 (1975).

11. N. F. MOTT and R. W. GURNEY, Electronic Processes in Ionic Crystals. Oxford, Clarendon Press (1940).

12. R. E. BURGESS, Proc. Phys. Soc. B 66, 430 (1953).

13. B. R. GOSSICK, Surface Science 18, 181 (1969); 21, 123 (1970); 25, 465 (1971).

14. H. K. HENISCH, Bull. Res. Council of Israel 4, 409 (1955).

15. M. A. LAMPERT and P. MARK, Current Injection in Solids. New York, Academic Press (1970).

16. C. POPESCU and H. K. HENISCH, Phys. Rev. B 11, 1563 (1975).

17. C. POPESCU and H. K. HENISCH (in publication).

For general reading:

A. G. MILNES and D. L. FEUCHT, Heterojunctions and Metal-semiconductor Junctions. Academic Press, London and New York (1972).

M. PEPPER (Hon. Editor), Metal-semiconductor Contacts. Inst. of Physics Conf. Series No. 22, London and Bristol (1974).

DISCUSSION

HEYNE : A) By what boundary condition(s) did you characterize your contact(s) ?
B) Did you study the influence of different types of contacts ?

HENISCH : A) We dealt with a semi-infinite system. On one side, therefore, the boundary conditions are those which characterize the undisturbed bulk material. on the other side, our X=0, we assume $J_p/J = 1$ and F = field = 0. The first implies maximum injection (just to find out what happens then), the second comes from p-n junction theory. There is always a point where F=0 when a forward voltage is applied to a junction. We simply found it convenient to call that point X=0. N_0 and P_0 had to be determined by reiterative calculations.

B) Different contacts would be characterized by different values of J_p/J. We have made no experiments of this kind. If J_p/J were smaller than unity, all the various effects would, of course, be smaller than those calculated by us. For $J_p/J=0$ (a very unlikely case), they would disappear.

MARK : What trap distributions are required to obtain the condition : (lifetime) < (ohmic relaxation time) to achieve the relaxation semi-conductor regime ?

HENISCH : No traps are needed to produce the relaxation regime in the sense $\Delta N < 0$. When traps are present, the criterion for majority carrier depletion under the influence of minority carrier injections is no longer $\tau_D/\tau_0 > 1$ but $\tau_D/\tau_0(M_0+1) > 1$ where M_0 is the normalized trap concentration. Changes of M_0 affect τ_D only in a minor way (via the carrier mobility), but they can affect τ_0 quite sensitively. The product $\tau_0(M_0+1) \approx \tau_0 M_0$ is therefore more nearly constant in practice then the above inequality condition immediately suggests.

When traps are present in increasing concentrations, the relaxation regime tends to be increasingly suppressed ; the system is then driven towards and into the lifetime regime. The most interesting aspect is then no longer $\Delta N < 0$ but F(=field) > F_∞ over a certain regime, because the latter implies an increase of total resistance.

MACDONALD : 1) Have you taken hot carrier effects into account ?
2) The Langevin theory of bimolecular recombination leads to a recombination time equal to the

dielectric relaxation time. How then is it possible for your parameter A ($\simeq \tau_D/\tau_o$) to be greater than unity in the trap free situation.

HENISCH : 1) No, there was no immediate reason for doing so. In the trap-free case, the maximum field is in the undisturbed bulk region, and can be controlled without difficulty to avoid hot electron effects at ordinary temperatures. In the trapping case, the highest field enhancement so far calculated by us is by a factor of only 3 or so, hence, no danger. However, a future calculation should certainly take high field effects into account.

2) I am not sure what the special assumptions of the Langevin treatment are. The form of the starting equation suggests that it is intended to apply to bimolecular recombination through a small number of neutral centers. In standard semiconductor theory, we actually distinguish between :

2a) direct electron-hole recombination by electron-hole encounters <u>without</u> centers, and

2b) Shockley-Reed type recombination through centers of arbitrary capture cross section for electrons and holes. In the steady state (but only then) this form of recombination is likewise bimolecular.

In the context of (a), it is customary to write the net recombination rate under excitation as :

$$R = \frac{np - n_e p_e}{\tau_o \, (n_e + p_e)}$$

where the subscripts denote equilibrium values. The concentration products reflect the applicability of Boltzmann statistics to electronic semi-conductors. The corresponding carrier lifetime τ_o, being defined for totally homogeneous field-free material, is then entirely independent of bulk transport considerations.

In the context of (b), the Shockley-Reed theory must be considered as a generalized Langevin analysis, except in so far as it is based on Fermi statistics. The arbitrary capture cross-sections for electrons and holes permit recombination in two stages, and then (in distinction from Langevin) allow for a net steady state charge storage in the recombination centers. The theory does not yield the suggested identity of τ_o and τ_D. Indeed, it is in no way concerned with the dielectric relaxation time τ_D, because it envisages (as above) recombination in a homogeneous medium, controlled by the capture cross-sections and again not by bulk transport considerations. Now, do the two standard continuity relationships for dn/dt and dp/dt call for any particular ratio of τ_D/τ_0 ? They contain (necessarily) equal recombination terms which drop out when dn/dt = dp/dt, which means that the continuity equations prevail for arbitrary values of the τ_D/τ_0 ratio.

It may be possible to envisage (as Langevin presumably did) a system in which recombination proceeds through very few centers, far apart, the situation being then transport controlled in a highly localized particle sense. However, it seems more than doubtful whether constants (like τ_D) derived from the statistics of bulk transport, can then be usefully applied as descriptive parameters. In any event, we know from experience that it is possible to obtain specimens of a given semi-conductor material with enormously different carrier lifetimes but only slightly different resistivities (and thus τ_D values).

ELECTROCHEMICAL COLORATION AND REDOX REACTIONS IN SOLID IONIC CONDUCTORS

Pierre FABRY and Michel KLEITZ

Laboratoire de Cinétique Electrochimique Minérale Associé au C.N.R.S. (E.R.A n° 384) - E.N.S. d'Electrochimie et d'Electrométallurgie de Grenoble - Domaine Universitaire - BP 44 - 38401 SAINT MARTIN D'HERES - France

Simple solid ionic crystals such as alkali halides are generally described as perfect crystals in which more or less mobile point defects are dissolved. With regard to the electronic distribution, these point defects constitute traps sometimes having energy levels located in the forbidden band. When a suitable voltage is applied to a cell involving such a material, it is observed that a change in the electronic trap filling, compensated by a change in trap concentrations, can occur. The phenomenon starts at one electrode and spreads through the bulk of the material. When it starts at the anode, it is in fact an electrochemical oxidation of the material and when it starts at the cathode, an electrochemical reduction. As described in chapter 1, this corresponds to an enrichment or an impoverishment in one of the basic crystal components. The process is commonly regarded as resulting from the ionic conductivity of the crystal. Sometimes it even serves as a qualitative test for the existence of such a conductivity[2,3]. It is frequently accompanied by the onset of a coloration of the material and for this reason is generally called electrochemical coloration. (In one case[1] the coloration was observed to suddenly occur in the whole crystal).

With solid ionic conductors having more complex structures the same observations were also made[4-9] and the phenomenon is certainly basically the same.

At first the electrochemical coloration of alkali halide crystals was interpreted as being the consequence of the dissolution of the alkali metal produced at the cathode of the cell. In 1937 Pohl[10] published an important synthesis of his co-workers results. At that time the coloration was regarded as resulting from

M. Kleitz and J. Dupuy (eds.), Electrode Processes in Solid State Ionics, 331–365.

the trapping of electrons on alkali ions and the propagation as proceeding by electron jumps from one alkali ion to another. This interpretation was revised when the concept of point defects was introduced. In 1950 Heiland[11] described a certain number of particularly significant experimental results obtained on KCl. He also developed a calculation method for the propagation rate. It is basically similar to that which was proposed by Karabascheff[12]. About 1960, with the development of MHD channels, electrodes in refractory oxides were considered. The electrochemical reduction, called blackening of such electrodes, which is a nuisance in this application, was studied[3,13-41] mainly with a view to preventing it. Casselton[13] published a synthesis of the results obtained in this field on stabilized zirconia. Since 1950 the investigated range has somewhat broadened. Results were published on the electrochemical reduction to a divalent state of rare earth ions in CaF_2[42-47], on the reversible electrochemical coloration of CuCl[48,49] on the electrochemical reduction of oxide systems[50-53,120]. However, the alkali halide crystals remain the subject of most studies[1,10-12, 54-78].

Changes in the electronic trap filling in ionic crystals, very often called coloration by analogy, are also obtained either by high energy radiation bombardment or chemically by bringing the material to equilibrium with an oxidizing or reducing system. Although possible in principle, this result cannot be obtained as easily as in semiconductors by a simple addition of an aliovalent dope. As a rule, this leads to changes in the concentration of complementary point defects and not necessarily to a change in the Fermi level position. Roughly speaking, bombardment gives rise to the formation of point defects through "mechanical" displacements of ions and bound electrons. No global chemical effect, no change in the overall stoichiometry for instance, are expected in this case. The process is carried out at a temperature sufficiently low to prevent the displaced ions from diffusing back to their original positions. The secondary point defects created are generally different from those which would exist under thermodynamical equilibrium. The electrochemical process, like the chemical process, results in a true state of thermodynamical equilibrium in experimental conditions. Comparisons of the results obtained by radiation bombardment and by addition (chemical or electrochemical) were performed on several materials[42,54,57,79]. The optical spectra are the same for the chemical and electrochemical colorations and are frequently different for the radiation coloration.

The comparison of the chemical and electrochemical colorations leads to a question unsolved to date : what are the first steps in electrochemical colorations? Two assumptions have been considered. In the case of reduction,one assumption[11] is : at the cathode-phase-boundary a system is created which has a reducing power similar to that of the reducing system which would be used in chemical coloration. According to this assumption, the two processes are not essentially different apart from the effect of the applied field on the diffusing species. The other proposed assumption[13] is : the potential applied to the electrode is sufficient to permit direct injection of electrons in the crystal. The chemical reduction then primarily results from the electric compensation which occurs in the material. Here, the elementary steps, particularly at the electrode, are certainly different. Further details on these approaches will be given later on.

Continuing this comparison with other related processes, it must be mentioned that electrochemical reactions other than those leading to the simple enrichment or impoverishment in one of the basic crystal components can also give rise to change in point defect concentrations. These are deposition or dissolution at the electrodes, of a doping ion. The electrochemical deposition of Cd, Hg, Co dissolved in the LiCl-KCl eutectic was observed by Hladik[80]. Experimental results obtained with the electrode C/AgCl/Cd^{2+} were also interpreted[81] as resulting from the deposition of cadmium metal on the graphite surface. In the same way, we can expect for instance, a Cd^{2+} ion injection in the crystal under anodic polarization of a Cd/NaCl electrode. The occurrence of such reactions obviously supposes a certain mobility of the corresponding ions. The reactions are more likely to be observed with cations in cationic conductors and with anions in anionic conductors. The Cl^- ions for instance are easily injected into stabilized zirconia[82]. Such reactions are well known in silica based materials[83-85]. They are somewhat similar to the traditional electrode reactions in the presence of a supporting electrolyte.

It must also be stressed that experimental studies on the dielectric breakdown of ionic crystals provide evidence, in the so-called high temperature region (usually above 300°C) of electrochemical coloration phenomena during the pre-breakdown period[86-88]. This has already been noted[1,59] ; however, due consideration to the phenomenon has rarely been given in the interpretation proposed[89]. According to the paper by Manfredi and Paracchini[90] the electrochemical coloration seems also to be closely related to the switching phenomena, at least in this type of material. Obviously it is certainly also similar in many respects to the phenoma described by Lampert[91]. The paper by Zhelev and Todorov[76], is implicitly based on such an assumption.

To avoid any confusion, comparison must finally be made with the experiments performed within the framework of the Wagner theory for measuring the electronic conductivity parameters[92]. If the electrochemical coloration is performed with a counter electrode which acts as a "reference electrode" there is, in principle, no difference : in both approaches the working electrode is regarded as making the Fermi level in the ionic crystal in contact, shift from its initial position to another position. There are slight formal differences. The electrochemical coloration is primarily intended to increase the number of entrapped electrons at point defects, and essential parameters for the investigators are the relative position of the trap levels and the Fermi level variation range. The departure from stoichiometry which accompanies the coloration is described as a variation in the concentrations of certain point defects resulting from the electric compensation of the entrapped electrons. In the electronic conductivity measurement, the investigators are primarily interested in the free-electron concentration variations. They do not enter into the detail of the electronic structure of the material and treat the departure from stoichiometry "macroscopically" from a thermodynamical viewpoint, in terms of anion and cation concentrations. This may result in different interpretations of transient phenomena as will become evident in the chapter dealing with the progression of the colored zone. In practice, the differences are more important. In electronic conductivity measurement, changes in stoichiometry are rather small and the steady state is quite rapidly reached. Investigators are mainly interested in the characteristics of this steady state. Electrochemical coloration generally implies larger deviations from stoichiometry and the relaxation times are very long. There, the investigators are principally interested in the transient behavior of the material. Another cause of important differences is the fact that "reference" electrodes are almost never used in electrochemical coloration experiments. Under these conditions, the final states may be totally different. It seems that sometimes a homogeneous composition of the material can be reached by electrochemical coloration ; that is, at least, an implicit assumption frequently made. In other cases a coloration occurs at both electrodes[64] and their effects cancel each other in the bulk of the material. This topic has not been thoroughly investigated so far.

1 - SPECIFIC PROPERTIES OF SOLID IONIC CONDUCTORS

In many respects solid ionic conductors, also called solid electrolytes, can be considered as semiconductors. However, they are characterized by some particular properties resulting from the presence of a relatively high concentration of mobile ionic species :

- Any current passing through the electrolyte is equivalent to a flux of matter. If the exchange velocities of matter with the phases in contact on the surfaces are not sufficiently rapid, this flux gives rise to an accumulation or a depletion of some chemical species in the neighbourhood of the corresponding surface. Then the current produces a chemical effect.
- The mobile ions play a dominant part in determining the electrostatic potential profiles within the material. In fact, they cons-

titute the majority carriers. This property is manifest especially during the electrochemical coloration process ; any electrons or holes which are injected tend to be compensated by a supply or a departure of a corresponding number of mobile ions. This effect which results in local change in crystal composition, is the basic principle of electrochemical coloration. Under these conditions, the band curvature in the macroscopic band schema of the material is principally determined by the properties of the mobile ions and no longer by those of the electronic carriers[93].

2 - CHANGES IN THE PROPERTIES OF AN IONIC CRYSTAL RESULTING FROM ITS COLORATION

On the hypothesis of local equilibrium, a description analogous to that proposed by Kröger[95] can be used to determine the local situation in the crystal. The basic principles and assumptions of this description are briefly recalled in what follows. Here we will use as an independent variable the electron chemical potential μ_e. It is the most appropriate parameter to characterize the degree of trap filling. Within a constant, it equals the distance between the Fermi level and the valence band edge.

To give an example of the information which can be deduced from this type of description, we will consider a simple schematic crystal KX, where the only point defects are the K vacancy V_K' (Kröger's notation) and the X vacancy $V_X^\bullet$. These point defects can trap electron holes and electrons, respectively to form V and F centers, noted here $V_K^\times$ and $V_X^\times$. They are not supposed to form any association. The conditions of local equilibrium are :

$$V_K' + V_X^\bullet \leftrightarrows \text{zero} \quad (1)$$

$$V_X^\bullet + e' \leftrightarrows V_X^\times \quad (2)$$

$$V_K' + h^\bullet \leftrightarrows V_K^\times \quad (3)$$

$$e' + h^\bullet \leftrightarrows \text{zero} \quad (4)$$

Accordingly, the corresponding concentrations (noted with squared brackets) are correlated as :

$$a = \ln[V_K'] + \ln[V_X^\bullet] \quad (5)$$

$$b = \ln[V_X^\bullet] + \ln[e'] - \ln[V_X^\times] \quad (6)$$

$$c = \ln[V_K^\times] - \ln[h^\bullet] - \ln[V_K'] \quad (7)$$

$$d = \ln[e'] + \ln[h^\bullet] \quad (8)$$

These concentrations are moreover linked by the electroneutrality equation which expresses the local electric compensation. In the case of a pure material it can be written :

$$[V_K'] + [e'] = [V_X^\bullet] + [h^\bullet] \quad (9)$$

According to the usual derivation, the existence domain of the material can be split into three intervals where this equation simplifies to a simple equality between two concentrations, the other concentrations being negligibly small when compared with them. The intervals are respectively characterized by the equalities :

$$[V_K'] = [h^\bullet] \quad (10)$$

$$[V_K'] = [V_X^\bullet] \quad (11)$$

$$[e'] = [V_X^\bullet] \quad (12)$$

As underlined by Andreev and co-workers[55] the space charges which can occur during the electrochemical coloration generally correspond to charge densities much smaller than those which appear in these equations. As a first approximation they can be neglected in the calculation of the point defect concentrations in the bulk of the material.

From the system formed by equations (5) to (8) and one of the three equations (10)(11) or (12), all the species concentrations can be calculated, in the corresponding intervals, as a function of $\ln[e']$ and plotted in a diagram. The parameter $\ln[e']$ is simply proportional to the chemical potential μ_e according to :

$$\mu_e = \mu_e^o + RT \ln[e'] \quad (13)$$

(the solid solution is supposed ideal)
μ_e^o is a constant, R the gas constant and T the absolute temperature.

The general shapes of the graph obtained in this way are depicted in fig. 1. In this figure we have supposed, as an example, that the initial state of the material is defined by a value $\mu_e(A)$ of the variable μ_e and that after coloring it reaches the state B. On the assumption of local equilibrium, the coloration process consists in a gradual shift from A to B.

A simple reading of the diagram gives the changes in concentrations which accompanied the coloration process. Obviously, if the formation of color centers is the main goal of the operation, the concentration of V_X^x centers are of primary interest.

For an understanding of the mechanism, the quantities :

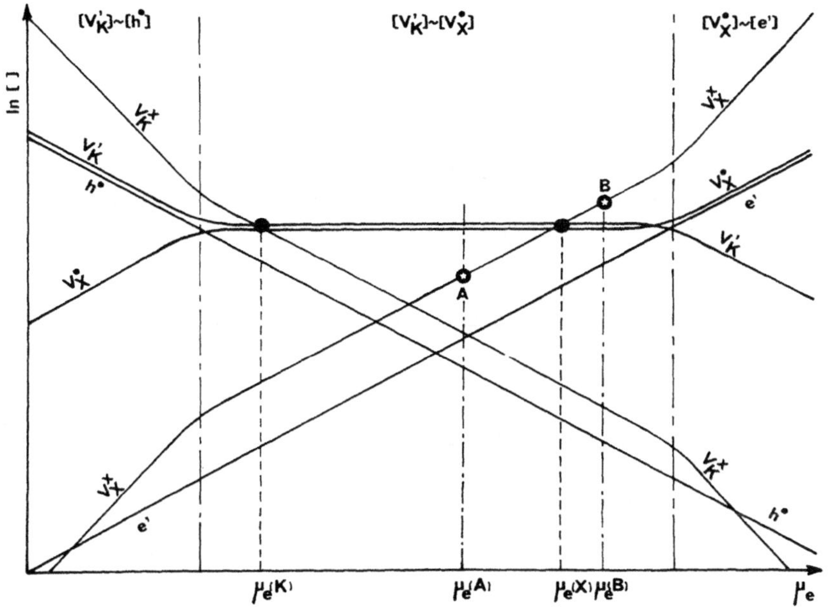

Fig. 1 - *Typical concentration diagram (see text)*

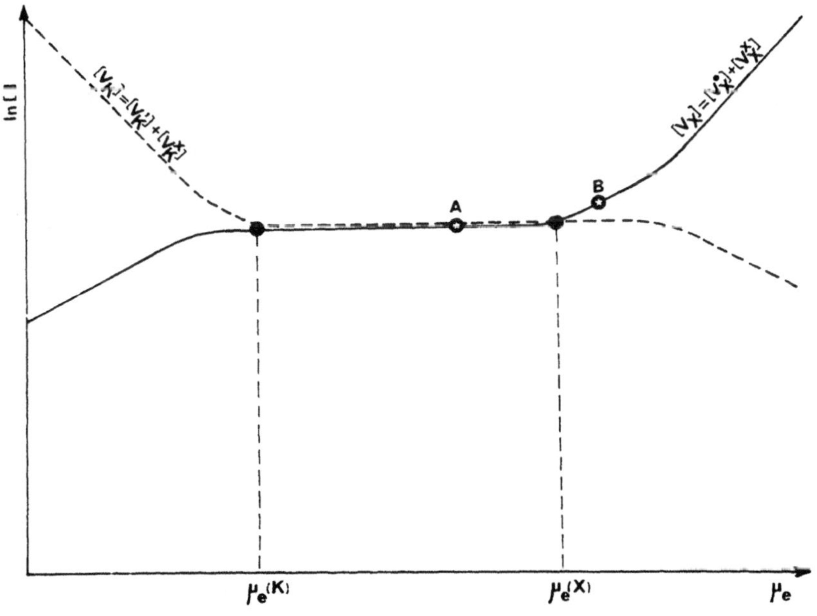

Fig. 2 - *Variations in vacancy concentrations which accompanied the coloration*

$$[V_K] = [V'_K] + [V^{\times}_K] \qquad (14)$$

and $$[V_X] = [V^{\bullet}_X] + [V^{\times}_X] \qquad (15$$

are particularly important. They represent the number of K or X particles which must move away (or in) for the coloration to occur. Their variations are represented in fig. 2. They show that the coloration is not simply accompanied by a change in stoichiometry but implies the displacement of definite particles.

From these curves we can also directly deduce the changes in the stoichiometry. They confirm, should it be necessary, that the coloration corresponds to a chemical oxidation or reduction of the material.

As emphasized in another paper[96], the variable μ_e can be viewed in certain simple situations, as equal, to within a constant and a multiplication factor, to the potential of an electrode in electric equilibrium with the material. So, to within the same multiplication factor, the variation $[\mu_e(A) - \mu_e(B)]$ equals the electrode potential variation which can be measured.

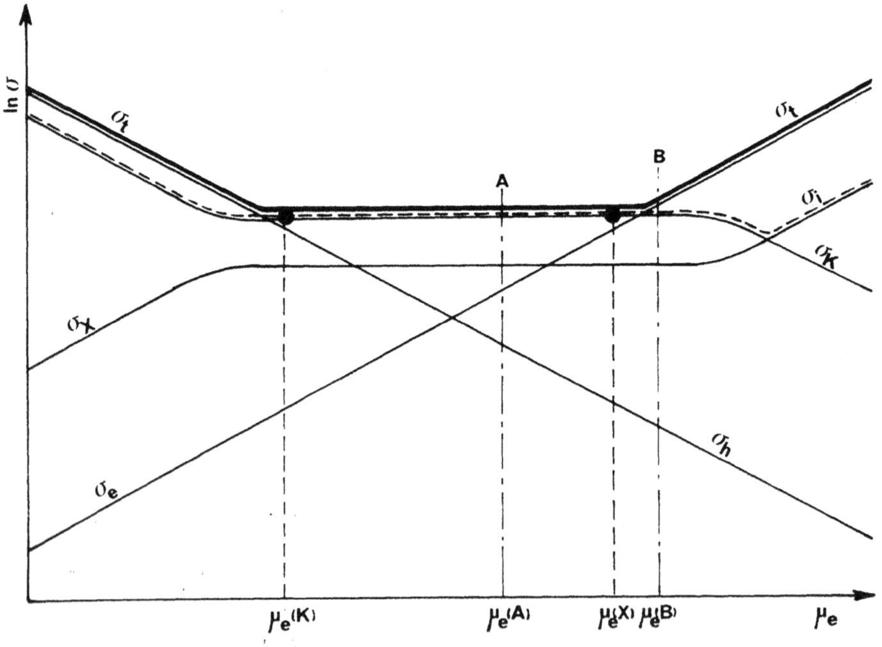

Fig. 3 - *Variations in conductivities which accompanied the coloration*

On the μ_e scale, two points deserve special mention : they are noted $\mu_e(K)$ and $\mu_e(X)$; they correspond to equal numbers of empty and occupied traps of one sort. They are called the equivalent points. If the degeneracy factors of the corresponding levels are one, at these points, the variable μ_e equals, to within a constant, the energy level of the corresponding traps.

From the diagram in figure 1 we can also, in principle, deduce the variations of the partial conductivities of the material. For a given particle p, its partial conductivity is related to its concentration [p] by the usual expression

$$\sigma_p = \tilde{u}_p z_p^2 F^2 [p] \tag{16}$$

where $\tilde{u}_p$ is its electrochemical mobility and z_pF its electric charge. As a first approximation, $\tilde{u}_p$ can be frequently assumed independent of the concentration. Under these conditions :

$$\ln\sigma_p = \ln[p] + C^t \tag{17}$$

and the variations of the conductivities are simply deduced from the initial diagram by translations of the relevant curves. A typical situation is depicted in figure 3. In this figure, besides the partial conductivities (σ_K, σ_X, σ_e, σ_h), the ionic conductivity

$$\sigma_i = \sigma_K + \sigma_X \tag{18}$$

and the total conductivity

$$\sigma_t = \sigma_K + \sigma_X + \sigma_e + \sigma_h \tag{19}$$

are represented. Such a diagram simply shows the variations in magnitude and nature of the conductivity which can be expected.

We can also deduce from it the ionic transport number t_i

$$t_i = \frac{\sigma_i}{\sigma_t} \tag{20}$$

The plot of this parameter (fig. 4) gives direct information on the limits between which the electrochemical coloration process can be easily performed : the coloration occurs more easily the greater the parameter t_i. If the ionic transport number is nil, no transport of matter occurs under usual experimental conditions, and thence no change in composition and no coloration.

In the extrinsic domain where the vacancy concentrations are not determined by the Schottky disorder (equation 1) but by the concentration of aliovalent impurities dissolved in the material, the situation is slightly more complex. If, for simplicity, we suppose that there is only one aliovalent impurity M, of valence 2,

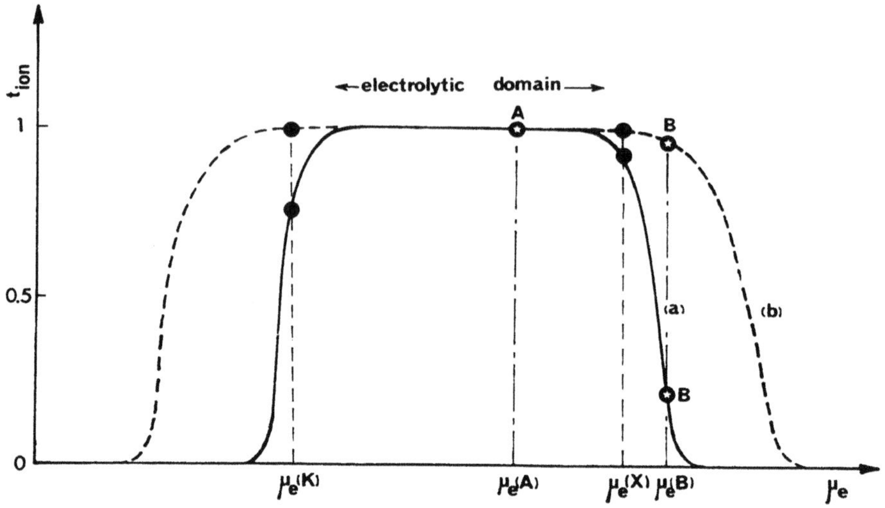

Fig. 4 - *Variations in the ionic transport number ; a) pure material b) M^{2+} doped material*

substituted for K^+ ions, then the electroneutrality equation is :

$$[V_K'] + [e'] = [M_K^\bullet] + [V_X^\bullet] + [h^\bullet] \tag{21}$$

The corresponding variations in concentration are represented in fig. 5. As regards the electrochemical coloration process, several particular properties must be noted :

- In general, the ionic conductivity of the doped material is greater than that of the pure material
- The concentration of $V_X^\bullet$ is much smaller
- In the typical situation where the aliovalent impurities have valencies that are stable in the investigated range, the interval where the ionic transport number is noticeably different from zero is larger than that of the pure material (cf. Fig. 4). The variation of μ_e, i.e. the shift of the Fermi level, that can be obtained by the electrochemical process, encompasses a larger interval
- The concentrations of vacancies as functions of μ_e show abrupt variations. In our simple example, they occur in the range where the electroneutrality equation simplifies into the equality :

$$[e'] = [M_K^\bullet] \tag{22}$$

- In the particular situation where one of the aliovalent impurities undergoes a change in valency in a range where its concentration is dominant in the electroneutrality equation (21), changes in the concentrations of X or K vacancies occur in the vicinity of the corresponding equivalent point. A typical situation is depicted in fig. 6. Generally it results in a change in the ionic and total

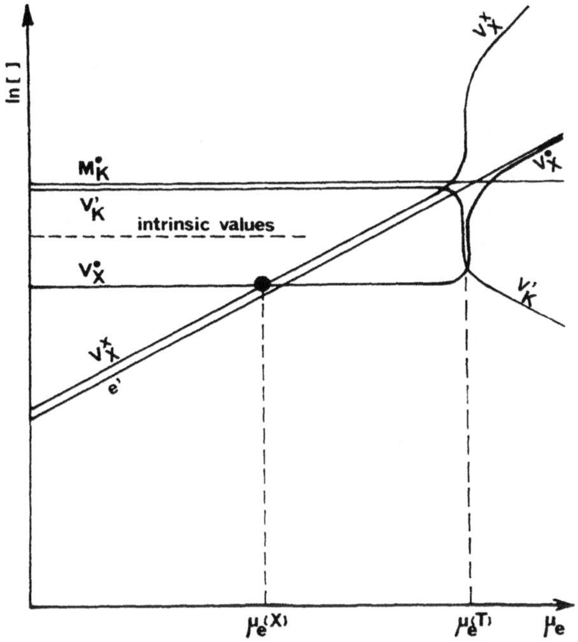

Fig. 5 - *Variations in vacancy concentrations (half diagram) in the case of a material substitutionally doped with M^{2+} ions*

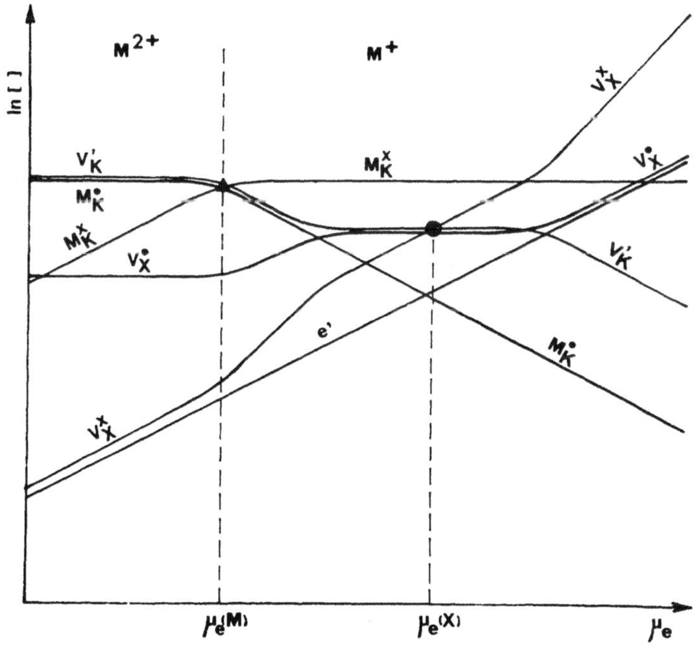

Fig. 6 - *Variations in vacancy concentrations (half diagram) in the case of a material substitutionally doped with M^{2+} ions undergoing a valency change*

conductivity of the material (Fig. 7 curve b). At this point, it must be emphasized that other "accidents" of this type may occur in the range which is usually called the electrolytic domain[97]. If the aliovalent impurities, are sufficiently concentrated, they may form a narrow impurity band. Then, in the vicinity of the equivalent point an additional electronic conductivity may occur as has already been suggested[94,98-100] (Fig. 7 curve c). For this reason, it is sometimes useful to utilize, in addition to the notion of electrolytic domain, that of redox stability interval[98] to characterize the intervals where no point defect undergoes a change in valency.

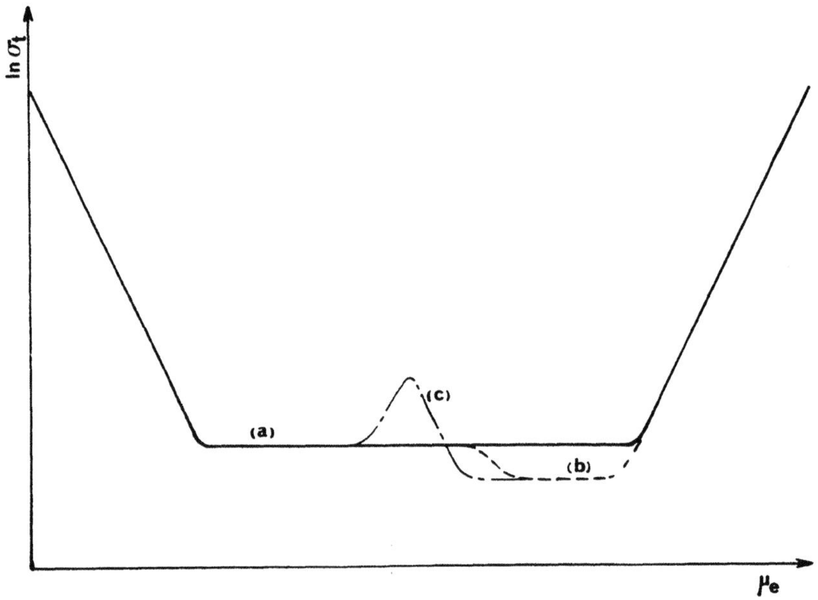

Fig. 7 - *Typical conductivity curves (see text).*

The published experimental results substantiate the preceeding deductions. Departure of matter and change in conductivities have been observed after electrochemical (and chemical) coloration.
- Relative variations in density of the order of 10^{-4}-10^{-5} were measured by Hacskaylo[54] on NaCl. Andreev and co-workers[55] have shown that similar variations occur in KBr ; their magnitude depends on the distance from the electrodes and on the degree of coloration. With stabilized zirconia losses in weight are more remarkable. Beranger[14] observed a loss of 4.8 % and Gokhshtein[15] of 3.75 %. These important variations induce change in the lattice parameters of the material[16,17].
- Several authors[56,72] showed that, in the presence of dissolved impurities (Pb^{2+}, Sr^{2+}, Zn^{2+}, Mn^{2+}) electrochemical coloring results in a noticeable reduction in ionic conductivity. In the same way Sootha and Mitra[58] observed a decrease in the total conductivity

of KCl doped with $MnCl_2$ after coloration. In all cases the concentration of free electrons was observed to increase with reduction. Jacquin and col.[18] measured a drastic change in stabilized zirconia where the ionic transport number decreases from 1 to almost zero. It was also shown, by several authors[68,101,102] that the increase in the conductivity of colored KCl is due to an increase in the number of free electrons. Durand and co-workers[103] recently confirmed that the reduction of a KCl crystal can reach a state where the electroneutrality condition(9)simplifies into the equality

$$[V_{Cl}^{\bullet}] = [e'] \qquad (23)$$

- It must be noted that Pershits and Pinsk[69] observed different influences of chemical and electrochemical coloration on the conductivity of potassium chloride samples. Another remark must be made regarding the reduction of the silver ion. It appears to be reducible to the Ag^- state in NaCl and KCl crystal[73] and then would occupy an anionic site.

3 - THE ELECTROCHEMICAL COLORATION PROCESSES

For the sake of simplicity we will confine the derivations to reduction reactions. The extension to oxidation reactions is obvious.

The electronic filling of a given type of trap in the solid ionic conductor is given by the usual relation :

$$\frac{N_c}{N} = \frac{1}{1 + \frac{1}{g} \exp \frac{W_T - W_F}{kT}} \qquad (24)$$

where N is the trap concentration
N_c the concentration of trapped electrons
W_T the level of the corresponding electronic state
g its degeneracy factor (here, we will only consider the simple case where $g = 1$)
W_F the Fermi level

This relation can also be written in a form :

$$W_F = W_T - kT \ln \frac{N - N_c}{N_c} \qquad (25)$$

which is analogous to the traditional expression of an electrode voltage

$$E = E^o + \frac{RT}{zF} \ln \frac{[Ox]}{[Red]} \qquad (26)$$

In the case of F center formation, for instance, the analogy applies for $[Ox] = [V_X^{\bullet}]$ (27) and $[Red] = [V_X^{\times}]$ (28)

Both formalisms can be used to describe electrochemical coloration. For a while we will adopt the electrochemical viewpoint. This will allow us to refer to situations that are well identified in traditional electrochemistry. This viewpoint was also adopted by Lanyi and Harasta[104] to describe the charging and discharging current on NaCl.

Preliminary reduction of the surface

As a rule when the resting potential of an electrode lies, on a 100 mV scale, far from the relevant E^o of a given reaction, it is customary to assume that the potential is determined by another oxidizing-reducing system. The details of the experimental results published on electrochemical coloration indicate that in many situations, the initial value of the electrode potential is indeed noticeably different from that observed when coloration occurs. It is reasonable under these conditions to assume that it is fixed, at first, by a certain system that we will call (B^+, B)

$$E = E^o_B + \frac{RT}{F} \ln \frac{[B^+]}{[B]} \qquad (29)$$

Species adsorbed on the surface of the ionic conductor are sometimes considered to play this role[96,105].

To make the electrode potential reach the values where a noticeable coloration occurs, it is then necessary to reduce the concentration of B^+ or (and) increase that of B. The passage of a current through the ionic conductor generally has such an effect, as initially mentioned (chemical effect). More generally, by analogy with traditional results in electrochemistry, we can expect a steady-state E(I) characteristic of the electrode shaped as indicated in figure 8a. It shows the existence of a limiting current I_ℓ. When this value is reached, the electrode reaction rate is high enough for all the B^+ particles arriving at the electrode surface to be immediately consumed. The surface tends to be completely depleted of B^+ particles and, according to formulae (29), the potential E can shift towards lower values.

During this shift, the electrode potential may cross potential ranges' specific to other reactions. Then, similar phenomena may occur in addition to the one just described. This situation has been well identified on stabilized zirconia electrolytes[98] (Fig. 8b). At first, the Fermi level on the surface of the material usually appeared to be fixed by surface states due to absorbed oxygen particles. They can be eliminated by passing a certain current I_a trough the electrode. When the electrode potential shifts towards more cathodic values it encounters the range where the traces of carbon dioxide molecules adsorbing on the surface can also undergo a reduction process. The same observations are then made ; a certain current $I_a + I_b$ is necessary to completely deplete the surface of all these species.

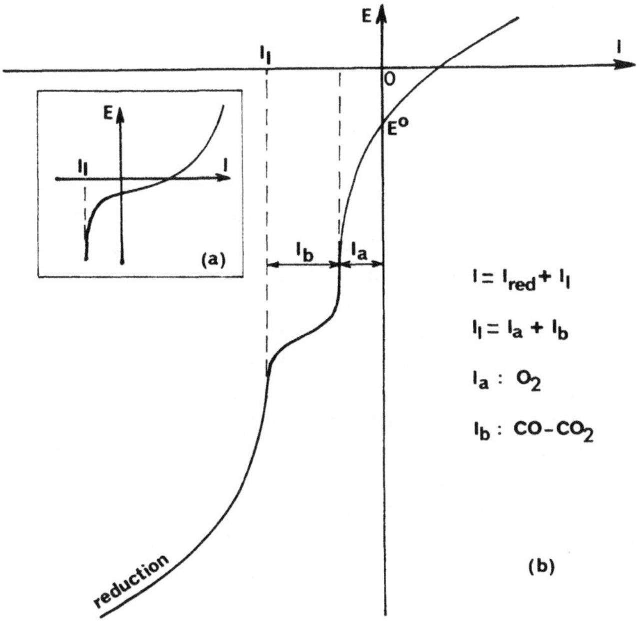

Fig. 8 - *Typical electrode characteristics (see text)*

In principle the simple volatilization of alkali metal from the surface of an alkali halide can give rise to the same phenomena. More complex reactions of this type have also been observed. Oxygen was shown to react at the surface of alkali halides to form oxides or mixed compounds[11,60]. Nitrogen reacts with zirconia to form nitrite or oxynitrite[19-22]. The material of the electrode itself can form complex compounds under certain conditions. This has been observed with platinum[11,17].

Generally the limiting currents I_ℓ are principally determined by the electrodes utilized. The more permeable the electrode to B^+ species, the higher the value $|I_\ell|$. With a platinum electrode working in air, the coloration of stabilized zirconia is only obtained for a current of the order of 100 mA/cm^2 [13], or even more[19,36]. When the coloration of the ionic conductor is the main object of the experiment, it is obviously of interest to select an electrode as impervious as possible, and the most inert environment for the electrode. Heiland[11] isolated the electrodes by covering them with supremax. Many authors embed the electrode in the solid electrolyte to avoid exchange of chemical species with the surrounding atmosphere.

In an ideal situation the supply rate of any B^+ species is nil. Then, it is simply necessary to eliminate the finite quantity of B^+ which is initially present on the surface. To do so, only a finite quantity of electricity needs to be passed through the electrode. The necessity of passing a certain quantity of electricity to appropriately reduce the surface has been experimentally shown by Sanchez and co-workers[59-61]. In practice, these authors applied a constant current during a certain time. They suggested, as other authors[11,65,77], that this leads to the formation of an intermediate layer containing a certain quantity of potassium. (In our description this could be viewed as an increase in the B concentration).

It must be stressed at this point that the general sketch of the previous derivation certainly applies to the different degrees of reduction which can be reached electrochemically, the ultimate step being, in some favourable cases, the formation of the metal. If the current passing through the electrode is too high, not only the first degree of coloration, but higher degrees are reached simultaneously. This point has been well confirmed on stabilized zirconia[94]. By adequately controlling the experimental conditions, only the first degree is obtained ; by passing a high current density a complete reduction with formation of zirconium metal is observed. So, the observation of potassium metal deposition during electrochemical coloration might not prove to be a prerequisite condition.

Progression of the colored zone

At the present time, there is some confusion in the literature on the choice of a mechanism which would explain the observed progression of the colored zones in the bulk of solid ionic conductors. Let us call T^+ and T the empty trap and the filled trap respectively. In some models the species T^+ and T are viewed as entities capable of diffusing as regular chemical species. In other models the T species in itself is not supposed to move. Its displacement results from its dissociation,

$$T \rightarrow T^+ + e^- \tag{30}$$

the displacement of the free electron and its subsequent entrapping to form another T species, some sites farther on. Then the progression rate is only determined by the mobility of the electrons and their life time. The displacement of color centers proper was studied by several authors[106-111]. It is generally admitted that the dissociation results in a higher diffusion coefficient.

As far as the formalism is concerned, in the first case the electrode is sometimes said to be "localized". The reduction of the traps occurs at the electrode interface ; the T species are created there and subsequently diffuse in the bulk of the material. In the second case, the electrode is viewed as "extended" in the sense that the electrode reaction proper (30) occurs in a diffused part of the material. At the electrode interface the corresponding charge transfer can be described as a simple injection of electrons. In some particular cases the coloration front appears as a rather sharp transition in the space separating a colored zone, which is almost entirely an electronic conductor, from the uncolored zone, which is an ionic conductor. According to the usual definition of an electrode[93], it must be localized at this transition. In this case it is called a "virtual" electrode[13].

As already stressed by Shamovskii and co-workers[112] another frequent cause of confusion is the erroneous utilization of the model proposed by Pohl and co-workers and recalled by Mott and Gurney[113]. The authors assumed that the propagation of the colored zone resulted from the drift of the electronic cloud under the sole effect of the electric field. They stressed that such an assumption was reasonable only when the electronic conductivity of the colored material remains small with respect to the initial ionic conductivity. Through the various papers which followed, this restrictive condition was progressively neglected and now, with the same assumption, the same calculations are applied to situations where the electronic conductivity becomes larger than the ionic conductivity by one order of magnitude[33]. Moreover, under the same conditions the electric field is frequently supposed to be simply equal to the ratio of the applied voltage to the sample length. This is very likely to be a rough approximation.

It is also frequently assumed that the free electron concentration is locally proportional to the $V_X^\times$ center concentration. Referring to a Kröger diagram (Fig. 1) this means that the colorations occurred in the central interval. According to recent results [101,103] it seems likely that certain colorations may have occurred over broader μ_e ranges involving a zone where the free electron concentration was proportional to the square root of the $V_X^\times$ center concentration.

In fact the situation may be even more complex than that. If we refer to the diagram shown in the first part, the coloration is necessarily accompanied by a change in the local concentration of point defects. In some cases the relevant process may be rapid, in others it may not. In a solid oxide electrolyte, for instance, the trapping of an electron in an oxygen vacancy implies the creation of a new vacancy, i.e. the departure of an oxide ion, according to the overall reaction :

$$\searrow 2\,e^- + V_O^{\bullet\bullet} + O_O \rightarrow 2\,V_O^{\bullet} + O^{2-}\nearrow \tag{31}$$

This process can certainly occur easily in a material which is a conductor, precisely by O^{2-} ions. In an alkali chloride, the formation of an F center corresponds to the overall reaction :

$$\searrow e^- + Cl_{Cl} \rightarrow V_{Cl}^\times + Cl^-\nearrow \tag{32}$$

It implies the supply of an electron and the departure of a Cl^- ion as previously shown on the Kröger diagrams. Frequently, the observed electronic conductivity in the colored zone is higher than the initial ionic conductivity, which is itself higher than the partial anionic conductivity. Then the moving away of the chlorine ions is likely to take over the role of the slow limiting process of the overall reaction.

General equations

The difficulty in solving the problem lies in the complexity of the equation system which describes the evolution of the material. This point has already been underlined in an earlier attempt [98] to rationalize the selection of the equations. The main sets of equations of this system are : (for simplicity we will consider a one-dimensional geometry)

- 1) The equations giving the local magnitude of the fluxes J_p and partial current density I_p of point defects

$$J_p = -[p]\tilde{u}_p \frac{\partial \mu_p}{\partial x} - [p]\tilde{u}_p z_p F \frac{\partial \phi}{\partial x} \tag{33}$$

$$I_p = z_p F J_p \tag{34}$$

where $[p]$ is the local concentration of the point defect
$\tilde{u}_p$ its electrochemical mobility
$z_p F$ its electric charge
μ_p its chemical potential
ϕ the electrostatic potential

Under ideal conditions μ_p can be simply expressed as a function of $[p]$ (cf for instance (13)) and J_p simplifies to :

$$J_p = -\tilde{u}_p RT \frac{\partial [p]}{\partial x} - [p]\tilde{u}_p z_p F \frac{\partial \phi}{\partial x} \tag{35}$$

- 2) The equations giving the local balances for the point defects

$$\frac{\partial [p]}{\partial t} = -\frac{\partial J_p}{\partial x} + \sum_q \nu_{pq} v_q \tag{36}$$

where q enumerates the local reactions involving the point defect p
v_q are their rates
ν_{pq} the coefficient of p in their expression

- 3) The Poisson equation :

$$\frac{\partial^2 \phi}{\partial x^2} = \frac{\rho}{\varepsilon} \tag{37}$$

As previously mentioned, the charge concentration ρ which appears in this equation is generally much smaller than the main point defect concentrations. Therefore as far as the calculation of these point defect concentrations is concerned we can state that :

- the local electroneutrality is maintained

$$\sum_p z_p [p] = 0 \tag{38}$$

As previously recalled, this equation can be frequently simplified into an equation involving only two concentrations.

- The total current density is a constant through the material

$$\sum_p I_p(x,t) = I \quad (39)$$

with

$$\frac{\partial I}{\partial x} = 0 \quad (40)$$

- 4) The equations expressing the limiting conditions. One is particularly interesting ; it concerns the current density at the electrode where the electrochemical coloration occurs. The current entering (or leaving) the ionic crystal can be regarded as ionic for a part I_i equal to the limiting current density I_ℓ (Fig. 8) and electronic for the complementary part I_e

$$I_i(0,t) = I_\ell \quad (41)$$

$$I_e(0,t) = I - I_\ell \quad (42)$$

A straightforward conclusion can be drawn from this condition. At the end of the electrochemical process if the material is approximately homogeneous, the state reached corresponds to an ionic transport number t_i :

$$t_i = \frac{I_\ell}{I} \quad (43)$$

This point can be easily located on a diagram similar to that shown in figure 4. In principle, this provides an easy way to determine the final state reached by the material. If the electrode is ideally blocked $I_\ell = 0$ and $t_i = 0$. Then the electronic conductivity in the final state is necessarily much greater than the ionic conductivity.

Coloration of a perfect KX crystal

As an example of utilization of such a set of equations let us consider the electrochemical reduction of the schematic crystal previously described. The selected experimental conditions are as follows :

- The material is substitutionally doped by M^{2+} ions having a stable valence in the considered range. The coloration is performed under conditions such that the initial and final states lie within its extrinsic conductivity domain where the electroneutrality conditions simplifies, as a first approximation, into the equality :

$$[V_K'] = [M_K^\bullet] = a \qquad (a = C^t) \quad (44)$$

- The mobility of the $V_{Cl}^\times$ center itself is supposed to be very low and is neglected: $\tilde{u}_{V_{Cl}^\times} = 0$ (45)

The only mobile particles to be considered are V_K', $V_{Cl}^\bullet$, e' and the local reactions are

$$\text{zero} \rightleftarrows V_{Cl}^\bullet + V_K' \quad (1)$$

$$V_{Cl}^\bullet + e' \rightleftarrows V_{Cl}^\times \quad (2)$$

Equation (39) here, is

$$I_K + I_{Cl} + I_e = I \qquad (46)$$

where I_K and I_{Cl} stand for the partial current densities of V_K' and $V_{Cl}^{\bullet}$. Reaction (2) can be reasonably supposed to be equilibrated. Thence

$$[V_{Cl}^{\bullet}][e'] = m[V_{Cl}^{\times}] \quad (m = C^t) \qquad (47)$$

Its rate v_2 can be expressed by the quantity of $V_{Cl}^{\times}$ produced by unit time: $v_2 = Q_{V_{Cl}^{\times}}$. Then, according to (36) and (45) the $V_{Cl}^{\times}$ balance equation gives

$$\frac{\partial[V_{Cl}^{\times}]}{\partial t} = Q_{V_{Cl}^{\times}} \qquad \frac{\partial[V_{Cl}^{\times}]}{\partial t} = v_2 \qquad (48)$$

On the other hand, in a perfect material, reaction (1) cannot occur in the bulk of the material. This would imply the formation of K and X interstitial particles which are not supposed to exist. Such a reaction can only occur on the surfaces where a unit crystalline cell can be created or consumed. Thence the corresponding rate v_1 is nil : $v_1 = 0$. (49)

In the local balance equation of V_K' this results in :

$$\sum_q \nu_{Kq} v_q = 0 \qquad (50)$$

Consequently, this equation implies, taking into account (44)

$$J_K = b' \quad (b' = C^t) \quad (52) \qquad \text{and} \qquad I_K = b \quad (b = C^t) \quad (53)$$

(in fact b equals the limiting current density I_ℓ). Then the expression of I_K (33) gives :

$$\frac{\partial \phi}{\partial x} = \frac{U}{L} \qquad (55)$$

where U is the applied voltage and L the sample length.
Besides that, combining equations (46) and (53) yields :

$$I_e + I_{Cl} = I - b \qquad (56)$$

At the beginning of the process a large part of the material is unaltered. The application of equation (56) to this part, gives, taking into account (55),

$$|I - b| = (\sigma_e^o + \sigma_{Cl}^o)\,\frac{U}{L} \qquad (57)$$

where σ_e^o and σ_{Cl}^o are the partial electronic and anionic conductivities under initial conditions.
The electron balance equation is, according to (48)

$$\frac{\partial[e']}{\partial t} = -\frac{\partial J_e}{\partial x} - \frac{\partial[V_{Cl}^{\times}]}{\partial t} \qquad (59)$$

In the reduction process $\frac{\partial[e']}{\partial t}$ is necessarily positive, and therefore

$$\frac{\partial [V_{Cl}^{\times}]}{\partial t} \leqslant -\frac{\partial J_e}{\partial x} \quad (60)$$

By integration of this condition one gets for the total number of color centers $\frac{dN_{V_{Cl}^{\times}}}{dt}$ produced in the material per unit surface of the section :

$$\frac{dN_{V_{Cl}^{\times}}}{dt} = \int_{x=0}^{x=L} \frac{\partial [V_{Cl}^{\times}]}{\partial t} dx \leqslant -\left[(J_e)_{x=L} - (J_e)_{x=0}\right] \quad (61)$$

According to (56) the variation of $|J_e|$ is necessarily smaller than $\frac{|I-b|}{F}$ and then

$$\frac{dN_{V_{Cl}^{\times}}}{dt} \leqslant \frac{1}{F} |I-b| \quad (62)$$

Combining with relation (57) gives,

$$\frac{dN_{V_{Cl}^{\times}}}{dt} \leqslant \frac{(\sigma_{Cl}^{o} + \sigma_e^{o})U}{FL} \quad (63)$$

This could indicate, in particular, that the greater the concentration $[V_{Cl}^{\bullet}]$ in the initial material (i.e. the higher σ_{Cl}^{o}) the higher the rate of coloration. The fact that the rate of coloration suddenly decreases as the temperature is lowered[62] could support this point : when the temperature is decreased, the material enters the extrinsic conductivity range where the concentration $[V_{Cl}^{\bullet}]$ rapidly decreases (compare diagram 1 and 5). This point has already been underlined by Tsikin and co-workers[71]. In the same way the chemical coloration rate was observed to sharply decrease in the extrinsic conductivity range[114].

In fact, as Andreev and co-workers[55] have emphasized the values $\frac{dN_{V_{Cl}^{\times}}}{dt}$ experimentally obtained are usually much greater than that compatible with condition (63). Moreover, the calculation previously developed does not show the increase in current generally observed.

Coloration of a KX crystal with dislocations

One way of reconciling the calculation and the experimental results is to suppose that the rate of reaction (1) is not nil. This implies that there are locations in the bulk where the reaction can occur. Tsikin and co-workers[55] suggested that these locations could be the dislocations. They have shown[63] that the dislocation density has in fact a large influence on the electrochemical coloration process. (Similarly the role of the grain boundaries in sintered zirconia during blackening was underlined by several investigators[17,25]). The reaction occuring at dislocations, as written in (1), can be regarded as a simultaneous production of a unit cell KX, a vacancy V_K' and a vacancy $V_{Cl}^{\bullet}$. The production of V_K' is not incompatible with the condition $[V_K'] = C^t$, it simply results in a local variation of the J_K flux (cf. eq. 70). As an approximation let us consider under these conditions the other limiting case : reaction (1) is now supposed to remain equilibrated due to a sufficiently high density of dislocations. The relevant condition is :

$$[V_{Cl}^{\bullet}][V_K'] = d \qquad (d = C^t) \tag{65}$$

taking into account (44), this gives :

$$[V_{Cl}^{\bullet}] = \frac{d}{a} \tag{66}$$

We will also suppose as an approximation that the amount of $V_{Cl}^{\bullet}$ vacancy brought by the flux J_{Cl} previously considered is negligible compared to the amount produced by reaction (1)

$$|J_{Cl}| << v_1 \tag{67}$$

According to conditions (66) and (67) the balance equation of $V_{Cl}^{\bullet}$ simply gives :

$$v_1 = v_2 \tag{68}$$

According to condition (44) the flux of V_K' can be simply written :

$$J_K = \frac{\sigma_K}{F} \cdot \frac{\partial\phi}{\partial x} \qquad (\sigma_K = C^t) \tag{69}$$

So,the V_K' balance equation is, taking into account (44) :

$$\frac{\sigma_K}{F} \frac{\partial^2\phi}{\partial x^2} = v_1 \tag{70}$$

thence, according to (68) and (48) :

$$\frac{\sigma_K}{F} \cdot \frac{\partial^2\phi}{\partial x^2} = \frac{\partial[V_{Cl}^{\times}]}{\partial t} \tag{71}$$

Conditions (47) and (65) give :

$$[V_{Cl}^{\times}] = \frac{d}{a.m} [e'] \tag{72}$$

which can be introduced in (71) :

$$\frac{am\sigma_K}{Fd} \cdot \frac{\partial^2\phi}{\partial x^2} = \frac{\partial[e']}{\partial t} \tag{73}$$

On our assumption, equation (46) is

$$- \sigma_K \frac{\partial\phi}{\partial x} + \tilde{u}_e RTF \frac{\partial[e']}{\partial x} - \tilde{u}_e F^2 [e'] \frac{\partial\phi}{\partial x} = I \tag{74}$$

The equations (73) and (74) with the appropriate limiting conditions determine the behavior of the material.

To our knowledge this system has been neither compared to experimental results nor solved. The purpose of its development was just to show how a typical problem can be handled within the framework of a solid state chemistry description.

Various approximations can be proposed depending upon the relative influence of the electrical field and the concentration gradient on the diffusion of electrons. In the case of a dominant concentration gradient, the equation system reduces to an expression analogous to the second Fick law :

$$\frac{\partial [e']}{\partial t} = \alpha\, D_e \,\frac{\partial^2 [e']}{\partial x^2} \tag{76}$$

where D_e is the electron diffusion coefficient ($\tilde{u}_e RT$) and α their life time $\left(\frac{[e']}{[V_{Cl}^{\times}]}\right)$

In the case of a dominant electric field and an electronic conductivity remaining much smaller than the ionic conductivity the following relation is obtained :

$$\frac{\alpha I \tilde{u}_e F}{\sigma_K} \cdot \frac{\partial [e']}{\partial x} = \frac{\partial [e']}{\partial t} \tag{77}$$

It shows that the concentration profile obeys a law of the type :

$$[e'] = [e']_o + f(x - vt) \quad \text{with } v = \left|\frac{\alpha I \tilde{u}_e}{F a \tilde{u}_K}\right| \tag{78}$$

Its shape is not altered during the progression. The progression rate v is proportional to the current and inversely proportional to the K vacancy concentration.

Electric-field-determined progression

More generally when the electric field can be assumed to be the dominant force acting on the mobile particles, important simplifications can be made in the resolution of the general equation system. According to formula (33) this condition can be written :

$$\left|\frac{RT}{z_p F}\,\frac{\partial \ln [p]}{\partial x}\right| << \frac{\partial \phi}{\partial x} \tag{79}$$

It must hold for every particle p and in all parts of the material. To get an idea of the orders of magnitude, the term $\frac{RT}{z_p F}\frac{\partial \ln [p]}{\partial x}$ can be approximated by the ratio

$$\frac{RT}{z_p F}\,\frac{\Delta \ln [p]}{\Delta \ell} \tag{80}$$

where $\Delta \ln [p]$ represents the total logarithm variation of the considered concentration between the initial material and the colored material. A reasonable upper limit for $\frac{RT}{z_p F} \Delta \ln [p]$ can usually be selected around 1V

$\Delta \ell$ is the approximate thickness of the transition zone which separates the unaltered zone from the colored zone.

Under condition (79), formula (34) simplifies to

$$I_p = - \tilde{u}_p z_p^2 F^2 [p] \frac{\partial \phi}{\partial x} \tag{81}$$

or more usually $I_p = -\sigma_p \frac{\partial \phi}{\partial x}$ with $\sigma_p = \tilde{u}_p z_p^2 F^2 [p]$ (82)

and formula (39) to :

$$- \sum_p \sigma_p \frac{\partial \phi}{\partial x} = I \tag{83}$$

with the usual definition of the transport number t_p one obtains :

$$t_p = \frac{\sigma_p}{\sum_q \sigma_q} \quad (84) \qquad I_p = t_p I \quad (85)$$

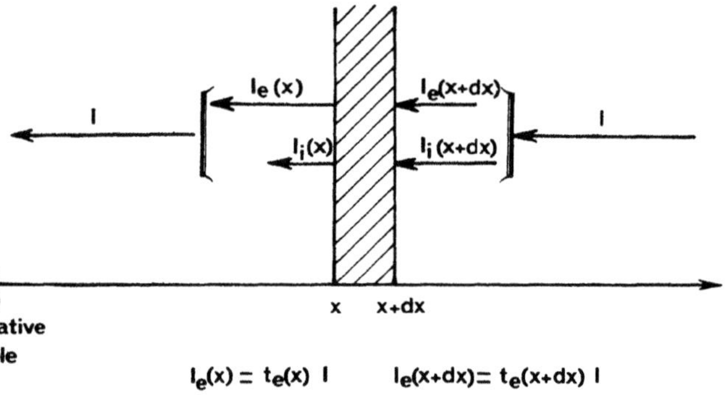

Fig. 9 - *(see text)*

Then the progression of the colored zone can be simply calculated from the sketch shown in figure 9. The electron balance in a layer located on an abscissa x and of thickness dx is simply :

$$\frac{I}{F} \cdot \frac{\partial t_e}{\partial x} = \frac{\partial [e_G]}{\partial t} \tag{86}$$

where $[e_G]$ stands for the overall local concentration of free and trapped electrons. Equations (83) and (86), with the appropriate limiting conditions completely describe the evolution of the system. As an example the situation which has been previously dealt with is described here by the equations :

$$-(a\tilde{u}_K + \tilde{u}_e[e']) \frac{\partial\phi}{\partial x} = \frac{I}{F^2} \quad \text{(equation 83)} \tag{87}$$

$$\frac{I}{F} \cdot \frac{\tilde{u}_e \, a \, \tilde{u}_K}{(a\tilde{u}_K + \tilde{u}_e[e'])^2} \cdot \frac{\partial[e']}{\partial x} = \frac{\partial[V_{Cl}^{\times}]}{\partial t} \quad \text{(equation 86)} \tag{88}$$

(The concentration of free electrons $[e']$ is supposed much smaller than the concentration of trapped electrons $[V_{Cl}^{\times}]$). When the electronic conductivity remains much smaller than the ionic conductivity :

$$a\tilde{u}_K >> \tilde{u}_e[e'] \tag{89}$$

equation (88) is identical to (77), if we take into account the definition of α. In general, equation (88) can be also written :

$$\frac{I}{F} \cdot \frac{\tilde{u}_e \, a \, \tilde{u}_K \, \alpha}{(a\tilde{u}_K + \tilde{u}_e[V_{Cl}^{\times}]\alpha)^2} \cdot \frac{\partial[V_{Cl}^{\times}]}{\partial x} = \frac{\partial[V_{Cl}^{\times}]}{\partial t} \tag{90}$$

From this expression we can deduce the progression rate v_C of a given concentration C in $V_{Cl}^{\times}$

$$v_c = -\left[\frac{\partial[V_{Cl}^{\times}]}{\partial t} \Big/ \frac{\partial[V_{Cl}^{\times}]}{\partial x}\right]_{[V_{Cl}^{\times}] = C} \tag{91}$$

$$v_c = \frac{|I|}{F} \cdot \frac{\tilde{u}_e \, a \, \tilde{u}_K \, \alpha}{(a\tilde{u}_K + C\,\tilde{u}_e\alpha)^2} \tag{92}$$

- v_c is a constant proportional to $|I|$
- it is the smaller, the higher the value of C. Therefore the front zone becomes more and more diffuse as it progresses in the material. The phenomenon is the more accentuated, the greater the conductivity ratio in the colored zone

$$\frac{\sigma_e}{\sigma_K} = \frac{\alpha\tilde{u}_e C}{a\tilde{u}_K} \tag{93}$$

- If this ratio remains noticeably smaller than 1, expression (92) can be simplified and calculated in accordance with the meaning of the different parameters. The usual expression is then obtained for the progression rate :

$$v = \alpha\tilde{u}_e F \frac{\partial\phi}{\partial x} \tag{94}$$

Virtual cathode

In the case of a progression rate limited by the electron supply rate, it is also directly calculable when the sample can be regarded as composed of three zones : a colored zone approximately homogeneous, an intermediate zone, and an unaltered zone. Then the mode of calculation sketched in fig. 9 can be applied to the intermediate zone. It gives :

$$-\frac{I}{F}\left[t_e(c) - t_e(u)\right] = \frac{dN(e_G)}{dt} \qquad (95)$$

where $t_e(c)$ and $t_e(u)$ are the electronic transport numbers in the colored and unaltered zone respectively and $\frac{dN(e_G)}{dt}$ the variation in the total number of free and trapped electrons in the intermediate zone. This formula is particularly useful when the colored material is predominantly an electronic conductor, i.e. when $t_e(c)$ is equal to 1.
If the usual assumptions hold: $t_e(u) \neq 0$, $N(e_G) \neq N_c$, where N_c is the number of color centers, considered as simply charged, then it simplies to :

$$-\frac{I}{F} = \frac{dN_c}{dt} \qquad (96)$$

This results in an average progression rate equal to :

$$v = \frac{|I|}{F\,C_\infty} \qquad (97)$$

where C_∞ is the color center concentration in the colored zone (the initial concentration is supposed to be nil). This formula has been verified in our laboratory on stabilized zirconia[116].

4 - EXPERIMENTAL CONDITIONS

Electrical conditions

Obviously it would be of great interest in many situations to perform the electrochemical coloration with a cell including a "reference electrode". Measuring the voltage between the working electrode and this electrode after interrupting the current gives valuable pieces of information on the coloration degree obtained[94]. The reference electrode which is supposed to fix the Fermi level in the vicinity of its contact in a stable position, can be either a third electrode specially devised for this purpose[115] or the counter electrode itself, if it is sufficiently reversible[94]. Unfortunately the interest of this procedure has been rarely recognized, and most experiments have been performed with simple-two-electrode cells.

As previously seen (relation 43) the current density which flows through the sample is an essential parameter which determines the degree of coloration reached in the material in contact with the electrode. The importance of this parameter has been stressed by Casselton[13].

To estimate the evolution of the electrochemical cell characteristics $U(I)$, we can, as a first approximation, refer to the simple expression :

$$U = E + RI \qquad (98)$$

where the electromotive force E can be approximated by the expression :

$$E = \frac{RT}{F} \ln \frac{[e']_1}{[e']_o} \qquad (99)$$

$[e']_1$ and $[e']_o$ are the free electron concentrations at the two electrodes. The effects of the space charges should also be taken into account in E ; they are frequently small when the process is progressing. The magnitude of the average local resistivity of the material can be estimated from the relevant diagram (Fig. 3) and the variation rate of R from the progression rate of the colored zone. In fact, defining an average resistivity in a non homogeneous material may imply a very large inaccuracy[98]. The utilization of formula (98) is only of interest in situations where the approximations previously described are appropriate. In the first case the term RI is well determined by equation (83). In the second case the resistance can be simply decomposed in accordance with the model :

$$R = R_c + R_i + R_u \qquad (100)$$

where R_e R_i R_u are the resistances of the colored, the intermediate and the unaltered zone. They can be expressed as a function of the length of the relevant zones. In simple cases this yields an expression of the type :

$$R = p + qvt \qquad (101)$$

where p and q are two constants and v the progression rate.

Size of the electrode

It is sometimes assumed that a point electrode is more favourable than a plate. Under this condition the local electric field is higher and the coloration, which is assumed to result from an electron injection, is facilitated. According to our description, the main parameters to be considered at the phase boundary are the current density and the fluxes of impurities which tend to adsorb on the crystal surface. A point electrode is obviously also more favourable for the obtention of high current densities. It favors a reduction of the impurity fluxes when it is more or less embedded in the ionic material. As was shown by Heiland[11] and many others, the electrochemical coloration can be easily obtained also on a large surface electrode provided its environment is sufficiently inert and the ionic conductivity of the crystal sufficiently high to allow noticeable current densities. In the literature quoted, various particular sizes and shapes are reported. They do not appear to be essential factors.

Shape of the coloration front

From a microscopic analysis it has been previously shown that the coloration front is more or less diffuse. A macroscopic observation also indicates that the coloration front may take particular shapes. The electrode size is obviously a determining factor in this regard. Gokhshtein[24] has demonstrated that a tongue shape is predictable when the colored material is much more conductor than the initial material. The current density is higher at the extremity of the tongue where the electric path is shorter and the phenomenon tends to accentuate. It must also be mentioned that reactions at the free surfaces of the sample may modify the shape of the colored zone. Volatilization of the metal excess in the alkali halides and reoxidation by oxidizing species present in the surrounding atmosphere are the more usual reactions of this type.

Observation techniques

Numerous properties, mostly optical and spectroscopic, have been used to identify the color centers produced by electrochemical coloration and their distribution in the material. Most of the measurements were performed on quenched samples. Their basic principles and techniques are detailed in an abundant literature[67]. In situ measurements, during the coloration process, are far less numerous.

- A direct optical observation of the colored zone is sometimes possible through a window in the experimental vessel. The most spectacular observations of this type were made on alkali halides. An apparatus based on this principle was even devised for teaching demonstrations[64].
- Some E.P.R. measurements have been carried out under working conditions[49].
- As evident in formula (101) interesting items of information are provided by the sample electric resistance. It is approximately determined from simultaneous measurements of U and I, more accurately by using the interruption technique, and in a more sophisticated manner, by superimposing a small a.c. current to the bias.
- More detailed information is obtained with appropriate probes located along the sample. This technique was extensively used by Casselton[13] on zirconia. It was also applied on several other materials[5,6,15,36]. A stringent requirement of this technique is the inertness of the surrounding atmosphere. There must be no noticeable electron electrochemical potential gradient between the bulk of the material, where the coloration is occurring, and the surface where the resulting Fermi level shift is measured. The sample must also be appropriately shaped to minimize the tendency to form a

tongue-shaped colored zone. Figure 10 shows a typical probe-potential-versus-time plot obtained in our laboratory. The stabilized zirconia sample was a square-section bar 3x3x12 mm. The probe formed of a platinum wire, was located 3 mm from the injecting electrode. Its potential was measured with respect to an air reference electrode. The curve clearly shows the arrival, at probe level, of two coloration stages characterized by two "waves". Such electronic probes measure the electron electrochemical potential variations $\Delta\tilde{\mu}_e$ which accompanied the passage of the coloration front

$$\Delta\tilde{\mu}_e = \Delta\mu_e - F\Delta\phi \qquad (102)$$

In practice, the measurement is especially sensitive when the chemical term $\Delta\mu_e$, generally smaller than one volt, is accurately measurable. This cannot be the case when cell voltages of several hundred volts are used . When the "waves" are not easily detectable because of too high cell voltage, the method may be still appropriate if the coloration is accompanied by a drastic electric resistance variation. The passage of the coloration front is then evidenced by a knee on the E(t) plot (cf. formulae 101).

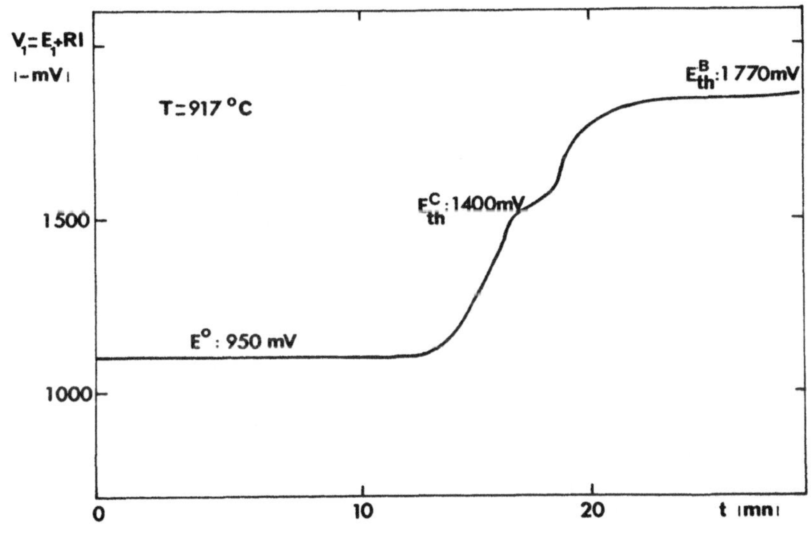

Fig. 10 - *Probe-potential variation during coloration of stabilized zirconia*

The method can be directly used for measuring the progression rate. It also provides information on the coloration stage that has been reached (cf. Fig. 10) and on the electric resistance variations[13].

-Jacquin and co-workers[18] assembled the electrochemical cell in a thermobalance. This allowed them to measure the losses in weight during the coloration progression simultaneously with the electrical

characteristics.
- Following a parallel route, Fouletier and co-workers[116] coupled the electrochemical cell with an oxygen pump and gauges and simultaneously measured the amount of matter being extracted from the sample. These last two methods are obviously applicable when the changes in composition correspond to a measurable quantity of matter.

5 - APPLICATIONS

So far, electrochemical coloration has not been put to use in many applications. Recently, black zirconia has been proposed as an "electrically controllable and renewable oxygen getter"[116-117]. It exhibits an oxygen affinity close to that of titanium and has a reasonable oxygen capacity. The getter is simply renewed by electrochemical reduction, i.e. by applying a suitable voltage between two electrodes deposited on it. Its electrical resistance varies sharply near the saturation point. So, measurements of the electric resistance between the same two electrodes provide an easy test of getter efficiency.

Other potential applications are based on the general band schema of an ionic conductor. As recalled in another paper[93], under open circuit conditions, the band edges tend to be "horizontal" and the material can be regarded as a rigid framework, unaltered by variations in the position of the Fermi level (within the limits which correspond to a predominant ionic conductivity). In an electrochemical cell, the measured voltage is then simply related to the Fermi level positions with respect to the band edges μ_e, in the vicinity of the electrodes, by the expression

$$E = a - \frac{1}{F}(\mu_e^1 - \mu_e^2) \qquad (a = C^t) \tag{103}$$

If the experimental conditions can be appropriately selected so that μ_e^2 is a constant, then the variations of E are direct measurements of the variations of μ_e in the material in contact with the other electrode . This property can be put to use to obtain information on the electronic structure of the material. A typical experiment is performed, first by electrochemically reducing a small zone of the material and then by recording the voltage during the chemical reoxidation by some oxidizing species. The curve obtained has a shape similar to that plotted in figure 10 (the only differences are due to the additionnal $F\Delta\phi$ term (eq. 102) and sometimes different local oxidation or reduction rates). It shows "waves" occurring at definite potentials. A simple calculation based on the previously developed description indicates that these waves must appear at the equivalent points (fig. 1 - 5 - 6) or at peculiar points (point $\mu_e(T)$ in fig. 5). The equivalent-point potentials thus obtained determine the redox stability intervals, the interest of which has been previously recalled. Such plottings were systematically carried out on a few materials [94,118]. Measurements were also performed over ranges wider than those delineated by the basic requirement of a predominant ionic conductivity. The equivalent-point-potentials measured under these conditions were certainly erroneous but they appeared not to depend upon the measuring conditions and to be characteristic of the material. They allowed us to determine the redox stability intervals over an extended range where the potential scale is unknown but fixed. The additional data are very useful in practice to predict the risks of errors in the measurements with solid state electrochemical cells[94].

Another potential application predicted by the previous model, could be of interest. A solid state cell in which a certain part of the ionic conductor has been colored generally exhibits an e.m.f and can be regarded as a battery. As shown by Heiland[11] the electrochemical coloration is reversible and such a battery can be discharged. This possibility has been demonstrated by Thiessen[119]. A more typical situation is obtained with an ionic conductor which can be electrochemically reduced or oxidized. When it is inserted between two electrodes, both reactions can occur simultaneously : oxidation at the anode and reduction at the cathode. According to our model the material can then be regarded as a double reservoir in which electrons can be stored at a certain level and holes at another, in two different parts of the material. The voltage of the corresponding cell simply equals the distance between the two levels to within the multiplication factor F^{-1} [93]. Its electrical capacity is determined by the number of electrons and electron holes which can be entrapped. The "electrode reactions" occurring at the phase boundary of such a cell simply imply an injection or extraction of electronic carrier and no transfer of matter. This would certainly obviate many difficulties encountered in more traditional batteries. The validity of the principle was demonstrated in our laboratory with KI. The voltage obtained was approximately 1.3 V. For practical purposes the KI cell is obviously not of interest since its electrical capacity is too small.

REFERENCES

1. M.N. Popova, G.B. Semushkin, A.N. Tsikin, Izvest. Akad. Nauk SSSR Ser. Fiz. 29 82 (1965)
2. C.J. Kevane, E.L. Holverson, R.D. Watson, J. Appl. Phys. 34 2083 (1963)
3. W.E. Danforth, Phys. Rev. 86 416 (1952)
4. S.P. Mitoff, R.J. Charles, J. Appl. Phys. 43 927 (1972)
5. J. Felsche, J. Lietz, Neues Jahrb. Mineral. Abh. 109 238 (1968)
6. J. Felsche, J. Lietz, Neues Jahrb. Mineral. Abh. 109 250 (1968)
7. H.L. Blood, S. Levine, N.H. Roberts, J. Appl. Phys. 27 660 (1956)

8. Z. Ogorelec, Fizika 3 41 (1971)
9. B. Collen, Acta Chem. Scand. 18 839 (1964)
10. P.W. Pohl, Proc. Phys. Soc. 49 3 (1937)
11. G. Heiland, Z. Physik 127 144 (1950)
12. N. Karabascheff, Z. Physik 118 718 (1941)
13. R.E.W. Casselton, J. Appl. Electrochem. 4 25 (1974)
14. G. Beranger, P. Desmarescaux, P. Lacombe, C.R. Acad. Sci. Paris 259 1961 (1964)
15. Ya P. Gokhshtein, R.A. Khaikin, High Temp. 7 965 (1969)
16. R.E.W. Casselton, J.S. Thorp, D.A. Wright, Proc. Brit. Ceram. Soc. n° 19 265 (1971)
17. R.E.W. Casselton, J. Penny, M.J. Reynolds, Trans. J. Brit. Ceram. Soc. 76 115 (1971)
18. M. Jacquin, M. Guillou, J. Millet, C.R. Acad. Sci. Paris 264 2101 (1967)
19. J.P. Loup, Z. Mihailovic, P. Morvan, C.R. Acad. Sci. Paris 261 109 (1965)
20. G. Beranger, P. Lacombe, Rev. Int. Hautes Temp. Refract. 3 235 (1966)
21. M. Jacquin, J. Millet, C.R. Acad. Sci. Paris 266 25 (1968)
22. A. Wilcockson, R.E.W. Casselton, J. Amer. Ceram. Soc. 53 293 (1970)
23. M. Guillou, J. Millet, M. Asquiedge, N. Busson, M. Jacquin, S. Palous, M. Pithon, A. Lecante, Rev. Int. Hautes Temp. Refract. 4 273 (1967)
24. Ya P. Gokhshtein, Dokl. Phys. Chem. 190 10 (1970)
25. L.N. Popova, L.M. Demidenko, D.N. Poluboyarinov, R.Ya Popil'skii V.S. Barunov, Tr. Mosk. Khim. Tekhnol. Inst. 63 132 (1969)
26. W.E. Danforth, F.H. Morgan, Phys. Rev. 79 142 (1950)
27. J. Besson, C. Deportes, M. Kleitz, Journées Piles à Combustibles, Bruxelles (june 1965)
28. R.E.W. Casselton, Int. Atom. Energy Agency Symp. MHD Electr. Power Gener. (july 1968)
29. R.E.W. Casselton, Electromotive Force Measurements in High Temp. Systems, Ed. C.B. Alcock, Inst. Mining Metal London p. 151 (1968)
30. S.K. Adams, R.E.W. Casselton, J. Amer. Cer. Soc. 53 117 (1970)
31. M. Guillou, J. Millet, S. Palous, Electrochem.Acta 13 1411 (1968)
32. M. Guillou, A. Lecante, J. Millet, Rev. Gen. Elec. 77 1051 (1968)
33. J. Millet, M. Jacquin, C.R. Acad. Sci. Paris 269(c) 917 (1969)
34. Ya P. Gokhshtein, A.A. Safonov, High Temp. 8 368 (1970)
35. Ya P. Gokhshtein, A.A. Safonov, Dokl.Acad.Nauk SSSR 195 109 (1970)
36. Ya P. Gokhshtein, A.A. Safonov, Zh. Fiz. Khim. 45 1593 (1971)
37. E.N. Belova, Ya P. Gokhshtein, A.A. Safonov, Zh. Fiz. Khim. 46 2672 (1972)
38. P.A. Tikhonov, A.K. Kuznetsov, E.K. Keler, M.D. Krasil'nikov, Fiz. Khim. Elektrokhim. Rasplavl. Solei Tverd. Elektrolitov n° 1 154 (1973)
39. J.L. Weiniger, D. Zemany, J. Chem. Phys. 22 1469 (1954)
40. J.S. Thorp, A. Aypar, J.S. Ross, J. Mater. Sci. 7 729 (1972)

41. D.A. Wright, J.S. Thorp, A. Aypar, H.P. Buckley, J. Mater. Sci. 8 876 (1973)
42. F.K. Fong, J. Chem. Phys. 41 2291 (1964)
43. F.K. Fong, R.C.A. Rev. 25 303 (1964)
44. H. Guggenheim, J.V. Kane, Appl. Phys. Letters 4 172 (1964)
45. V.F. Pisarenko, V.V. Popov, Vop. Radiofiz Spektrosk 3 80 (1967)
46. D.E. Brodie, J.W. Coburn, R.R. Haering, Can. J. Phys. 48 759 (1970)
47. K. Kubo, J. Phys. Soc. Japan 21 1046 (1966)
48. M. Regis, C.R. Acad. Sci. Paris 269(c) 1590 (1969)
49. M. Regis, J. Phys. Chem. Solids 33 1997 (1972)
50. C.J. Kevane, Final Rept. Feb. 28 (1962) Unclassified A.D.274510 Armed Serv.Techn. Inf. Agency
51. I.V. Vinokurov, Z.N. Zonn, V.A. Ioffe, Sov. Phys. Solid State 9 2659 (1968)
52. C.D. Cox, P.C. Eastman, J. Can. Ceram. Soc. 38 75 (1969)
53. B. Henderson, A.C. Tomlinson, J. Phys. Chem. Solids 30 1801 (1969)
54. M. Hacskaylo, G. Groetzinger, Phys. Rev. 87 789 (1952)
55. G.A. Andreev, G.B. Semushkin, A.N. Tsikin, Soviet Phys. Solid State 9 2564 (1968)
56. Ya N. Pershits, A.N. Il'in, Sov. Phys. Solid State 10 719 (1968)
57. Ya N. Pershits, V.A. Andrusich, Fiz. Tverd. Tela 13 280 (1971)
58. D.G. Sootha, V. Mitra, Physica (Utrecht) 46 531 (1970)
59. N. Paramo, C. Sanchez, J. Phys. Soc. Japan 30 1106 (1971)
60. M.T. Montojo, C. Sanchez, Solid State Comm. 14 485 (1974)
61. M.T. Montojo, C. Sanchez, J. Phys. Chem. Solids 35 1437 (1974)
62. V.F. Pisarenko, Soviet Phys. Dokl. 2 391 (1957)
63. V.D. Belyaev, G.B. Semushkin, A.N. Tsikin, Soviet Phys. Solid State 10 921 (1968)
64. H.N. Hersh, L. Bronstein, Am. J. Phys. 25 306 (1957)
65. Y. Uchida, Y. Nakai, J. Phys. Soc. Japan 8 795 (1953)
66. Y. Uchida, Y. Nakai, J. Phys. Soc. Japan 9 928 (1954)
67. J.H. Schulman, W.D. Compton, Color Centers in Solids, Pergamon Press N.Y. (1963)
68. Ya N. Pershits, A.N. Il'in, Sov. Phys. Solid State 11 631 (1969)
69. Ya N. Pershits, M.G. Pinsk, Sov.Phys. Solid State 11 2714 (1970)
70. V.G. Landau, G.B. Semushkin, A.N. Tsikin, Sov. Phys. Solid State 11 1951 (1970)
71. G.B. Semushkin, A.N. Tsikin, Sov.Phys. Solid State 12 492 (1970)
72. V. Topa, M. Yuste, Phys. Status Solidi A 3 131 (1970)
73. V. Topa, S. Ioan, B. Velicescu, G. Mitroaica, Rev. Roum. Phys. 18 571 (1973)
74. R.C. Tyagi, R. Singh, A.K. Gupta, J. Phys. C 3 769 (1970)
75. C. Ruiz Mejia, Rev. Mex. Fis. 19 319 (1970)
76. V. Zhelev, G. Todorov, Phys. Status Solidi A 3 195 (1970)
77. Yu I. Gritsenko, Phys. Status Solidi A 7 K113-K116 (1971)
78. I. Besoi, An. Univ. Timisoara, Ser. Stiinte Fiz. Chim. 9 97 (1971)
79. P. Görlich, H. Karras, R. Lehmann, Phys. Status Solidi 3 98 (1963)
80. J. Hladik, Physics of Electrolyte tome 2 p. 867, Ed. J. Hladik Acad. Press N.Y. (1972)

81. J. Dupuy, J. Bert, 24th Meeting ISE (1973)
82. C. Deportes, private communication
83. G. Hetherington, K.H. Jack, M.W. Ramsay, Phys. Chem. Glasses 6 6 (1965)
84. T. Dunn, G. Hetherington, K.H. Jack, Phys. Chem. Glasses 6 16 (1965)
85. J.K. Higgins, F.V. Roue, J. Appl. Electrochem. 3 201 (1973)
86. G.C. Walther, L.L. Hench, Phys. of Electronic Ceramics Part A; L.L. Hench, D.B. Dove Edit. Marcel Kekker (1971)
87. J.R. Hanscomb, K.C. Kao, J.H. Calderwood, O'Dwyer, P.R. Emtage Proc. Phys. Soc. 88 425 (1966)
88. J.J. O'Dwyer, The Theory of Electrical Conduction and Breakdown in Solid Dielectrics, Clarendon Press Oxford (1973)
89. J.W. Hinze, J.A. Baker, J.W. Patterson, J. Electrochem. Soc. 121 1328 (1974)
90. M. Manfredi, C. Paracchini, Phys. Status Solidi A 21 427 (1974)
91. M.A. Lampert, Phys. Rev. 103 1648 (1956)
92. J.B. Wagner, this book
93. M. Kleitz, P. Fabry, E. Schouler, this book
94. P. Fabry, M. Kleitz, C. Deportes, J. Solid State Chem. 6 230 (1973)
95. F.A. Kröger, The Chemistry of Imperfect Crystals, North-Holland Publ. Comp. Amsterdam (1964)
96. M. Kleitz, P. Fabry, E. Schouler, Fast Ion Transport in Solids W. van Gool Edit. North Holland (1973)
97. J.W. Patterson, J. Electrochem. Soc. 118 1033 (1971)
98. M. Kleitz, Thesis Grenoble France (1968)
99. M. Gouet, Thesis Ing. Doct. Paris VI France (1972)
100. L.G. Harrison, M. Prasad, J. Chem. Soc. Faraday Trans. 1-70 471 (1974)
101. R.E. Seevers, A.R. Scott, J. Phys. Chem. Solids 31 729 (1970)
102. D. Durand, Thesis Lyon France (1974)
103. D. Durand, G. Chassagne, J. Serughetti, J. Phys. (Paris) Colloq. (9) 465 (1973)
104. S. Lanyi, V. Harasta, Czech. J. Phys. 23 1084 (1973)
105. D.O. Raleigh, Phys. Stat. Solidi 4 215 (1971)
106. Kong Sop Song, J. Phys. Chem. Solid 31 1389 (1970)
107. P.V. Sastry, J. Phys. Soc. Japan 26 73 (1970)
108. V.I. Gotlib, S.A. Bolsun, K. Svarcs, Latv. PSR Zinat. Akad. Vestis, Fiz. Tech. Zinat. Ser. 1 25 (1972)
109. D.I. Vaisburd, A.A. Vorob'ev, A. Yu Gutman, Izv. Vyssh. Ucheb. Zaved. Fiz. 13 159 (1970)
110. G.C. Kuczynski, H. Ichinose, B. Swaroop, Proceedings British Ceram. Soc. 1 129 (1964)
111. G.C. Kuczynski, J.J. Byun, Phys. Status Solidi B 50 367 (1972)
112. L.M. Shamovskii, A.A. Dunina, M.I. Gosteva, Soviet Phys. Solid State 2 2252 (1960)
113. N.F. Mott, R.W. Gurney, Electronic Processes in Ionic Crystals Oxford University Press (1948)
114. L.M. Samovskii, L.I. Rybakova, M.I. Gosteva, Dokl. Akad. Nauk SSSR 91 67 (1953)

115. J. Hladik, Physics of Electrolyte tome 2 p. 839, Ed. J.Hladik Acad. Press N.Y. (1972)
116. J. Fouletier, M. Kleitz, Vaccum 25 307 (1975)
117. M. Kleitz, P. Fabry, J. Fouletier, A.N.V.A.R n° 73 11572 (30 Mars 1973)
118. A. Pelloux, P. Fabry, C. Deportes, C.R. Acad. Sci. Paris 276 241 (1973)
119. K. Thiessen, Monatsber. Deut. Akad. Wiss. Berlin 12 587 (1970)
120. S.N. Koikov, A.N. Tsikin, A.S. Shakirova, Sov. Phys. Solid State 9 2777 (1968)

DISCUSSION

VOÏNOV : β-Al_2O_3 is also known to darken when electrolysed but this darkening is generally attributed to sodium dendrite growth along grain boundaries. Could it be that colour centers are also injected in β-Al_2O_3 ?

FABRY : Very little is known about the electronic structure of β-alumina. To our knowledge no point defect capable of entrapping an electron or hole in β-Al_2O_3 has been reported in the literature.

ETSELL : Could you tell me what caused the initial plateau (at $\sim$ 1.4 V) in your first I-V curve and, secondly, am I correct in assuming that you have not detected zirconium metal near the cathode since, under conditions of thermodynamic equilibrium which should be approached at the low current densities you used its formation is not possible ?

FABRY : At the present time we don't know exactly which point defect is involved in the plateau at 1.4 V.

The formation of zirconium metal appeared possible only under high current densities on sintered samples. The metal formed at the grain boundaries.

PIZZINI : Did you observe any change of mechanical properties (embrittlement) during coloration ? Reoxidation could in the case of embrittlement, restore also the mechanical properties.

FABRY : The change of mechanical properties of black zirconia was studied by J.M. Farley and co-workers (J. Mater. Sci. 7 475 (1972)). For coloured alkali halides see for instance ref. 67. If the reduction of stabilized zirconia is performed under low current densities the phenomena is reversible during numerous cycle (ref. 116). After reduction under high current densities zirconia is destroyed during reoxidation or becomes very brittle.

HIGH TEMPERATURE FUEL CELLS AND ELECTROLYSERS

B.C.H. Steele

Imperial College, London, S.W.7

1. INTRODUCTION

Electrochemical energy convertors have many attractive features of which the most important is their ability to avoid the Carnot limitation inherent in all heat engines. A variety of fuel cells have been devised and adequately described in relevant books and reviews[1-4], and the success of the Pratt and Whitney engineered hydrogen/oxygen fuel cell in the Apollo manned lunar exploration programme is well known. More recently the proposals for a 'hydrogen economy'[5] as a means of distributing energy from nuclear power complexes, and to supplement the diminishing resources of hydrocarbon fuels has provided a stimulus to the development of new types of electrolysers. Both these applications of electrochemical energy converters will be considered in the present paper, but discussion will be limited to high temperature systems incorporating solid oxide electrolytes, and the other principal type of high temperature fuel cell incorporating molten carbonate electrolytes is not included.

The maximum theoretical efficiency of a fuel cell is given by the relationship,

$$\mathcal{E}_{max} = \frac{W_{rev}}{\Delta H} = \frac{\Delta G}{\Delta H} = \frac{n\,F\,V_e}{\Delta H}$$

where the symbols ΔG, ΔH, n, F, have their usual meaning and V_e refers to the reversible equilibrium cell voltage. However the actual operating voltage (V_i) will be lower than V_e due to polarisation losses (η_a, η_c) at both electrodes and to resistive losses (IR) within the electrolyte phase,

M. Kleitz and J. Dupuy (eds.), Electrode Processes in Solid State Ionics, 367–386.

$$\text{i.e., } V_i = V_e - \eta_a - |\eta_c| - IR$$

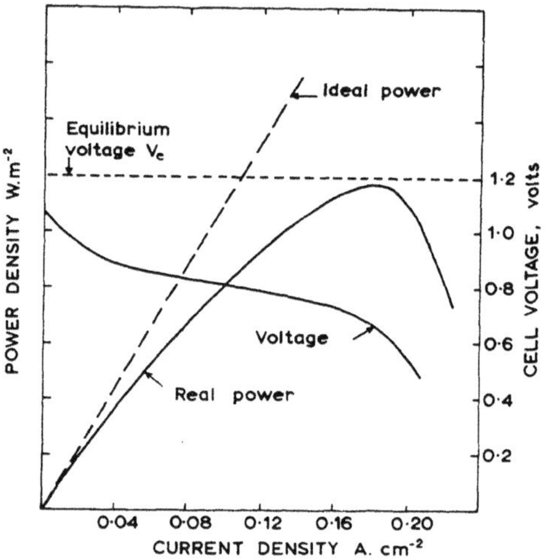

Fig. 1. Voltage-current and power-current relationships for hydrogen-oxygen fuel cell.

Figure 1 depicts a typical curve showing the relationship between the operating voltage (V_i) and current density. Also included in the diagram are curves for the ideal power(V_e x I) and actual power (V_i x I) output, and it is immediately obvious that a real fuel cell produces maximum power at a particular current density. The relationship between power output and current density is thus an important design parameter.

When an electrochemical energy converter is being operated as an electrolyser the various losses necessitate an increase in the voltage required to operate the cell according to the expression,

$$V_i = V_e + \eta_a + |\eta_c| + IR,$$

and thus the overall efficiency of the electrochemical process is again decreased.

The following sections summarise the various factors, including, properties and selection of cell component materials, cell geometry and fabrication, and electrode kinetics, that can influence the performance of electrochemical energy converters.

2. MATERIALS PROPERTIES AND SELECTION

a) Ionic Conductivity of Solid Electrolytes

The properties of solid oxide electrolytes have been extensively reviewed[6], and figure 2 contains a summary of the ionic

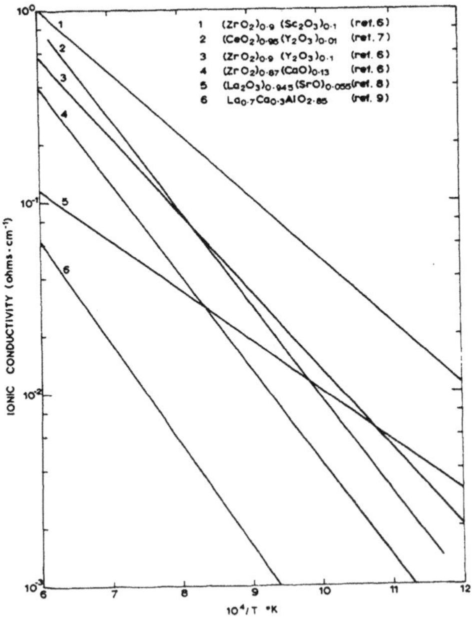

Fig. 2. Ionic conductivities of selected oxide electrolytes as a function of temperature.

conductivity of selected oxide electrolytes as a function of temperature. The electrolyte material must satisfy several criteria. In addition to possessing high oxygen ion conductivity and low electronic conductivity the fabrication techniques (see section 2c) usually require the electrolyte to be sintered to at least 95% theoretical density at relatively low temperatures ($< 1500^{\circ}C$) to produce dense strong impermeable components. In addition the isothermal ionic conductivity should be independant of oxygen partial pressure over the wide range encountered which may be as much as 20 orders of magnitude. Oxide electrolytes which best satisfy these criteria are based on zirconia solid solutions and most of the research and development on high temperature cells has favoured the use of yttria-stabilized zirconia (YSZ). Although scandia and ytterbia stabilized zirconia have higher conductivities the cost and availability of these oxides ensure that yttria is the preferred oxide. Both calcia[11] and yttria[12] stabilized zirconia possess the unfortunate characteristic of exhibiting a slow increase in resistivity with time. This 'ageing' phenomenon is associated with a complex ordering process within the highly defective crystal lattice but it appears that with a sufficiently fine grain size[13] the ageing rate can be reduced to a level which does not seriously impair the long time performance of the cell. For maximum ionic conductivity the optimum amount of yttrium oxide to be added to the zirconium oxide host lattice is in the range 8 - 10 mole %, the precise value being selected according to the expected temperature of operation of the cell.

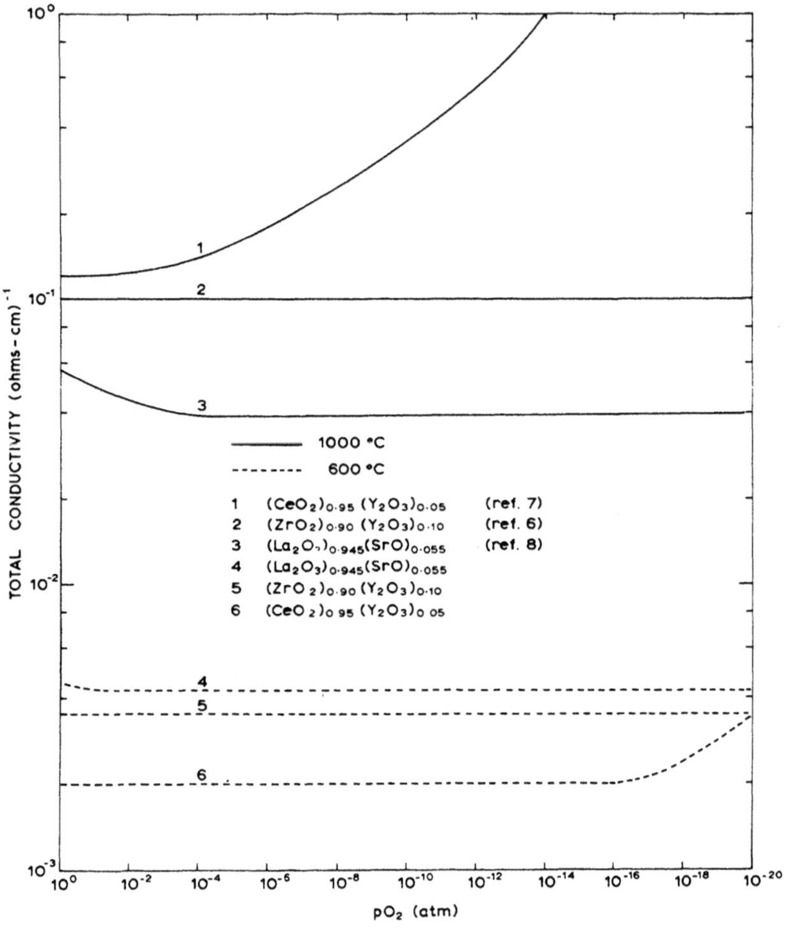

Fig. 3. Total conductivities of selected oxide electrolytes as a function of oxygen partial pressure.

Other oxide solid solutions have been examined as potential solid electrolyte materials. Ceria solid solutions[7,14] for example, possess high oxygen ion conductivities but the relative ease of reduction of the Ce^{4+} ion ensures that the ionic transference number (t_{ion}) only approaches unity over a very narrow range of oxygen partial pressure ($1 > pO_2$ atm.$> 10^{-2}$) at 1000°C. This property is clearly evident in figure 3 which depicts selected values of t_{ion} as a function of oxygen partial pressure. It is worth noting incidentally, that in contrast to the high temperature behavior, t_{ion} for $(CeO_2)_{0.95}(Y_2O_3)_{0.05}$ at 600°C remains at unity over an extensive range of oxygen partial pressures. The ionic conductivity of trivalent oxides such as Sc_2O_3, Y_2O_3, Sm_2O_3, La_2O_3, doped with CaO or SrO, has also been measured[8,15] to assess their application as fuel cell electrolytes. However, with one exception the values for ionic conduction at 1000°C are two to three orders of magnitude too low. The exception are the La_2O_3 - SrO solid solutions, and conductivity data for the composition $(La_2O_3)_{0.945}$ $(SrO)_{0.055}$ are included in figures 2 and 3. At 900°C this composition has a conductivity comparable with $(ZrO_2)_{0.87}$ $(CaO)_{0.13}$ and at lower temperatures (~600°C) the conductivity exceeds that of YSZ. The disadvantage of this material for fuel cell operation is the relative lack of stability of the individual oxides towards water vapour and carbon dioxide which can complicate the fabrication procedure. It is very interesting, however, to note the relatively low value of 0.55 eV for the activation energy for migration of oxygen ions in this material compared with the values of 0.75 - 1.00 eV normally associated with zirconia solid solutions.

Perovskite solid solutions such as $La_{0.7}Ca_{0.3}AlO_{2.85}$ have also been examined[9,15] but the ionic conductivity results are not encouraging (see fig. 2) and there is no evidence to suggest that alternative perovskite solid solutions will provide improved solid electrolytes for fuel cell operation compared to the existing zirconia based materials.

b) Electrode materials

The selection of appropriate electrode material is principally influenced by the oxygen partial pressure expected to be in contact with the electrode. For the reducing conditions imposed by the fuel at the anode in a fuel cell or at the cathode in an electrolyser then it is possible to use metallic electrodes. Taking into consideration such factors as cost, vapour pressure, ease of fabrication, then it appears that the most favoured metals are Co and Ni. There is one disadvantage associated with using Co and Ni in that each has a coefficient of thermal expansion substantially larger than YSZ, and thus tends to become detached from the electrolyte during thermal cycling. Fortunately this can be overcome[16] by incorporating the metal in a porous matrix of YSZ, a procedure which also inhibits the agglomeration of the fine metallic particles. These metallic electrodes can also be used in electrolysers as depolarised anodes[16] where the atmosphere is still reducing due to the presence of CO/CO_2 mixtures.

It has proved much more difficult to select satisfactory electrode materials for the oxidising environments prevailing at the cathode in fuel cells and at the anode in electrolysers. The noble metals are too expensive and it is generally agreed that the only practical class of materials to consider for the relevant electrodes are electronically conducting oxides.

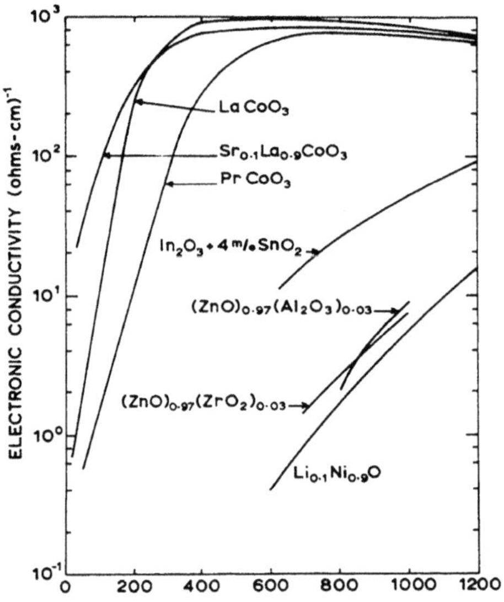

Fig. 4. Electrical conductivity as a function of temperature for various oxides proposed for use as cathodes (Taken from ref. 18).

Electrical conductivity data for the principal oxides that have been examined are summarised in figure 4, and it is apparent that several materials exhibit electronic conductivities of the required magnitude. However, only Sn doped In_2O_3 is without any obvious limitations. Li-doped NiO is subject to Li loss and thus rapid deterioration of electrical properties at typical operating temperatures. Al - and Sr - doped ZnO are unsatisfactory at temperatures in the vicinity of 1000°C when used in conjunction with zirconia based electrolytes due to rapid attack of the electrolyte by ZnO. La CoO_3 and Sr-doped La CoO_3 also react with the zirconia electrolyte to form electronically insulating layers at the electrode-electrolyte interface. Finally Pr CoO_3 has a thermal expansion coefficient some 2.5-3 times larger than that of zirconia and thus tends to become detached during thermal cycling. Although Sn-doped In_2O_3 apparently suffers none of these disadvantages there is no published information regarding long term cell performance, and it could be that this material is also subject to limitations. At present, however, it appears to be generally regarded as the best electrode material available for use in oxidising environments.

In the assembly of multi-cell devices certain designs (see section 2c) require a suitable material for a series connector between, for example, the metallic anode and oxide cathode of a fuel cell. In addition to being compatible with both the anode and cathode this intermediary material must also be capable of satisfying additional stringent requirements. It must be stable in both an oxidising environment and also in the reducing atmosphere imposed by the fuel gas because of the high permeability of nickel and cobalt to hydrogen; it must possess good electrical conductivity over an extensive range of oxygen partial pressure; and it should not exhibit any phase changes throughout the range of temperatures encountered during the operation of the cell. It appears that the most promising materials for this application are spinels such as $Co\,Cr_2O_4$ doped with a few per cent of Mn, or TiO_2 containing additions of Cr, Mn, Nb, or W.

c) Cell design and fabrication

The geometrical arrangement of high temperature fuel cells is dictated by the need to keep resistive losses to a minimum. For example at 1000°C YSZ has a resistance of about 10 ohm-cm., and so for current densities of 0.5 A/cm^2 the ohmic losses will be represented by (5 x t) volts, where t is the thickness of the electrolyte in cms. Thus if impermeable electrolyte layers of 0.01cm (100μm) thickness could be fabricated then the associated resistive losses would only be 50 mV. Two main approaches have been adopted to permit the fabrication of thin electrolyte components, and these are depicted schematically in figure 5. In the first design a continuous tube of porous substrate material (e.g. alumina, or stabilized zirconia) forms the structural element upon which the fuel cell components are constructed. Anode, electrolyte, cathode, insulating and connecting materials are successively deposited onto this substrate with appropriate masking between each step to produce a number of series connected cells as depicted. The advantage of this type of arrangement is that very thin films ($<$100 μm) of the cell component materials may be used since the porous substrate provides the mechanical support. An alternative approach is to make the anode-electrolyte-cathode assembly self-supporting. Successive deposition of the components is carried out on a metallic mandrel (e.g. aluminium) which is subsequently removed by leaching to produce the arrangement shown in figure 5(b). With this type of assembly the electrode and electrolyte layers must be thick enough to provide overall mechanical stability. At present it is difficult to decide which of these two types of multi-cell construction

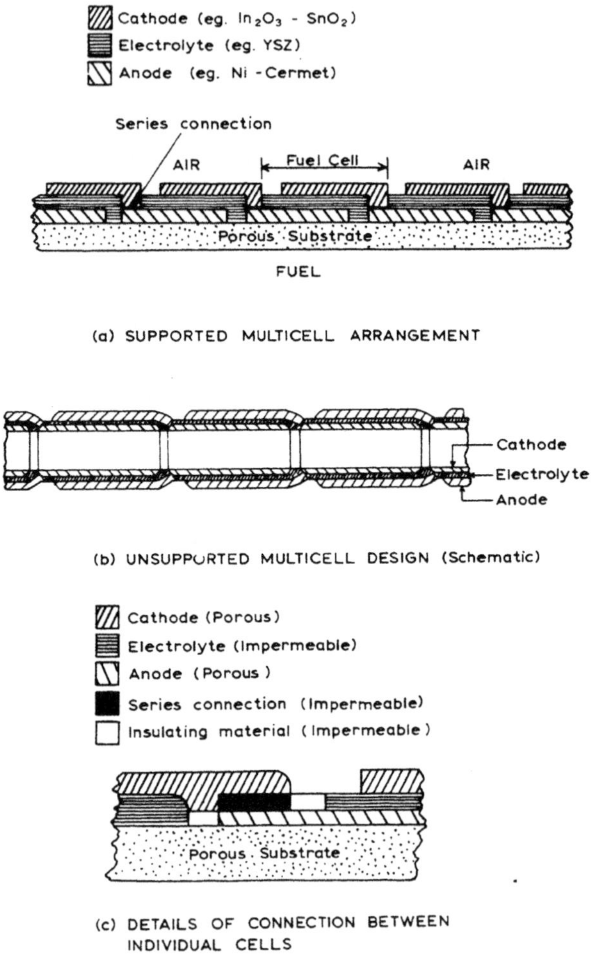

Fig. 5. Design features of multi-cell assemblies.

represents the best compromise for practical fuel cell designs. The supported type allows the use of very thin layers but is subject to additional polarisation losses arising from restricted mass transfer through the porous substrate. On the other hand the self supported design requires thicker component layers and thus material costs are higher and electrolyte resistance larger.

The fabrication of the multi-cell assemblies is usually accomplished by a plasma-spraying procedure, and preparation of the powders into an appropriate form is an essential pre-requisite for this technique. It is usual to produce agglomerates of the fine powder by either spray-drying or sol-gel processes. Two particular features of the current fabrication technology should be mentioned. The first concerns the preparation of the Ni-YSZ cermet electrodes, which involves mixing NiO with YSZ in the desired ratio followed by spray-drying of the powder. After plasma spraying the NiO in the electrode is reduced in H_2 at about 400°C to produce the Ni-YSZ electrode. The use of very fine particulate NiO, rather than Ni powder, results in much finer Ni grains within the reduced electrode thus improving the electrode kinetics. The second feature concerns the sintering of the YSZ electrolyte. As mentioned earlier the deposition of successive layers of cell components requires that the electrolyte powder after plasma-spraying should be capable of being sintered into an impermeable layer at temperatures below about 1400°C to avoid excessive reaction and interdiffusion of the various cell components. This is usually accomplished by the addition of a sintering agent such as cobalt oxide (~1 mol %). This addition enables YSZ to be sintered to about 95% theoretical density at temperatures as low as 1250°C. The addition of the cobalt does not measurably contribute to the electronic conductivity of the electrolyte down to oxygen partial pressures as low as 10^{-20} atm. Iron oxide has also been used as a sintering agent but the iron can react with the carbon atoms in the fuel gas (cf. reaction of CO with blast furnace refractories) to produce a gradual decline in cell performance with time.

It can be concluded that the fabrication of high temperature fuel cells and electrolysers involves a high level of technology which with the present designs will always be expensive. The complexity of a typical assembly can be appreciated by examination of the junction between units of a multi-cell design which is shown to an enlarged scale in figure 5(c).

d) Cell performance

Optimisation of the cell performance for either fuel cell or electrolyser operation is obviously dependent upon the particular criteria adopted. For example it might be required to design a fuel cell for maximum power output per unit volume, whereas minimisation of energy cost could well be the overiding consideration for an electrolyser. Electrochemical, electrical, and thermal parameters are obviously all inter-related, and typical analyses for optimisation are available in the literatures[17,18,19].

Information from prototype fuel cells using porous oxide cathodes (e.g. SnO doped In_2O_3) and porous Ni-YSZ anodes suggests that the predominant polarisation losses arise from gas-phase polarisation, constriction resistance, and of course the electrolyte resistance for assemblies incorporating thick electrolyte layers. Providing the pores are sufficiently large ($>1\ \mu m$) to avoid viscous drag effects, then gas phase polarisation usually arises due to depletion of the relevant electro-active species at electrodes situated at the down-stream end of the gas stream in multi-cell assemblies. The diffusion flux of the species down the pore will be reduced, concentration polarisation will develop, and eventually a mass transport limiting current density may be observed.

A typical expression[18] for the mass transport limiting current density (J_L) resulting from the interdiffusion of O_2 and N_2 in a porous air electrode is given by:

$$J_L = \frac{4FcpD}{st} \ln \frac{1}{1-Xo_2}$$

where F is the Faraday constant, c is the molar density of oxidant, D is the interdiffusion coefficient for O_2 and N_2 mixtures, p, s, t represent the fractional porosity, tortuosity, and thickness, respectively for the electrode, and Xo_2 is the mole fraction of O_2 in the gas phase adjacent to the electrode. Taking D at 900°C as 2.5 cm^2/s, and assumming p and s have values of 0.25 and 1.5 for an electrode with a thickness of 0.15 mm (150μ), the value of J_L for air (21% O_2) is 26.5 A/cm^2. When the oxygen content is reduced to 5% O_2, the same electrode parameters produce a value of J_L equal to 5.9 A/cm^2. This calculated value for the limiting current is of the same order of magnitude as that observed experimentally ($\sim$1.0 A/cm^2) for similar porous electrodes.

In addition to these features of a porous electrode, there is another factor that can introduce electrical losses which arises from the fact that a porous electrode is not in continuous contact with the electrolyte. The result of this discrete contact is that ions and electrons will be restricted in their flow to regions of electrode/electrolyte/gas phase contact, the so-called triple-point contact (T.P.C.). This introduces a resistive loss which is termed a 'constriction' resistance and is additional to the cell resistance that would be expected on the basis of bulk resistivities and geometry. This type of polarisation will be discussed further in section 3.

Ultimately the performance of high temperature fuel cells and electrolysers has to be compared with alternative power producing systems and low temperature electrolysers. It does appear that the present performance of high temperature fuel cells requires to be significantly improved in order to compete with existing power producing systems. The same is true for electrolysers. Both existing low temperature electrolysers and novel systems incorporating solid polymer electrolytes[20] at present offer many advantages compared to high temperature electrolysers. It does appear therefore that the principal application of high temperature devices incorporating solid electrolytes are likely to remain in the field of oxygen monitoring[21], small electrochemical pumps[21], and selected de-oxidation processes in metal extraction[22] and refining[23].

3. Kinetics of platinum electrodes on oxide electrolytes

Some general comments were made in section 2(d) relating to polarisation losses at electrodes in prototype multi-cell devices. A more detailed analysis of the kinetics associated with platinum electrodes is now presented, as nearly all the published data on experimental high temperature electrolysers refers to systems incorporating platinum electrodes. This type of electrode is also often employed in commercial high temperature oxygen monitors although of course platinum electrodes would not be used in commercial fuel cells and electrolysers.

It is generally accepted that the anodic oxygen reaction at porous platinum electrodes is not rate-controlling and so available information relates to the cathodic reaction. Examination of selected investigations[24-30] confirms that the choice of experimental conditions and electrode morphology often controls whether the subsequent rate limiting mechanism is associated with gaseous phase mass transfer (i.e. oxygen condensation flux) or with the adsorption and subsequent charge transfer process. In this connection the scanning electron micrographs reported by Pizzinni[24,25], are particularly interesting as they clearly indicate that the electrode morphology is very dependant upon the mode of preparation, and also confirm that the structure may change with time as the small active crystallites of platinum agglomerate to produce large grains. It should also be noted that interpretation of the results can be complicated by the fact that many commercial platinum pastes contain a silicate flux to promote adhesion but which may also block pores and reduce the triple-phase contact area (T.P.C.).

A variety of possible processes have been proposed as the rate limiting step in the cathodic oxygen reaction, e.g.

a) gaseous diffusion within the pores of the electrode,
b) adsorption and dissociation on the surface of the platinum or solid electrolyte,

c) surface diffusion of adsorbed species to the triple-phase contact area (T.P.C.)
d) dissolution of the adsorbed species into the metallic or electrolyte phase,
e) bulk and/or grain boundary diffusion of the dissolved species in the metallic or electrolyte phase,
f) charge transfer across the electrode/electrolyte phase boundary.

Investigations designed to identify the rate controlling mechanism usually determine such as parameters as electrode resistance (R_{el}) or the limiting current (J_L) as a function of temperature and oxygen partial pressure, and conclusions of the selected investigations are summarised in Table I.

For macroporous electrodes (pore diam. $> 1\,\mu m$) and point electrodes there is some measure of agreement between investigators and the following pattern emerges for the cathodic reduction of oxygen at platinum electrodes.

High pO_2	Low pO_2
$J_L \propto pO_2^{\frac{1}{2}}$	$J_L \propto pO_2$
$R_{el} \propto pO_2^{-\frac{1}{2}}$	$R_{el} \propto pO_2^{-1}$

A dependence of $pO_2^{\frac{1}{2}}$ is usually interpreted as the influence of atomic oxygen and implies that a step involving this species is rate limiting, whereas a process involving the concentration of molecular oxygen would be directly proportional to pO_2. At low oxygen partial pressures, therefore, it is quite likely that gas phase mass transfer is the rate controlling step arising from interdiffusion in boundary layers and pores. A similar oxygen partial pressure dependence at high oxygen partial pressure observed by Etsell and Flengas[28] which has been explained[24] by suggesting that these investigators used an electrode structure invorporating very fine pores ($< 0.01 u$) thereby curtailing the rate of oxygen mass transfer. Using a thick foil platinum electrode Casselton[30] also reported a direct proportionality between limiting current and oxygen partial pressure. This was explained by proposing that singly charged oxygen molecules (O_2^-) are a necessary precursor to the formation of other mobile oxygen species which diffuse along the interface between the platinum foil and electrolyte.

Interpretation of the $pO_2^{\frac{1}{2}}$ dependence is much more controversial. Baurle[24] suggests that the adsorption - dissociation step is rate controlling in the formation of atomic oxygen in the temperature range 400-800°C, where as Pizzinni[24] concludes from examination of the relevant activation energies that surface

Table I

Rate limiting steps for the cathodic oxygen reaction at platinum electrodes

Investigator (refs)	Electrode Morphology	Oxygen Argon Pressure Range (atm)	$J_L \propto pO_2^n$ Relationship	Rate limiting Mechanism
Pizzini (24, 25)	Micro and macro-porous	$10^0 - 10^{-3}$ $10^{-3} - 10^{-5}$	$n = \frac{1}{2}$	$>$ 700°C Chemi-sorption $<$ 700°C Surface diffusion Mass transport
Kleitz (26)	Point electrode	$10^0 - 10^{-3}$ $10^{-3} - 10^{-5}$	$n = \frac{1}{2}$ $2/3 < n < 1$	Adsorption-dissociation Mass Transport
Baurle (27) A.C. Impedance	Dense and semi-porous	$10^0 - 10^{-5}$	$n = \frac{1}{2}$	Oxygen dissociation OR electron transfer
Etsell (28)	Micro-porous	$10^0 - 10^{-3}$	$n = 1$	Mass transport through pores
Brook (29)	Foil and paste	$10^0 - 10^{-3}$	$n = \frac{1}{2}$	Bulk diffusion of atomic oxygen
Casselton (30)	Thick disc	$10^0 - 10^{-3}$	$n = 1$	Interfacial diffusion of O_2^- species

diffusion (ΔE = 35 kcal/mole) of atomic oxygen on the platinum is the limiting step at temperatures below 700°C. The suggestion that surface diffusion is the slow step is obviously more tenable than the proposal of Brook et al[29] that bulk diffusion of atomic oxygen is rate controlling as it is impossible to account for the magnitude of the limiting currents reported by these workers using available oxygen diffusion and solubility data in platinum[31]. At higher temperatures (700-1000°C) Pizzinni[24] reports a relatively low activation energy (8-16 kcal/mole) for the rate controlling mechanism which is interpreted as arising from the chemisorption step. Evidence for this interpretation is provided by the observation[25] that the limiting currents at temperatures higher than 700°C obtained with oxygen-nitrogen mixtures are reduced when water-vapour is present in the gases. Water is strongly chemisorbed on the platinum thus decreasing the concentration of active sites available for oxygen-reduction.

The results reported by Fabry and Kleitz[26] for point electrodes obeying the $pO_2^{\frac{1}{2}}$ dependance at low current densities are important. Platinum, silver, and gold were investigated as metals and different ZrO_2 based solid solutions as electrolytes. The overpotential activation energies determined from the experimental results were a function of both the nature of the metal and also the composition of the electrolyte. These investigators conclude that the oxygen exchange step between the surrounding gas and the electrode micro-system is slower than the actual Faradaic electrode reaction especially at lower oxygen partial pressures. Because of the influence of both metal and electrolyte dopant it is probable that both the oxygen capacity and work function of the metal are involved together with the rate at which oxygen exchanges with the surface of the oxide electrolyte. In this connection it is interesting to note that the deliberate addition of Fe_2O_3 to a solid electrolyte[26] did not significantly influence the electrode kinetics, and it does appear that the most important factor influencing surface exchange coefficients on fluorite oxide solutions is the concentration of anion vacancies and not the concentration of electronic defects, a conclusion already made by Floyd[32] from his isotopic exchange measurements on ZrO_2, ThO_2, and CeO_2 solid solutions. Claudel[33] has noted that the concentration of anion vacancies influence the catalytic behaviour of ceria, and Grabke[34] also observed that the kinetics of oxygen exchange on transition metal oxides (e.g. FeO, NiO) were a function of the concentration of point defects. However the fact that the type of both metallic electrode and oxide electrolyte appear to influence the electrode kinetics suggests another interpretation namely that of reaction between the noble metal and oxide electrolyte.

It is well known[35] that excellent metal-ceramic seals can be produced by placing refractory oxides (e.g. Al_2O_3, ZrO_2 solid solutions) in contact with platinum foil at elevated temperatures (1000-1200°C) in oxidising environments. It is difficult, however, to understand the nature of the metal-ceramic bond that is produced under these conditions as available thermodynamic data indicates that there should be no reaction between the noble metals and the refractory oxides under these circumstances. In contrast, when reducing conditions prevail (i.e. low oxygen partial pressures) then a reaction is to be expected due to the formation of a very stable intermetallic compound. The unusually high stability of these transition metal alloys was predicted by the Brewer-Engel correlation[36,37], and involves the transfer of electrons from metals having internally paired d-electrons unavailable for bonding (e.g. platinum group metals) to other metals having empty d-orbitals (e.g. Group IV and V transition metals).

Thermodynamic data[38,39] are available for the alloy $ZrPt_3$ and so it is possible to determine the conditions under which platinum and zirconia would be expected to react, as follows.

$$Zr + O_2 \longrightarrow ZrO_2$$

$$\Delta G^o = -262{,}160 + 44.64\ T\ \text{cal/mole}$$

$$Zr + 3\ Pt \longrightarrow Zr\ Pt_3$$

$$\Delta G^o = -108{,}120 + 12.60\ T\ \text{cal/mole}\ ZrPt_3$$

$$ZrO_2 + 3\ Pt \longrightarrow Zr\ Pt_3 + O_2$$

$$\Delta G^o = +\ 154{,}040 - 32.04\ T\ \text{(cal)}$$

$$\mu_{O_2} = -154{,}040 + 32.04\ T\ \text{(cal.)}\ (ZrO_2,\ Pt,\ ZrPt_3\ \text{at unit activity})$$

and $E = 1.671 - 3.476 \times 10^{-4}\ T$ (Volt)

T (°K)	uO_2 (cal)	pO_2 (atm)	E (V)
1000	-122,000	2.1×10^{-27}	1.323
1273	-113,253	3.5×10^{-20}	1.228

It can be concluded therefore that at 1273°K, a cathodic overpotential of 1.228V (with reference to an electrode maintained at an oxygen partial pressure of 1 atm.) should be sufficient to initiate a reaction between the platinum electrode and the zirconia based electrolyte*

*The assumption of unit activity for ZrO_2 introduces only a very small error into the calculation as the zirconia-yttria fluorite solid solution only exhibits minor deviations from ideal behaviour.

This conclusion is relevant because it does provide an explanation for the increase in current which is observed when cathodic overpotentials in excess of about 1 volt are imposed during the operation of electrolysers[21,28,40]. The usual explanation for this observation is to assume that the zirconia electrolyte is being reduced according to reactions of the type:

$$O_o^x\ (ZrO_2) \longrightarrow \tfrac{1}{2}O_2 + V_o^{\circ\circ} + 2e^1,$$

which produce an electronic contribution to the total conductivity. The presence of significant electronic conductivity at the electrolyte surface could effectively increase the area of the triple phase contact (T.P.C.) region which in turn would permit a higher current density. Whilst this process undoubtedly occurs at high overpotentials ($>$1.5 V) and eventually produces blackening of the electrolyte, there is no definite evidence to confirm that it is the most significant process at moderate overpotentials (1.0 - 1.5 V). In fact, as already mentioned the deliberate addition of small quantities of transition metal oxides (e.g.Fe_2O_3)[26], designed to introduce a small but significant electronic transference number, has relatively little effect on the electrode behaviour. Moreover gaseous exchange measurements do not suggest that availability of electrons is likely to be the predominant factor in determining the rate of the overall reduction process.

It is probable, however, that the formation of $ZrPt_3$ with its interesting electronic structure and high affinity for oxygen, could produce an enhanced catalytic activity for the oxygen reduction process. If this is indeed the situation then the initial increase in current observed at cathodic overpotentials greater than 1V could be attributed to the catalytic behaviour of $ZrPt_3$ produced by reaction of the zirconia electrolyte and platinum electrode. Some support for this proposal is provided by the investigation of Pancharatnan et al[41] who studied the catalytic decomposition of nitric oxide using zirconia based electrolytes incorporating platinum or gold electrodes. The dissociation rate of nitric oxide could be increased by at least a thousandfold at cathodic overpotentials greater than 1V and this observation was attributed to the formation of active sites ($Vo^{\circ\circ} + e^1$) at the surface of the zirconia electrolyte. However the formation of the intermetallic alloy $ZrPt_3$ (or $ZrAu_3$) could also provide a satisfactory explanation for the enhanced catalytic decomposition of nitric oxide.

The reaction between platinum and zirconia based electrolyte obviously merits further investigation and is a further example of the propensity of most materials to react when brought into contact at elevated temperature. It was this type of behaviour that produced many of the technological problems associated with the development of commercial high temperature fuel cells and electrolytes and still provides the incentive for developing systems that can operate at ambient or moderate temperatures.

References

1. H.A. Liebhafeky and E.J. Cairns, 'Fuel Cells and Fuel Batteries' (John Wiley, New York, 1968)
2. M.W. Breiter, 'Electrochemical Processes in Fuel Cells' (Springer-Verlag, Berlin, 1969)
3. G. Samstede, Ed., 'Electrocatalysis to Fuel Cells' (University of Washington, Press, Seattle, 1972)
4. F.T. Bacon and J.M. Fry, Proc. Roy. Soc. A, 334, 427,1973
5. D.P. Gregory, Scient. Am. 228, 13, 1973
6. T.H. Etsell and S.N. Flengas, Chem. Rev., 70, 339, 1970
7. H.L. Tuller and A.S. Nowick, J. Electrochem. Soc. 122, 255, 1975
8. F. Vanbaelinghem, 'Elaboration et Proprietes Electrochimiques des Solutions Solides La_2O_3-SrO', Thesis, University of Grenoble (Nov. 1973). To be published in J. Applied Electrochem
9. T. Takahashi and H. Iwahara, Energy Conversion, 11, 105, 1971
10. T. Takahashi, p.989 in 'Physics of Electrolytes' Vol 2., Edit. J. Hladik, (Academic Press, London, 1972)
11. R.E. Carter and W.L. Roth, p.125 in 'Electromotive Force Measurements in High Temperature Systems', Edit. C.B. Alcock, (Institution of Mining and Metallurgy, London, 1968)
12. W. Baukal, Electrochim. Acta, 14, 1071, 1969
13. R.E. Carter, U.S. Patent 3,351,496
14. T. Kudo and H. Obayashi, J. Electrochem. Soc. 122, 142, 1975
15. B.C.H. Steele, B.E. Powell and P.K.R. Moody, Proc. Brit. Ceram. Soc., No.10, 87, 1968
16. H.S. Spacil and C.S. Tedmon, J. Electrochem. Soc. 116, 1627, 1969
17. E.F. Sverdrup, C.J. Warde, and R.L. Eback, Energy Conversion 13, 129, 1973
18. C.S. Tedmon, H.S. Spacil and S.P. Mitoff, J. Electrochem. Soc. 116, 1170, 1969
19. H.S. Spacil and C.S. Tedmon, J. Electrochem. Soc. 116, 1618, 1969
20. Proceedings 8th Intersociety Energy Conversion Engineering Conference, University of Philadelphia, 1973
21. J. Fouletier, H. Seinera, and M. Kleitz, J. Appl. Electrochem. 4, 305, 1974
ibid 5, 111, 1975
ibid 5, 177, 1975
22. B. Marincek, Schweiz Arch. Angew. Wiss. Tech. 33, 395, 1967
23. K.E. Oberg, L.M. Friedman, W.M. Boorstein, and R.A. Rapp, Met. Trans. 4, 75, 1973
24. S. Pizzini, p.461, 'Fast Ion Transport in Solids', Edit. W. van Gool, (North Holland, Amsterdam, 1973)
25. S. Pizzini, M. Bianchi, P. Colombo and S. Torchio J. Appl. Electrochem., 3, 153, 1975

26. P. Fabry, M. Kleitz, Electroanal. Chem. and Interfacial Electrochem. 57, 165, 1974
27. J.E. Baurle, J. Phys. Chem. Solids, 30, 2657, 1969
28. T.H. Etsell and S.N. Flengas, J. Electrochem. Soc. 118, 1890, 1971
29. R.J. Brook, W.L. Pelzmann and F.A. Kroger, J. Electrochem. Soc. 118, 185, 1971
30. R.E.W. Casselton, J. Appl. Electrochem, 4, 25, 1974
31. L.R. Vehlo and R.W. Barlett, Met. Trans. 3, 65, 1972
32. J.M. Floyd, 'Oxygen Transport in Oxides of Fluorite Structure' Ph.D. Thesis, University of London, 1969
33. M. Breysse and B.M. Claudel, J. Catel. 27, 275, 1972
34. H.J. Grabke, Ann. New York Acad. Sci. 213, 110, 1973
35. H.J. de Bruin, A.F. Moodie and C.E. Worble J. Mat. Sci. 7, 909, 1972
36. L. Brewer, p.221, 'Electronic Structure and Alloy Chemistry of the Transition Elements' Ed. P.A. Beck, (Interscience, New York, 1963)
37. L. Brewer and P.R. Wengert, Met. Trans. 4, 83, 1973
38. P.J. Meschter, and W.L. Worrell, Proc. 4th Int. Conf. Chem. Thermodym. (Monpellier, France, 1975) 3, 233
39. P.J. Meschter and W.L. Worrell, Met. Trans., to be published
40. C.B. Alcock and S. Zador, J. Appl. Electrochem, 2, 289, 1972
41. S. Pancharatnam, R.A. Huggins and D.M. Mason, J. Electrochem. Soc. 122, 869, 1975

DISCUSSION

KROGER : The conclusion by Professor Brook that oxygen diffusion through Pt was rate-limiting was based on the observation that the current density was progressively reduced as the thickness of the platinum electrode was increased by successively adding additional layers of platinum foil. It is of course possible that the predominant oxygen flux was via grain boundaries.

The suggestion that at high voltage oxygen may enter directly into the zirconia electrolyte when it has become electronically conducting was deduced from considerations of the dependence of the conductivity upon oxygen partial pressure.

STEELE : I have to repeat my comments in the paper that it is impossible to obtain the limiting currents mentioned by Brook (ref. 29) using available data for oxygen diffusion and solubility in platinum. In addition to the data of Vehlo and Barlett (ref. 31) there are many studies including our own (ref. 32) where isotopic exchange measurements have been conducted in the presence of platinum reaction tubes. Even at 1000°C the flux of ^{18}O into the platinum is negligible. Whilst grain boundaries would increase the flux,it is difficult to believe that the increase would be the several orders of magnitude which is required. In the absence of detailed microscopical examination it is difficult to formulate an alternative explanation.

I must emphasise that I agree that oxygen can probably enter the zirconia electrolyte directly at larger cathodic overpotentials, and my remarks about current enhancement due to the possible electrocatalytic behaviour of $ZrPt_3$ are more relevant to the moderate overpotential (1-1.5 V) regime.

ETSELL : As further verification of the feasibility of $ZrPt_3$ formation at high cathodic overpotentials, large current increases are frequently reported at significantly lower potentials than would be predicted by simply calculating the expected rise in electronic current from transport number data. It might prove interesting to compare cathodic I-V curves for several rate determining steps which can be changed by varying temperature, the cathodic gas mixture and electrode structure. Unlike reaction control the proposed mechanism should not apply under conditions of mass transport control.

PIZZINI : Regarding the catalytic activity of $ZrPt_3$, I would like to comment that it is well known that as soon as you replace noble metals by oxides with known catalytical activity, e.g. ($LaCoO_3$, Pr_2NiO_4) you immediately observe a reduction of the cathodic overvoltage.

STEELE : I agree with preceding comments and it is obvious that further work is necessary to evaluate the role of $PtZr_3$ and other intermetallic alloys involving the noble metals.

THE APPLICATION OF SOLID STATE IONICS TO BATTERIES

R. M. Dell

A.E.R.E. Harwell, Didcot, Oxon, OX11 ORA.

1. INTRODUCTION

Solid state ionics, that is solids which possess unusually high diffusion coefficients and conductances for specific ions, have assumed considerable importance in recent years for battery research. This stems from the discovery in 1966/7 of the alkali silver halides KAg_4I_5 and $RbAg_4I_5$ (1,2) and the beta alumina family of oxides (3). The halides, which are pure Ag^+ ion conductors, have resistivities as low as 5-10 ohm-cm at room temperature. They have been employed as solid electrolytes in small primary batteries, especially for applications where long life and freedom from self discharge are important. Sodium beta alumina has a comparably high conductivity at 300-400°C and is of interest as electrolyte for the sodium/sulphur battery. Several reviews on the development of solid electrolyte batteries have been written (4-6).

There are two quite distinct applications for solid state ionics in batteries. In addition to the electrolyte application, for which the electronic conductivity must be extremely low to avoid short circuiting the cell internally, there is also the possibility of using a solid state ionic as a cathode in which the diffusing cation dissolves to form an intercalation compound. For this application the solid should, ideally, have high values of both the ionic and electronic conductivities. These ionic cathodes may be distinguished from the more conventional cathodes which undergo reaction with a change of phase (e.g. PbO_2, NiOOH), although the distinction is not entirely clear-cut and intermediate situations may exist.

M. Kleitz and J. Dupuy (eds.), Electrode Processes in Solid State Ionics, 387–410.

Batteries which utilise solid state ionics are of three general types:

(1) All-solid-state batteries in which the anode, electrolyte and cathode are solids. A typical example of this battery type is the $Ag/RbAg_4I_5/RbI_3$ cell. Generally, solid state batteries are small primary or reserve batteries which operate at ambient temperatures.

(2) Solid electrolyte batteries with a liquid metal anode and/or a liquid cathode. These batteries can be ambient temperature systems (e.g. using a Na/Hg amalgam anode), although more generally the interest is in high temperature batteries for large scale applications, such as the $Na_{(1)}/\beta Al_2O_3/S_{(1)}$ battery.

(3) Ionic cathode batteries. Ionic cathodes are being developed mostly for use with liquid electrolytes, in particular for lithium batteries.

This "teaching" paper sets out to survey broadly the application of solid electrolytes and ionic cathodes to battery systems and to highlight recent developments. References to the literature are selective and the paper is by no means a comprehensive review of the subject.

2. SOLID ELECTROLYTES

If a solid state ionic compound is to find application as a battery electrolyte it must, generally speaking, meet three criteria:

(1) Its specific resistivity should not exceed ~20 ohm cm. This criterion is particularly important for large, rechargeable batteries (e.g. traction batteries) where high power levels are involved and electrical efficiency is a key parameter, but it may be less important for small, low power batteries, particularly those designs which utilise ultra-thin electrolyte layers.

(2) The transport number for ionic conduction should exceed 0.95 and, preferably, 0.99. Any electronic conduction will short-circuit the cell and lead to self-discharge on standing. This is especially undesirable for batteries which are required to have a long shelf life.

(3) The kinetics of ion exchange at the two electrolyte/electrode interfaces should be rapid, in order that significant currents may be passed without polarisation taking place. For high power batteries, current densities in the range 50-500 mA cm^{-2} are commonly required.

Solid state ionics which have potential for use as electrolytes in electrochemical devices may be divided conveniently into groups according to the temperature range at which their ionic conductivity has the desired value, viz:

(i) High temperature electrolytes (600-1100°C). These are mostly oxides or fluorides in which conduction is by anion migration through a vacancy mechanism, although Li^{+} ion conductors are known also.

(ii) Medium temperature electrolytes (200-400°C). These include the classical silver salts (αAgI, AgBr, etc) as well as the recently discovered oxide electrolytes with layer or tunnel structures (e.g. β alumina).

(iii) Low temperature electrolytes (0-100°C). This class is composed of certain complex salts of silver and copper halides which have been discovered since $RbAg_4I_5$.

A further useful classification of solid electrolytes is based upon whether the fast ion-conduction process is three-dimensional throughout the crystal lattice, as in the high temperature electrolytes and the silver salts, or is confined to two-dimensional layers, as in the beta alumina family of compounds, or to one-dimensional tunnels, as in the hollandite materials such as $K_xMg_{x/2}Ti_{(8-x)}O_{16}$.

2.1 High Temperature Electrolytes

The high temperature electrolytes are mostly oxides of composition MO_2 based upon the fluorite structure. The best investigated is "calcia stabilized zirconia" (CSZ) which consists of a solid solution of 12-15% CaO in ZrO_2. The addition of calcia transforms ZrO_2 from the monoclinic to the cubic (fluorite) structure and also introduces anion vacancies for charge compensation. Conduction is by $O^{=}$ ion diffusion through anion vacancies and ZrO_2-CaO has a resistivity of ~30 ohm-cm at 950°C. Trivalent cations may also be used to stabilise ZrO_2 with resistivities at 950°C of ~12 ohm-cm for ZrO_2-Y_2O_3 and ~6 ohm-cm for ZrO_2-Yb_2O_3 or ZrO_2-Sc_2O_3 (Figure 1). Stabilized zirconia is of interest as an electrolyte for fuel cells, but no battery applications have been proposed and the temperature of conduction is too high to be of real interest.

Other oxide electrolytes with the fluorite structure are ThO_2-Y_2O_3, employed in galvanic cells to monitor low oxygen potentials, and CeO_2 doped with Y_2O_3 or rare earth oxides. Fluorite-type electrolytes based upon CeO_2 have generally been less favoured than those based upon ZrO_2 or ThO_2 as Ce^{4+} is more readily reduced to Ce^{3+}, resulting in the onset of electronic conduction under highly reducing conditions. Recent studies of doped CeO_2 have, however, shown that CeO_2-5% Y_2O_3 has a substantially higher conductivity than ZrO_2-CaO[7] and remains purely ionic down to $Po_2 = 10^{-13}$ atm at 600°C, while CeO_2-10% Gd_2O_3 has a resistivity of 10 ohm-cm at 750°C, which is comparable to that of ZrO_2-CaO at 1100°C[8].

Another interesting family of high temperature solid oxides with the fluorite structure is that based upon Bi_2O_3[9]. This oxide, which has a monoclinic structure and is an electronic conductor at low temperatures, undergoes a phase transition at 730°C to the cubic, fluorite structure. It then has one-quarter of the oxygen lattice sites vacant and is an ionic conductor with resistivity <1 ohm-cm. Solid solutions of B_2O_3-WO_3 have been investigated and the composition $3B_2O_3$-WO_3 shown to have the fluorite structure at all temperatures and to remain an ionic conductor down to 400°C[10].

The fluorite-type oxides are not the only possible $O^=$ ion conductors. Perovskite-type oxides with anion vacancies can be formed by substituting M^{2+} for M^{3+} or M^{3+} for M^{4+}. The ionic

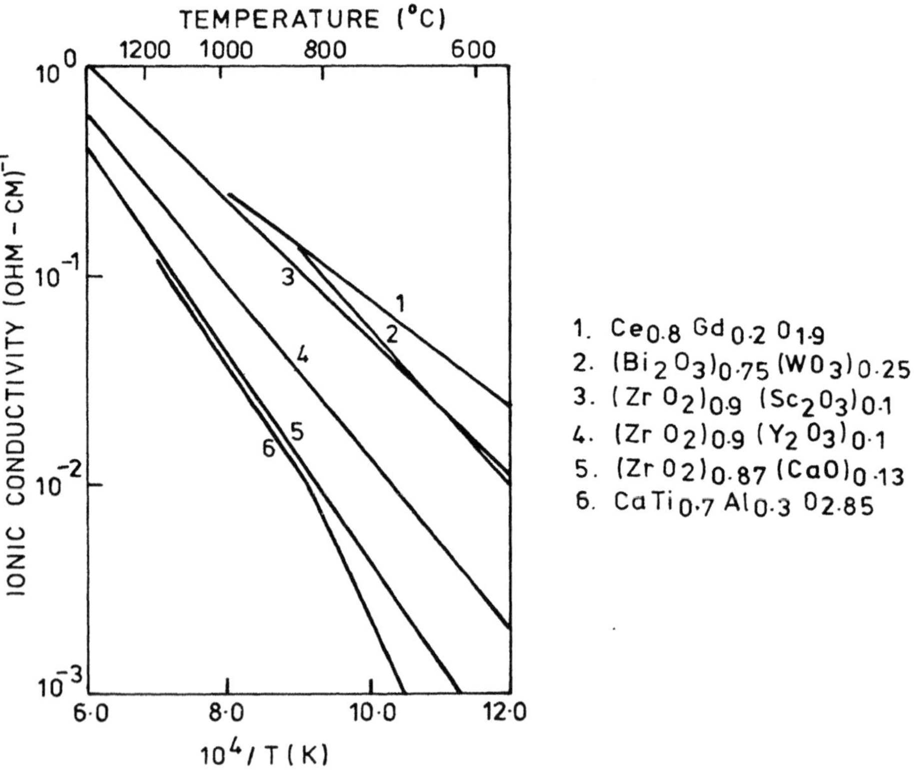

Figure 1 Ionic Conductivities of Fluorite and Perovskite type Oxides

conductivities of $LaAlO_3$ with La^{3+} partially substituted by Ca^{2+} or Ba^{2+}, and $CaTiO_3$ (and $SrTiO_3$) with Ti^{4+} partially substituted by Al^{3+} have been measured by Takahashi and Iwahara(11). The highest ionic conductivity observed, for $CaTi_{0.7}Al_{0.3}O_{3-x}$, was only comparable to that of ZrO_2-CaO.

Most theoretical treatments of ionic conduction in fluorite-type oxides have assumed a point defect model and vacancy diffusion. This is thought to be unrealistic for a material containing, typically, 10-15% vacancies on the oxygen sub-lattice; rather, a co-operative jump mechanism might be expected. Whittingham has discussed recently(12) a model which involves the jump of a single ion along a string of vacancies. This implies a co-operative diffusion mechanism for the fluorite structure, in some ways analogous to diffusion in the ReO_3-type shear structures.

High temperature solid oxide electrolytes are generally of greater interest to fuel cells than batteries and so will not be discussed further here. It may be noted that by operating at high temperatures it is possible to circumvent the catalytic activation problems normally encountered with the air cathode, at the expense of introducing materials compatibility problems. In view of the attractions of air-depolarised batteries this is of considerable interest in principle, but the practical problems of devising batteries based upon high temperature oxide electrolytes are formidable.

A quite different class of high temperature electrolytes are the lithium ion conductors. Several lithium salts (Li_2SO_4, Li_2WO_4 and the double salts $LiAgSO_4$ and $LiNaSO_4$) have been shown to possess high temperature modifications with resistivities below 1 ohm-cm(13). Work in this field has been reviewed by Pizzini(14). Unfortunately, with decreasing temperature these structures are thermodynamically unstable and revert to their low temperature phases in which the Li^+ ion is relatively immobile. This makes them of limited value as potential battery electrolytes.

Nevertheless some preliminary tests on the cell $Mg/Li_{1.72}Mg_{0.14}SO_4/MnO_2$ at 650°C gave interesting results(15) and it was suggested that low temperature Li^+ solid electrolytes could be produced by using three or more component systems.

2.2 Silver and Copper Halides

Silver iodide, the prototype of all the silver solid electrolytes, undergoes a phase transition at 140°C. Above this temperature the α phase, which has a body-centred cubic structure, has a resistivity of ~1 ohm-cm, while below the transition temperature βAgI has a resistivity of 10^4-10^6 ohm-cm. Unlike the high temperature oxide electrolytes, silver iodide is a stoichiometric

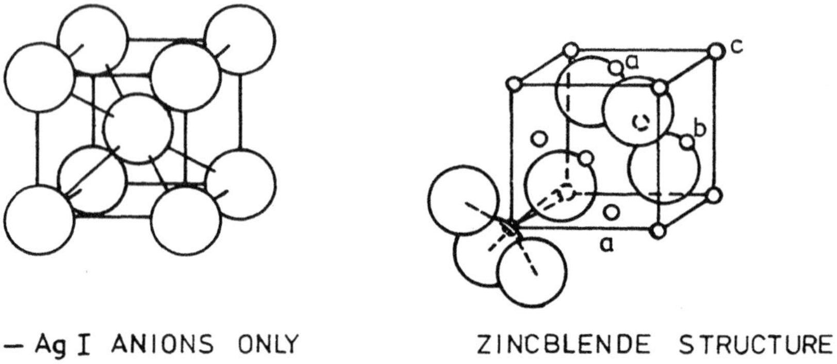

Figure 2 Structures of αAgI and Silver Mercuric Iodides

compound and its high cation conductivity cannot be ascribed to vacancy diffusion. Rather, in αAgI the body centred cubic packing of I^- ions provides a relatively open structure through which the Ag^+ ions can diffuse with low activation energy (Figure 2). The Ag^+ ions are arranged in a semi-random manner among the excess of available sites (tetrahedral, octahedral and trigonal bipyramidal), subject to the condition that nearest-neighbour sites cannot be occupied simultaneously. Each Ag^+ ion may be regarded as occupying a region of the lattice rather than a single site and other ions are excluded from this region(15). The mechanism of silver ion diffusion in αAgI has been reviewed recently by Whittingham(12).

A number of other silver salts, for example Ag_3SI and Ag_2HgI_4, show similar phase transitions, with their high temperature forms possessing high Ag^+ ion conductivities. The situation as it existed in 1970 is summarised in the well-known diagram published by Heyne(17) (Figure 3). The lesser conductivity of the high temperature phase of Ag_2HgI_4 is due to the fact that it does not have the b.c.c. structure of αAgI and αAg_3SI, but has a randomly ordered rearrangement of the low temperature f.c.c. (zinc-blende) structure (Figure 2).

$RbAg_4I_5$ is remarkable in that the α iodide phase, with its associated high ionic conductivity, extends down to room temperature. In principle $RbAg_4I_5$ disproportionates below 27°C to Rb_2AgI_3 and AgI, but this reaction is so slow that, in practice, the conductive compound may be employed as an electrolyte in a dry atmosphere at room temperature. The MAg_4I_5 structure exists with M = K, Rb or NH_4. Analogous bromides do not exist. The copper halide KCu_4I_5 is reported stable over a narrow temperature range (257-332°C)(17).

Since 1970 a number of other silver-iodide based electrolytes have been reported. These may be regarded as anion substituted- or cation-substituted derivatives of αAgI. Takahashi and co-workers

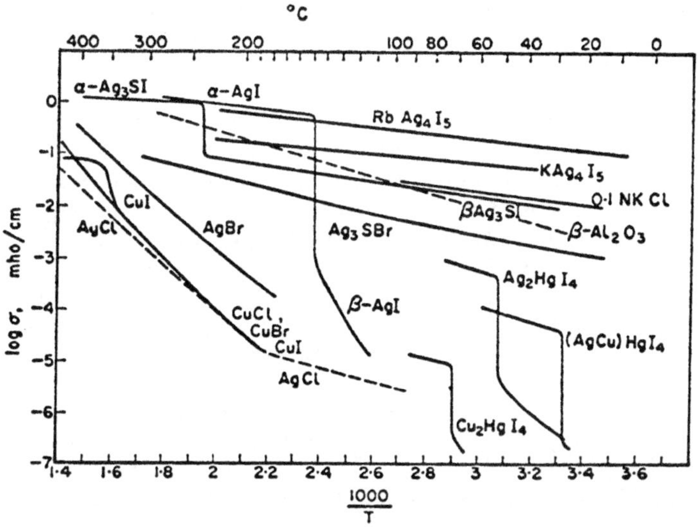

Survey of the conductivity curves of some solid electrolytes.

Figure 3

investigated the systems $AgI-Ag_3PO_4$ [18], $AgI-Ag_4P_2O_7$ [18] and $AgI-Ag_2WO_4$ [19] and in each system discovered a compound of comparatively high ionic conductivity at room temperature (Table 1). The stability and thermodynamic properties of these anion-substituted iodides have been reviewed by Owens [20]. Recent studies of the systems $AgI-Ag_3AsO_4$ and $AgI-Ag_3VO_4$ have revealed high ionic conductivity compounds $Ag_7I_4AsO_4$ and $Ag_7I_4VO_4$ [21], analogous to $Ag_7I_4PO_4$. Similarly, the systems $AgI + Ag_2XO_4$ (X = Cr, Mo, W)

Table 1

Anion Substituted Silver Iodide Electrolytes

System	Compound	Ionic Conductivity at 25°C $(\Omega\ cm)^{-1}$	Decomposition Temp (°C)
$AgI-Ag_2S$	Ag_3SI	0.01	*
$AgI-Ag_3PO_4$	$Ag_7I_4PO_4$	0.019	79°
$AgI-Ag_4P_2O_7$	$Ag_{19}I_{15}P_2O_7$	0.09	147°
$AgI-Ag_2WO_4$	$Ag_6I_4WO_4$	0.047	293°
$\alpha AgI-Ag_2SO_4$	Solid solution	0.05	-

* Phase transition βAg_3SI to αAg_3SI at 235°C

gave $Ag_6I_4CrO_4$, $Ag_6I_4MoO_4$ and $Ag_6I_4WO_4$ as compounds of high ionic conductivity(22). It is notable that all these conducting compounds contain 80 mole % AgI.

The cation-substituted silver iodides of high conductivity, apart from KAg_4I_5, $RbAg_4I_5$ and $NH_4Ag_4I_5$, are mostly organic quaternary ammonium silver iodides of varying formulation, the simplest being tetramethyl ammonium silver iodide represented variously as $((CH_3)_4N)Ag_6I_7$ and $((CH_3)_4N)_2Ag_{13}I_{15}$ (20). There are also similar heterocyclic salts, e.g. pyridinium silver iodide $(C_5H_5NH)Ag_5I_6$. The crystal structures of these compounds are discussed by Geller (23). Their conductivities are generally comparable to those of the anion-substituted phases.

Recently, Geller has reported a further pyridinium silver iodide $(C_5H_5NH)_5Ag_{18}I_{23}$ whose structure is such that it is a two dimensional solid electrolyte (24). In accordance with this two dimensionality, its specific conductance is an order of magnitude less than that of $(C_5H_5NH)Ag_5I_6$.

Comparatively little work has been carried out on the copper (I) analogues of the silver iodides, although for battery applications there is an economic incentive to replace silver by copper. As Figure 3 shows, CuI behaves analogously to AgI although the transition temperature is higher and the conductivity of the α phase is less than for αAgI. A similar comparison may be made between Cu_2HgI_4 and Ag_2HgI_4. Both α CuI and α Cu_2HgI_4 have a f.c.c. arrangement of anions, with the cations distributed at random on the tetrahedral sites. It is this random distribution which gives rise to the high cation conductivity in CuI, while in Cu_2HgI_4 the presence of the Hg^{++} ions impedes Cu^+ ion diffusion resulting in a lower conductivity. A detailed study of disorder phenomena in the low temperature phases of these compounds has been made (25).

Until recently all attempts to prepare a copper salt which was highly conducting at room temperature had failed. Takahashi and Yamamoto have recently discovered a class of such salts which have a Cu^+ ion conductivity of up to 0.035 $(ohm\text{-}cm)^{-1}$ at 20°C (26). These are double salts of the cuprous halides (80-94 mole %) with the NN'-dialkyl (or dihydro)-triethylene diamine dihalides (20-6 mole %). There is evidence that the compositions are true compounds and not mixtures of the constituents. Interestingly, the salts are not confined to the iodides, but include chlorides and bromides as well, e.g. $3(C_6H_{12}N_2.2HCl)17CuCl$, $3(C_6H_{12}N_2.2C_2H_5Br)$ 17CuBr.

2.3 Beta Alumina and other Cation Conducting Oxides

Since the original discovery of fast ion transport in beta

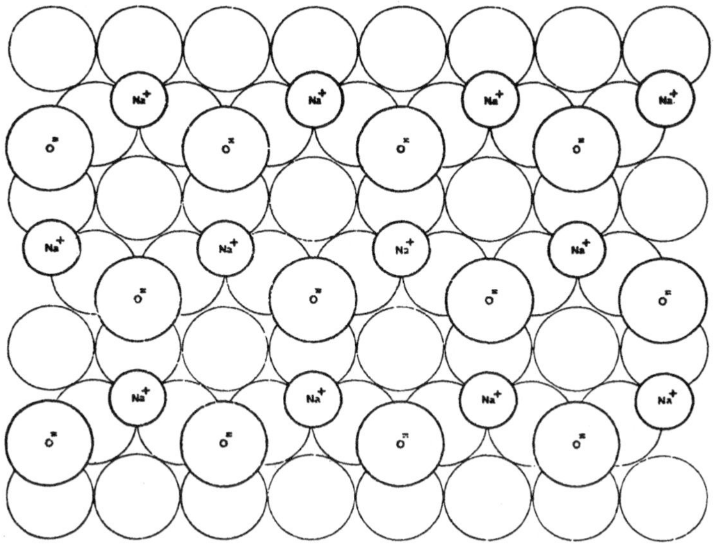

Figure 4 Conducting Plane of Beta Alumina

alumina, leading to high ionic conductivities and rapid cation exchange reactions[3], there has been a world-wide interest in this family of compounds formed between alumina and the alkali oxides. In addition to the alkali ions the monovalent cations Ag^+, Tl^+ and NO^+ also form beta aluminas, although the sodium compounds are the only ones of technical significance as potential battery electrolytes.

Three different hexagonal crystallographic structures are known, designated β, β'' and β''' alumina respectively[27,28]. These are all built-up on a common structural principle in which blocks of spinel structure, four or six anion layers thick in the crystallographic c direction, are separated from each other by layers perpendicular to the c direction which contain the mobile alkali cations together with immobile $O^=$ ions bridged between two Al^{3+} ions. The arrangement of sodium and oxygen ions in the conducting phase of βAl_2O_3 is shown in Figure 4.

The three structures are shown along the c axis in Figure 5, where it is seen that the spinel blocks are 11.3 Å thick in β and β'' alumina (4 anion layers) and 15.9 Å in β''' (6 anion layers). The principal difference between β and β'' alumina is that in the former the conduction plane is a plane of symmetry and the unit cell consists of two spinel blocks related by a two-fold screw axis (c = 22.6 Å), while in β'' Al_2O_3 the conduction plane is not a mirror plane and the unit cell contains three spinel blocks

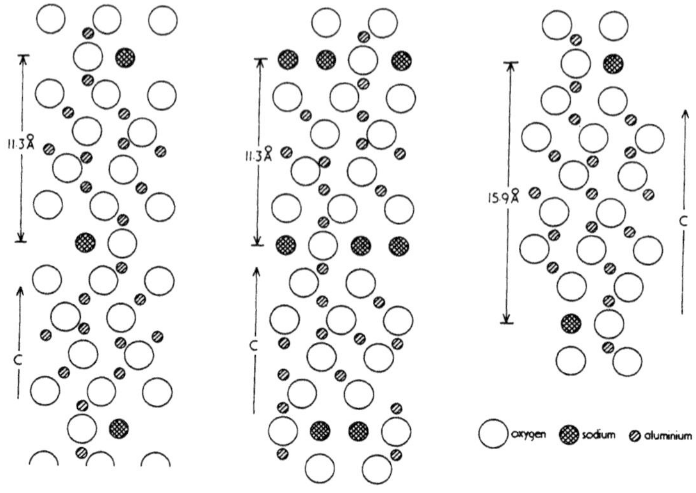

(a) β - Alumina (b) β" - Alumina (c) β‴ - Alumina

PROJECTIONS OF STRUCTURES OF β-ALUMINAS ON TO 1$\bar{1}$20 SIZES OF ATOMS APPROXIMATELY TO SCALE.

Figure 5

related by a three-fold screw axis (c = 33.9 Å).

The ideal compositions of the sodium phases and their unit cell dimensions are given in Table 2. Sodium βAl_2O_3 generally contains substantially more soda than is indicated by the idealised formula $NaAl_{11}O_{17}$; there are presumed to be corresponding Al^{3+}

Table 2

Ideal Compositions and Lattice Constants of Sodium βAl_2O_3

Phase	Ideal Composition	Lattice Constants (Å)	
β	$Na_2O.11Al_2O_3$	a = 5.59	c = 22.6 (2 x 11.3)
β''	$Na_2O.5.33Al_2O_3$	a = 5.59	c = 33.9 (3 x 11.3)
β'''	$Na_2O.4MgO.15Al_2O_3$	a = 5.62	c = 31.8 (2 x 15.9)

vacancies in the spinel blocks. Sodium β'' alumina has been shown to be metastable in the pure form $Na_6Al_{32}O_{51}$ [27], but is stabilised by the addition of MgO. An X-ray structural determination has been carried out on a single crystal of $Na_2O.MgO.5Al_2O_3$ [29]. The β''' phase invariably contains MgO or other additives, though not usually in the proportion indicated by the idealised formula.

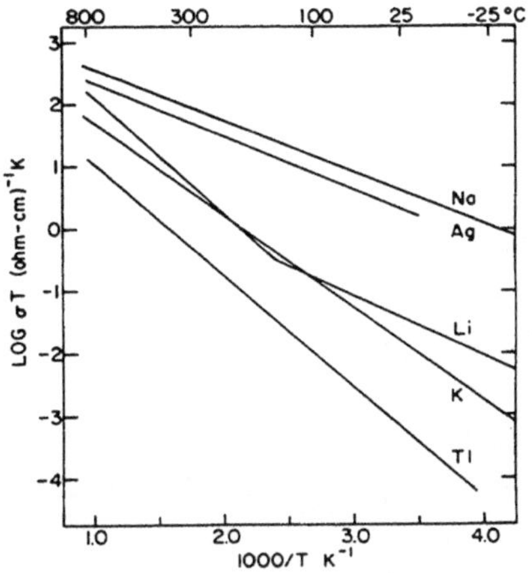

Figure 6 Ionic Conductivities of Substituted Beta Aluminas

Beta alumina ceramic electrolyte is generally a mixture of sodium β and β'' alumina in varying proportions. The ionic resistivity decreases linearly with increasing proportion of β'' phase from 16 ohm-cm for β at 360°C to 2.5 ohm-cm for β''[30]. Most of the other beta aluminas can be prepared from the sodium salts by cation exchange, either directly or via silver β alumina as an intermediary. Self diffusion coefficient measurements[3] and ionic conductivity measurements[31] (Figure 6) are in agreement in showing that sodium is the most mobile of the monovalent cations. Recently, A.C. conductivity methods have been employed to analyse the impedance of poly-crystalline sodium β alumina and it was shown that the electrical properties are determined more by the grain boundaries than by the grain interiors[32]. It follows that much of the voltage drop during current flow appears across the grain boundaries, giving rise to very large electric fields in this region. Electrical breakdown at grain boundaries, leading to incipient cracking, may be one cause of failure in ceramic electrolytes. However, it has been demonstrated that beta alumina is capable of transferring large amounts of electrical charge (i.e. 1700 Ah/cm^2) without failure[33].

Some interesting information on the crystal structure of ceramic grade sodium β alumina has stemmed from electron microscopy using the fringe imaging technique[34]. The existence of intergrowths of β and β'' alumina within a single crystal has been clearly demonstrated, the domains of each structure being oriented with the conduction planes parallel to each other (Figure 7). Spacings of 16Å, corresponding to laminae of the β''' phase have been seen in addition to the more usual 11Å spacing.

Figure 7 Oriented Intergrowths in Beta Alumina (Electron Microscope Lattice Imaging)

A number of studies have also been made of impedence effects at the sodium/βAl_2O_3 interface[35], polarisation at the βAl_2O_3/ sodium sulphide interface[36] and Na^+ transport between a liquid and beta alumina[37]; these topics will be discussed by other speakers at the summer school.

The beta aluminas are not the only oxides which show high ionic conductivity at moderate temperatures. Goodenough and colleagues[38] have investigated the meta-antimonates $MSbO_3$ (M = Li, Na, K, Rb, Tl and Ag) which crystallise in cubic, disordered phases. The SbO_3 octahedra form a rigid subarray through which pass <111> tunnels in which the M^+ ions are randomly distributed. Preliminary measurements suggested that the ionic conductivity of $NaSbO_3$ is comparable to that of $Na_2O.11Al_2O_3$.

Other complex oxides which are known or expected to exhibit high ionic conductivities at moderate temperature as a result of possessing tunnel structures are the hollandites, $K_xMg_{x/2}Ti_{(8-x/2)}O_{16}$ and $K_xAl_xTi_{(8-x)}O_{16}$[39], the tungstates ($K_2W_4O_{13}$), niobates and vanadates(31). Little work has yet been reported on these compounds.

3. SOLID ELECTROLYTE BATTERIES

For a battery to be of practical use it must be capable of sustained current drain without polarising. The magnitude of the current density required depends upon the application, ranging from a few micro-amps cm^{-2} for heart-pacemakers to ~0.1-0.2 A cm^{-2} for traction batteries. Thus it is in battery applications that the surface interfacial problems of solid electrolytes and electrode polarisation phenomena assume real significance. When a solid anode and/or a solid cathode is employed in conjunction with a solid electrolyte, electrode polarisation is almost inevitable at high current densities. Solid state batteries are therefore generally confined to small primary cells with low current densities, while for high current-density batteries which are capable of repeated recharge cycling it is desirable to employ liquid or gaseous electrodes in conjunction with a solid electrolyte. This explains the wide-spread interest in the sodium/sulphur battery.

A solid state battery which operates at ambient temperature may be represented as shown:

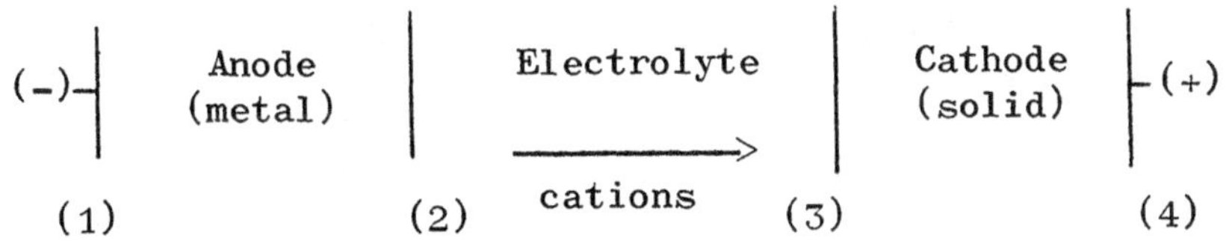

Four distinct cases may be identified:

<u>Case 1</u> The cathode is a good electronic conductor but poor ionic conductor. The reaction product will be deposited at interface (3) and may impede further reaction, leading to rapid polarisation.

<u>Case 2</u> Cathode is a good ionic conductor but poor electronic conductor. The reaction product will form at interface (4) - the current collector - and the cell will operate well provided that the product is itself an electronic conductor. Alternatively, an electronic conducting phase such as a metal or graphite may be incorporated in the cathode.

<u>Case 3</u> Cathode is a poor ionic and electronic conductor. This incorporates the worst features of cases 1 and 2.

<u>Case 4</u> Cathode shows good ionic and electronic conduction. This is the most favourable case, as is recognised by the current interest in the development of ionic cathodes.

3.1 Silver and Copper Electrolyte Batteries

Developments in the field of solid state batteries up to 1972 were reviewed by Liang[6]. Early interest in the $Ag/Ag_3SI/I_2$ cell diminished when it was realised that Ag_3SI was not stable with respect to iodine and reacted as shown:

$$I_{2(g)} + Ag_3SI_{(s)} \longrightarrow 3AgI_{(s)} + S_{(s)}$$

This cell was superceded by the Ag/$RbAg_4I_5$/RbI_3 system which can deliver moderately high currents but is only a low energy density cell. As mentioned previously, $RbAg_4I_5$ is thermodynamically unstable below 27°C, but the kinetics of decomposition are slow enough, in the absence of moisture, that the cell operates satisfactorily at low temperature. However, the phase change is significant with respect to the cell discharge product formed; above 27°C the cell product is $RbAg_4I_5$ itself, which serves to extend the electrolyte, while below 27° the poorly conducting phase Rb_2AgI_3 is formed. This problem at low temperatures may be partially overcome by incorporating $RbAg_4I_5$ in the cathode structure, in addition to the graphite current collector, so as to facilitate ionic flow during discharge.

The Ag/$RbAg_4I_5$/RbI_3 cell has been employed in military applications, especially where long shelf life is required. Among its advantages are claimed

1. Reliable performance over wide temperature range.

2. Ease of fabrication.

3. Resistance to mechanical shock and vibration.

4. Low cost automated manufacture.

5. Long shelf life - limited by slow diffusion of iodine through the electrolyte along grain boundaries.

Recent research has shown that the new silver electrolytes of mixed anion structure, e.g. $Ag_6I_4XO_4$ where X = Cr, Mo, W (vd. Section 2.2) are more stable towards moisture and are not subject to iodine migration through the electrolyte to the anode[19,22]. Similar advantages are claimed for $Ag_7I_4AsO_4$ and $Ag_7I_4VO_4$[21]. For this reason pure iodine may be used as cathode in place of the complex RbI_3. A typical cathode consists of iodine mixed with graphite, to facilitate electronic conduction, and powdered electrolyte to aid ionic conduction. Cells of the type Ag, $Ag_6I_4WO_4$/$Ag_6I_4WO_4$/I_2, $Ag_6I_4WO_4$, C give an open circuit voltage of 687 mv at 21°C, which agrees with that calculated from ΔG^o for the Ag/I_2 reaction. From these observations it seems likely that $RbAg_4I_5$ will be superceded by the new mixed anion electrolytes.

Another class of cathode which has been employed with solid state silver cells are the organic charge transfer complexes[40]. These are two component systems consisting of an electron donor (typically, polycyclic aromatic compounds) and an electron acceptor (such as a halogen). Their resistivity is several powers of ten less than that of either component alone. The best studied of these charge transfer complexes is the perylene-iodine system which has a resistivity as low as 20 ohm-cm. When used in the $Ag/RbAg_4I_5/P-I_2$ cell a stable discharge curve at 0.2 mA was obtained with high cathodic utilisation (65%). The cell also showed promising recharge behaviour. It seems likely that further work will be done on these charge transfer complexes.

Comparatively little work on copper batteries has been reported. The substitution of silver by copper in $RbAg_4I_5$ does not exceed 0.34 wt % and cells of this electrolyte with copper anodes behaved in an unstable fashion [41]. Preliminary cell measurements have also been reported using NN' dimethyl triethylene-diamine dibromide - cuprous bromide in conjunction with copper anodes [61]. With charge transfer complex cathodes (Br_2-perylene, iodine-perylene) the cells were unstable as the halogen oxidised the CuBr (or CuI) reaction product to the Cu^{++} state. With a Cu_2S cathode a more stable behaviour was observed.

3.2 Beta-Alumina Electrolyte Batteries

The principal battery which is under development with a beta-alumina electrolyte is the sodium/sulphur system which operates at 300-400°C and has liquid electrodes. This is described in my second lecture to the summer school. The interest in this battery stems from its projected application as a traction battery for vehicles and as a load levelling device for power stations. For such applications the energy and power requirements are many orders of magnitude larger than for the solid state batteries described above and current densities of $\sim$200 mA/cm^2 are commonplace. Liquid electrodes are therefore necessary to avoid electrode polarisation effects. The practical problems of developing these high temperature, high power level batteries are formidable and a substantial world-wide effort is involved.

Beta alumina has also found application as electrolyte in small, primary batteries operating at ambient temperatures, even though its specific conductance is only $\sim 10^{-3}$ (ohm-cm)$^{-1}$. Such batteries are clearly low power devices, but their significance lies in very long life with freedom from self-discharge and reliability. This is an important criterion for applications such as heart pace-makers and electronic watches. Research in this field has been undertaken mostly at the General Electric Company (USA) by Will and Mitoff who have employed sodium amalgam anodes and halogens, air or water as cathodes [42]. Some current/voltage curves for these cells are shown in Figure 8. The amalgams, which are sodium-rich, are two phase (liquid + solid) at all temperatures above 21.5°, the freezing point of the eutectic composition at

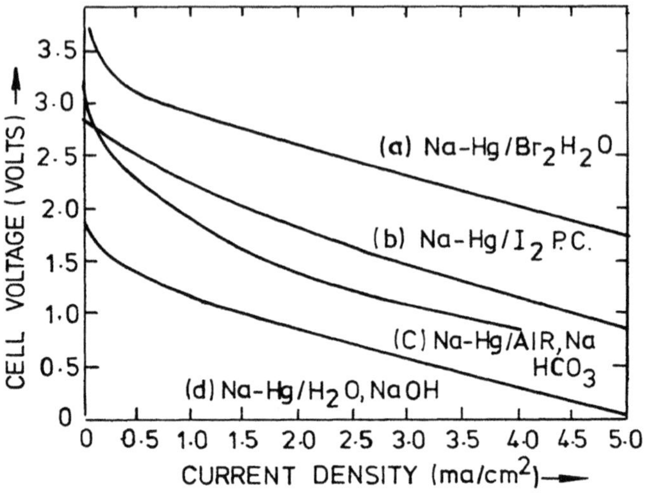

Figure 8 Current/Voltage Curves for Sodium-Mercury Amalgam Primary Cells with various cathodes at 25°C.

39.8% Na. The cathodes studied were aqueous solutions of bromine and iodine, air saturated $NaHCO_3$ solution and NaOH solution. General Electric has now announced the successful development of heart-pacemaker batteries of this type with bromine cathodes, giving an open-circuit voltage of 3.6 V, which are expected to have an operating life of ten years.

4. IONIC CATHODE BATTERIES

Solid state ionic cathodes are mixed ionic/electronic conductors, generally with layer or tunnel structures in which ionic diffusion is rapid. The phenomenon of mixed conduction in solids has been discussed in theoretical terms by Heyne(43) and a review of suitable compounds with tunnel and layer structures is included in a paper by Whittingham(12).

4.1 Cathode materials

Several different classes of potential ionic cathode materials have been identified:

Inorganic Bronzes

The well known tungsten and vanadium oxide bronzes of general formula M_xWO_3 and $M_xV_2O_5$, where M = Li, Na and $0 < x < 1$, possess tunnel structures likely to exhibit high ionic mobility and are also good electronic conductors[44]. These compounds generally exhibit a moderately wide range of stoichiometry with respect to the alkali metal without change of phase. For example, sodium tungsten bronze Na_xWO_3 has two non-stoichiometric phases, a tetragonal phase with $0.2 < x < 0.5$ and a cubic (perovskite) phase with $0.45 < x < 1$. The tetragonal phase exhibits a fair mobility for Na^+ ions but the cubic phase is a poor ion conductor[45]. Recently thermodynamic studies have been carried out of $Na_xV_2O_5$ and Na_xWO_3[46,47] allowing estimates to be made of the partial molar free energy of solution of Na in the bronzes. The corresponding voltages of the cell Na/electrolyte/Na_xWO_3 were in the range 2.43-2.60 V depending upon the value of x.

Transition Metal Chalcogenides

The di- and tri-chalcogenides of the transition metals Ti, Zr, Nb, Ta, Mo and W are assuming interest as possible ionic cathodes. The structures of these and other chalcogenides have been reviewed by Flahaut[48]. Certain of these compounds can be intercalated readily by electron donor molecules such as ammonia, pyridine etc. Thus TiS_2 and TaS_2, which are layer compounds with the $Cd(OH)_2$ structure, can absorb substantial amounts of ammonia by intercalation between the layers resulting in expansion of the unit cell[12].

Graphite complexes

Graphite is well known as a material which intercalates the alkali metals. Less well known is the fact that graphite reacts with fluorine to form a compound $(CF)_n$ which is essentially an intercalation compound, the fluorine atoms being situated between the carbon layers[49]. The spacing between the layers expands from 3.35 to 5.8 Å during reaction. This fluorographite provides a potential ionic cathode, particularly for use in lithium batteries where the cell voltage is ~2.8 V. During electrochemical discharge it seems likely that the lithium diffuses into the cathode to form a Li-F-C complex.

A further class of graphite interstitial compounds suitable for use as ionic cathodes has the general formula $M_xC_\alpha M'X_n$, where

M is an alkali metal

M' is a transition metal (Cr, Fe, Mn) in a high valence state

X is an electronegative non-metal (O,F.Cl)

These compounds were reported by Armand[50] who described the preparation and properties of C_8CrO_3. Alkali metals were then further inserted into the lattice either chemically or electrochemically. The cell Na/beta alumina/C_8CrO_3 was shown to have an open circuit voltage as high as 3.9 V. Other possible insertion compounds are CrO_2F_2, MnO_3F, CrF_6, Mn_2O_7.

Charge Transfer Complexes

These are discussed in section 3.1.

Iron-doped Beta Alumina

There is some interest in iron-doped beta alumina for use as an ionic cathode in conjunction with a beta alumina electrolyte. The obvious attraction of this system is that interfacial problems are largely eliminated and the ionic diffusivity in the cathode should approximate to that in the electrolyte. The advantages of an all-ceramic battery were recognised some while ago[51], but the observed power density was too low for practical battery applications, although it was suggested that the device might find application as a capacitor.

More recently, a cell of the type $Na_{(1)}$/β alumina/$Na_2O.11(Fe_xAl_{2-x}O_3)$ has been investigated and shown to have an OCV of ~2.4 V[52]. Surprisingly, at 120°C currents as high as 40-50 mA/cm^2 could be drawn.

Lithium Batteries

Ionic cathodes have been investigated most widely in the context of lithium batteries. In part this is because of the small size of the Li^+ ion, and therefore its ease of diffusion and intercalation, and in part because there was already substantial interest in lithium batteries on account of their high energy densities.

The development of organic electrolyte lithium batteries up to 1970 was reviewed by Jasinski[53]. Up to that time, and indeed since, most cathodes employed were simple transition metal oxides, sulphides or halides mixed with graphite to provide electronic conduction. Typical examples are MoO_3, CuS, MnO_2, PbO_2[54,55]. It has generally been assumed that the cathodic reduction is a conventional solid state reaction with phase change:

$$\text{e.g.}\quad MoO_3 + 2Li \longrightarrow MoO_2 + Li_2O$$

although Campanella and Pistoia have pointed out[54] that the observed OCV for the Li/organic electrolyte/MoO_3 cell is in excess of that expected from thermodynamic data for the above reaction. This is suggestive either that the electrolyte participates in the cathodic reaction to form a complex product or that an intercalation compound may be formed. Further experimental work at Imperial College with TiS_2 cathodes and K_xWO_3 cathodes, using organic electrolytes, has provided evidence for the formation of alkali metal intercalation compounds[56]. Lithium is particularly easy to diffuse into non-stoichiometric oxide or sulphide cathodes with layer or tunnel structures. Research has been conducted at the Bell Laboratories into the electrochemical behaviour of lithium cells with transition metal chalcogenides as cathodes[60]. Compounds studied included $NbSe_3$, NbS_3, $TaSe_3$, TaS_3, TiS_3. All these systems were found to be rechargeable with OCV's of 2.4-2.8 V and preliminary data suggested that they were suitable candidates as cathodes for high energy batteries. Further work is required on these materials to establish the relationship between ionic mobility and crystal structure and also whether the organic electrolyte enters into the reaction.

The electrochemical behaviour of the fluorographite $(CF)_n$ intercalation cathodes has also been investigated in lithium/organic electrolyte cells. These electrodes have very high theoretical energy contents (e.g. 3280 Wh/kg compared to 650 Wh/kg for MoO_3) and in a practical lithium battery energy densities up to about 330 Wh/kg may be expected[57]. Open circuit voltages of 3.1-3.3 V were observed, below the theoretical value of 4.66 V. There are, however, several problems associated with these cathodes[58]: they are non-conductors electronically, the electrochemical reaction rate is slow, resulting in low current densities and the need for high surface area electrodes, and, finally, the electrodes may be polarised by the reaction products.

Further information on the electrochemical performance of practical lithium cells, including those with $(CF)_n$ cathodes, is given in a battery review paper[59].

5. CONCLUSION

From a number of papers published in the past three years it is clear that the application of solid state ionics to battery systems of all types is a dynamic field of research which is growing rapidly. Satisfactory high conductivity ionic electrolytes are now available for small primary batteries operating at ambient temperatures and for large rechargeable batteries at 300-400°C. The remaining problem areas in the ionic electrolyte field are mostly associated with electrode polarisation phenomena, although there is still a need to discover an alkali metal conductor with a room temperature conductivity better than beta alumina and approaching that of the silver salts.

The field of ionic cathodes is still in its infancy and, indeed, it is at present often unclear whether a particular cathode material undergoes classical chemical reaction or takes up the diffusing cation in an intercalation process. Further work on this topic is to be expected.

References

1. J. N. Bradley and P. D. Greene, Trans. Farad. Soc. 62, 2069 (1966), 63, 424 (1967)

2. B. B. Owens and G. R. Argue, Science N.Y. 157, 308 (1967)

3. Yung-Fang Yu Yao and J. T. Kummer, J. Inorg. Nucl. Chem. 29, 2453 (1967)

4. M. N. Hull, Energy Conversion, 10, 215 (1970)

5. B. B. Owens, Adv. Electrochem. and Electrochem. Eng. 8, 1 (1971)

6. C. C. Liang, "Fast Ion Transport in Solids", (Ed. W. van Gool), North Holland (1973) P.19

7. H. L. Tuller and A. S. Nowick, J. Electrochem. Soc. 122, 255 (1975)

8. T. Kudo and H. Obayashi, J. Electrochem. Soc. 122, 142 (1975)

9. T. Takahashi, H. Iwahara and Y. Nagai, J. Appl. Electrochem. 2, 97 (1972)

10. T. Takahashi and H. Iwahara, J. Appl. Electrochem. 3, 65 (1973)

11. T. Takahashi and H. Iwahara, Energy Conv. 11, 105 (1971)

12. M. S. Whittingham, Electrochimica Acta. 20, 575 (1975)

13. A. Kvist and A. Lunden, Z. Naturfors. A21, 1509 (1966)

14. S. Pizzini, J. Appl. Electrochem. 1, 153 (1971)

15. B. Head and A. Lunden, Power Sources Symposium (Brighton 1974) Paper 31

16. H. Rickert, Z. Phys. Chem. B24, 418 (1960)

17. L. Heyne, Electrochim. Acta. 15, 1251 (1970)

18. T. Takahashi, S. Ikeda and O. Yamamoto, J. Electrochem. Soc. 119, 477 (1972)

19. T. Takahashi, S. Ikeda and O. Yamamoto, J. Electrochem. Soc. 120, 647 (1973)

20. B. B. Owens, "Fast Ion Transport in Solids" (Ed. W van Gool) North Holland (1973) P.593

21. B. Scrosati, F. Papaleo and G. Pistoia, J. Electrochem. Soc. 122, 339 (1975)

22. G. Chiodelli, A. Magistris and A. Schiraldi, Electrochim. Acta. 19, 655 (1974)

23. S. Geller, "Fast Ion Transport in Solids" (Ed. W. van Gool) North Holland (1973) P.607

24. S. Geller, P. M. Skorstad and S. A. Wilber, J. Electrochem. Soc. 122, 332 (1975)

25. G. W. Herzog and H. Krischner, "Reactivity of Solids" (Ed. J. S. Anderson, M. W. Roberts and F. S. Stone) Chapman and Hall (1972), P.140

26. T. Takahashi and O. Yamamoto, J. Electrochem. Soc. 122, 83 (1975)

27. Y. Le Cars, J. Théry and R. Collongues, Rev. Int. Haut Tempér et Réfract 9, 153 (1972)

28. M. Bettman and L. L. Terner, Inorg. Chem. 10, 1442 (1971)

29. M. Bettman and C. R. Peters, J. Phys. Chem. 73, 1774 (1969)

30. J. L. Sudworth, M. D. Hames, M. A. Storey, M. F. Azim and A. R. Tilley, Power Sources 4, (Ed. D. H. Collins) Oriel Press (1973) P.1

31. M. S. Whittingham and R. A. Huggins, "Reactivity of Solids" (Ed. J. S. Anderson, M. W. Roberts and F. S. Stone), Chapman and Hall (1972), P.125

32. R. W. Powers and S. P. Mitoff, J. Electrochem. Soc. 122, 226 (1975)

33. Y. Lazennec, C. Lasne, P. Margotin and J. Fally, J. Electrochem. Soc. 122, 734 (1975)

34. D. J. M. Bevan, B. Hudson and P. T. Moseley, Mat. Research Bull. 9, 1073 (1974)

35. R. D. Armstrong, T. Dickenson and J. Turner, Electroanal. Chem. and Interfacial Electrochem. 44, 157 (1973)

36. R. D. Armstrong, T. Dickenson and R. Whitfield, Electroanal. Chem. and Interfacial Electrochem. 32, App. 9 (1971)

37. M. Voinov and H. Tannenberger, Electrochim. Acta. 19, 959 (1974)

38. H. Y-P. Hong, J. A. Kafalas and J. B. Goodenough, J. Solid State. Chem. 9, 345 (1974)

39. J. Singer, H. E. Kautz, W. L. Fielder and J. S. Fordyce, "Fast Ion Transport in Solids" (Ed. W. van Gool) North Holland (1973) P.653

40. B. Scrosati, M. Torroni and A. D. Butherus, Power Sources 4, (Ed. D. H. Collins) 453 (1973)

41. B. Scrosati, G. Pistoia, M. Lazzari and L. P. Bicelli, J. Appl. Electrochem. 4, 201 (1974)

42. F. G. Will and S. P. Mitoff, J. Electrochem. Soc. 122, 457 (1975)

43. L. Heyne, "Fast Ion Transport in Solids" (Ed. W. van Gool) North Holland, (1973) P.123

44. P. G. Dickens and M. S. Whittingham, Quart. Rev. 22, 30 (1968)

45. P. G. Dickens, M. Jewess, D. J. Neild and J. C. W. Rose, J. C. S. Dalton, 1973, P.30

46. T. A. Ramanarayanan and W. L. Worrell, J. Electrochem. Soc. 121, 1530 (1974)

47. M. S. Whittingham, J. Electrochem. Soc. 122, 713 (1975)

48. J. Flahaut, Int. Rev. Sci. Inorg. Chem. (Series 1) Vol. 10, 189 (1972)

49. M. S. Whittingham, J. Electrochem. Soc. 122, 526 (1975)

50. M. B. Armand, "Fast Ion Diffusion in Solids" (Ed. W. van Gool) North Holland, 1973, P.665

51. K. O. Hever, J. Electrochem. Soc. 115, 830 (1968)

52. J. H. Kennedy and A. F. Sammells, J. Electrochem. Soc. 121, 1 (1974)

53. R. Jasinski, Coll. Czech. Chem. Commun. 36, 1079 (1971)

54. G. Campanella and G. Pistoia, J. Electrochem. Soc. 118, 1905 (1971)

55. F. W. Dampier, J. Electrochem. Soc. 121, 656 (1974)

56. B. C. H. Steele, Mat. Res. Bull (in press)

57. H. F. Hunger and G. J. Heymach, J. Electrochem. Soc. 120, 1161 (1973)

58. W. Tiedemann, J. Electrochem. Soc. 121, 1308 (1974)

59. D. Linden, N. Wilburn and E. Brooks, Power Sources 4, 483 (1973)

60. J. Broadhead, Power Sources 4, 469 (1973) Power Sources Symposium (Brighton 1974) Paper 41

61. M. Lazzari, R. C. Pace and B. Scrosati, Electrochim. Acta. 20, 331 (1975)

DISCUSSION

FARRINGTON : You identified "K Hollandite" as a conductor of K^+ ions. I am not aware of any conclusive evidence that K^+ ions are actually conducted through the Hollandite structure under d.c. conditions. Our results (1) reveal significant ion shifting at high frequencies but a d.c. ionic conductivity of $< 10^{-8}\ \Omega^{-1}\ cm^{1}$ between 25°-150°C in single crystal samples.
(1) G.C. Farrington, "Electrochemical Properties of K Hollandite" extended Abstracts, Electrochemical Society, May 1975.

DELL : The evidence for K^+ ion conduction in Hollandites was provided by Singer et al.(39). Using an a.c. screening method, they calculated values of $\sigma_{a.c(298K)}$ of up to $10^{-2}\ (\Omega cm)^{-1}$ for potassium magnesium hollandites. There was however a wide spread in reported values.

EICHINGER : The reason why intercalation compounds (especially alkali intercalation compounds) are not widely used until now seems to be, that they are reacting with the common organic solvants. Even solvants which are "stable", versus lithium metal are reacting with intercalation compounds (example PC). This reaction may be due to the fact no protective film formation occurs at an intercalation compound, whereas e.g. a film of protective Li_2CO_3 on a lithium surface is formed in PC.

DELL : There seems little doubt that the organic electrolyte plays a role but I am not too clear whether it reacts with the lithium intercalation complex as suggested by Dr. Eichinger, which might be deleterious, or whether the electrolyte plays an essential role in the intercalation reaction itself, serving to open up the layer structure and so allow the lithium to enter. Possibly either effect can occur with different systems.

DOTY : Since you chose to mention G.E. Na/β-alumina/Br_2 pacemaker cell, I feel that the Li/I_2 complex battery developed by Dr. A.L. Schneider at Catalyst Research Corp. should also be mentioned. Although there have not been very many publications about this system, Dr. Schneider has presented papers on it at Brighton, last year, and also at the various U.S. Power Sources Conferences. This battery is probably one of the few really commercially available batteries. It has been used in several thousand pacemakers over the last 2 1/2 years and is finding increasing usage each year.

RECHARGEABLE SOLID ELECTROLYTE TRACTION BATTERIES

R. M. Dell and L. S. Evans

A.E.R.E. Harwell, Oxfordshire, U.K.

1. Introduction

In recent years there has been a rapidly developing interest in the scientific phenomena of fast ion transport in solids[1] and of electrode processes in solid state ionics, the subject of the present summer school. This interest has arisen in part from a desire to understand the basic processes involved and in part from an appreciation of the role which super ion conductors and electrode processes may play in useful electrochemical devices. Among the latter, much attention is being directed towards the development of batteries based upon solid electrolytes.

Solid state batteries are of two general types, viz: small primary batteries, frequently employing a derivative of silver iodide as the electrolyte, and much larger, rechargeable batteries for use in traction applications and in bulk electricity storage. Research into rechargeable solid electrolyte batteries has centred around the sodium/sulphur system which has a sodium beta alumina electrolyte and liquid electrodes at the operating temperature of $350 \pm 50^{o}C$. This battery is under intensive development in U.S.A., France, Germany, Japan and the United Kingdom. Its attractive features include a high energy density and the ready availability and low cost of the essential components. Relatively little development work has been undertaken to date on other solid electrolyte, rechargeable batteries.

In considering future possible battery systems, a gulf is apparent between the viewpoints of the solid state scientist, interested in the fundamental transport properties of solids and electrode processes but knowing little about battery applications,

M. Kleitz and J. Dupuy (eds.), Electrode Processes in Solid State Ionics, 411–430.

and the battery design engineer to whom the idea of a solid electrolyte and liquid electrodes is novel. The object of this paper is to attempt to bridge this gap by drawing together some of the scientific, engineering and operational aspects involved in the development of a commercially viable traction battery for use in road transport and showing how these interact in a complex and iterative fashion. The paper is based upon the experience of the authors and their colleagues in attempting to develop sodium/sulphur batteries. As the summer school is composed predominantly of scientists already knowledgeable about fundamentals, the emphasis will be on battery engineering and technology. In this way it is hoped to draw attention to the many constraints which must be borne in mind if an exciting new battery concept, as seen from the laboratory, is to be developed into a practical traction battery.

2. The Sodium/Sulphur Battery

The principle of the sodium/sulphur cell was first demonstrated in 1967 by the Ford Motor Co. in USA[2].

The cell is based upon a liquid sodium anode and liquid sulphur cathode, separated by a beta alumina ceramic-type electrolyte which is an electronic insulator, but through which sodium ions diffuse rapidly at 300-400°C. During discharge the reaction $2Na + 5S \rightarrow Na_2S_5$ leads to an open circuit voltage (OCV) of 2.08V. Continued reaction beyond Na_2S_5 results in the formation of lower polysulphides in the range Na_2S_4 - Na_2S_3 (OCV 1.78V), after which solid Na_2S_2 separates out. The polysulphides are liquid at cell operating temperature, but are immiscible with liquid sulphur. In the first stage of discharge, $S \rightarrow Na_2S_5$, two liquids co-exist (2 phase region); beyond Na_2S_5 discharge continues to Na_2S_3 as a single ionic liquid of moderate conductivity. During recharge the polysulphide dissociates and sodium ions diffuse backwards through the ceramic membrane to the anode where they are re-converted to sodium metal.

As liquid sulphur has a high resistivity and a low solubility for sodium ions it is necessary to incorporate an electron conductor in the sulphur electrode. This takes the form of a graphite or carbon felt, impregnated with liquid S, which extends from the ceramic electrolyte to the outer metal container. Current is collected from this outer metal container (positive) and passes to the liquid sodium electrode (negative) through the external load.

The theoretical energy of the $2Na + 3S \rightarrow Na_2S_3$ reaction is as high as 790 Wh/kg of reactants, but when allowance is made for the mass of the electrolyte, current collectors, container etc., practical cell energy densities of 250-350 Wh/kg are predicted (cf. 20-40 Wh/kg for Pb/acid). These are in the range of interest for vehicle traction batteries, even though the energy density will be degraded further by the assembly of cells into battery containers and by the mass of associated electrical connectors.

The feasibility of designing and building a Na/S traction battery capable of driving an electric vehicle was demonstrated in 1972 when the U.K. Electricity Council successfully operated an urban-type delivery van powered by a 50 kWh experimental sodium/sulphur battery (Figure 1)[3]. This battery, which had a peak power output of 29 kW, was constructed of 960 cells arranged in 40 modules of 24 cells each. Figure 2 shows the modules being prepared for assembly into the heated and insulated battery housing. A second experimental vehicle battery, built in 1973, consisted of 864 cells of similar design in 24 modules each containing 36 cells. Each of these batteries gave the van a maximum speed of 55 km h^{-1} on the level and a range without stops of 125 km.

The cells from which the battery was constructed are shown in Figure 3. The external casing, of stainless steel, contained a vertical beta-alumina ceramic electrolyte tube. The sodium metal anode was contained within this tube and in the connecting sodium reservoir located above the cell body. As the cell discharged, sodium ions passed through the electrolyte tube into the sulphur cathode which was contained in the surrounding narrow annulus. The cathode compartment was filled with compressed graphite felt. A mechanical, gasket-type seal was employed to isolate the sodium and sulphur from each other and from the atmosphere. Each cell had a capacity of 52.5 Wh and an energy density of 160 Wh/kg. More than 2000 of these cells were manufactured.

The performance of the experimental vehicles was impressive and confirmed the technical feasibility of the sodium/sulphur battery for traction applications. At the same time it was recognised that for commercial viability it would be necessary to improve the volumetric energy density of the battery and to custom-design it for a particular vehicle, as well as extending the lifespan and reliability of the cells. These developments involve problems of cell and battery design and further work in Britain has been directed towards these goals.

3. The Design Problem

The central problem which confronts all design engineers is to design a product which meets a stated performance and cost specification, within constraints imposed by the laws of science, the properties of materials and known engineering practice. The particular difficulty in the present instance is that the performance specification for vehicle traction batteries tends to be

Figure 1 Experimental Na/S battery-powered van

Figure 2 Sodium/sulphur battery modules as used in van

complex and comprehensive, especially when one is concerned with a battery operating at 350°C, while the fundamental scientific and materials constraints inherent in the sodium/sulphur battery concept are also fairly severe. In this situation it is necessary that the scientist and design engineer work together closely as a team to solve some difficult problems.

The performance specification and the acceptable cost of a battery depend, of course, on its application. Here we are concerned with traction batteries, primarily for use in road vehicles, and we have concentrated our attention upon urban buses and light delivery vehicles. In order to define the battery performance targets it is first necessary to consider the operational problem from the point of view of the traffic manager, who specifies the journey profile which the vehicle must be capable of making, and then to select a particular design of vehicle in order to interpret this duty cycle in terms of energy and power requirements and the permissible mass and volume of the battery. When this was done, we arrived at a specification of battery performance for each class of vehicle (Table 1).

These figures relate specifically to the Leyland National bus, for use on city suburban routes throughout the day without recharging, and to the British Leyland 3.5 tonne (gross) van, used for urban goods deliveries. The railway application is for a battery electric railcar for short journeys on non-electrified lines. The energy density and power density figures relate to the complete battery including cells, containment box and insulation. These are ideal "target" values and a lesser specification may well prove acceptable for many applications.

Comparing the bus and van batteries it may be seen that, apart from an obvious difference of size, they are similar as regards the target energy densities and cycle life, but differ principally in the key parameters of energy/power ratio and the time available overnight for recharge. Since stored energy is determined by the mass of sodium and sulphur present, while peak power is determined by the maximum permissible current density across the β alumina electrolyte during discharge multiplied by its active surface area, it will be clear that the E/P ratio must be reflected in cell design. The bus battery will have a larger mass of stored reactants per unit area of electrolyte. Although this gives a lower production cost per kWh of stored energy, it leads to a problem on recharge. Compared to the van, three times as much current must flow across each cm^2 of electrolyte, with only half the time available for recharge, i.e. the electrolyte in the bus battery needs to withstand up to six times the current density of the van battery during recharge without overheating or failure. This may well lead to differences in electrolyte specification for the two applications.

Table 1

Battery Performance Targets

		City Bus	Delivery Van	Railcar[+]
Vehicle gross weight	tonnes	15	3.5	36
Range	km/day	240	165	270
Peak power	kW	150	50	250
Stored Energy	kWh	450	50	425
Energy/Peak power ratio	h	3	1	1.7
Battery weight	tonnes	2.2	0.25	2.84
Battery energy density	Wh/kg	200	200	150
Battery power density	W/kg	70	200	88
Recharge time available	(minimum) hours/day	6	12	6
Battery life	cycles	>1000	>1000	>1000

[+]Data kindly supplied by British Rail

There are further operational criteria to be met which are less readily quantifyable than those in the table. These are:

(1) Geometrical considerations

Traction batteries must meet a target as regards volumetric energy density which is even more stringent than that of gravimetric energy density. Furthermore, if an existing vehicle is to be used (rather than designing the vehicle around the battery) there are geometrical constraints imposed by the structure of the vehicle body and the spaces available in which to accommodate the battery. For example, the floorheight, which is a particularly important parameter in bus design, limits the cell length which is acceptable if the battery is to be housed under the floor.

(2) Reliability

This is a key parameter, both because of the inconvenience of changing failed cells and because the overall economics of electric vehicle operation depend critically upon the maintenance costs being less than those of internal combustion engined vehicles. A battery life of at least three years is looked for. Also included under reliability are factors such as the incidence of cell failure and the ability to cool the battery to room temperature and reheat without problems resulting from thermal expansion effects.

(3) Ease of maintenance

When cell failure does occur, it will be necessary to obtain access to the interior of the battery in order to change individual cells or groups of cells, preferably without cooling down first. This poses a difficult design problem.

(4) Safety

Molten sodium and sulphur at 300-400°C are inherently reactive materials, especially when exposed to air or water. The uncontrolled sodium/sulphur reaction itself, which would take place in the event of electrolyte fracture, is rapid and highly exothermic. It is therefore necessary to design cells which are intrinsically safe under normal operating conditions and can withstand likely accident situations[1].

The overall design problem then is to produce a battery which meets these criteria within constraints imposed by science, materials and production engineering and to an acceptable cost. In the next section we consider these technical constraints in more detail and their interaction with the battery specification outlined above. Detailed cost considerations are outside the scope of the present paper.

4. Technical Constraints

The technical problems facing the Na/S battery designer may be considered, loosely, as scientific or engineering in nature as listed below:

Technical Problem Areas

Scientific	Engineering
1. Structure and resistivity of β Al_2O_3	1. Sealing of cells
2. Ceramic fabrication problems	2. Optimisation of cell design
3. Electrolyte failure mechanism	3. Battery design
4. Physical chemistry of molten S/Na_2S_x mixtures	4. Heat transfer and insulation
5. Electrochemistry of sulphur electrode	5. Safety criteria
6. Corrosion phenomena	6. Production engineering

Each of these problem areas gives rise to its own particular technical constraints which the design team must take into account. Furthermore, many of these constraints are interactive with each other and with the performance specification outlined above. Some of these relationships will be mentioned as we discuss each problem briefly.

4.1 Scientific Problem Areas

Structure and Resistivity of β Al_2O_3

Sodium beta alumina is not a single compound but exists as two distinct phases (designated β and β''), each of variable composition. The idealised formulae are

β Na_2O 11 Al_2O_3 (5.24 wt % Na_2O)

β'' Na_2O 5.33 Al_2O_3 (10.24 wt % Na_2O)

In practice, the β phase is invariably richer in soda than suggested by the idealised formula, while the β'' phase is meta-stable in the absence of added impurity ions (vd. phase

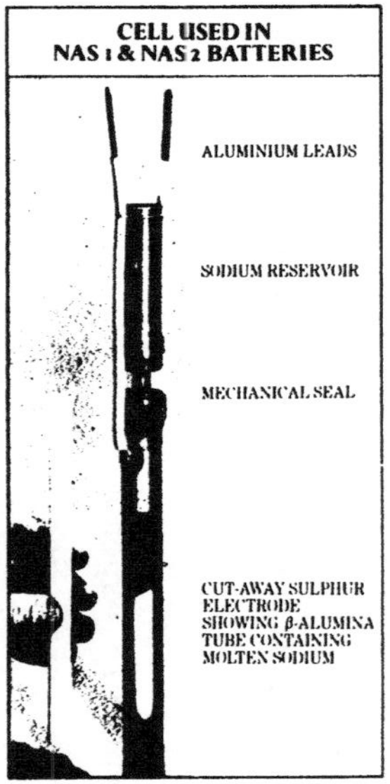

Figure 3

Sodium/sulphur cell

(U.K. Electricity Council)

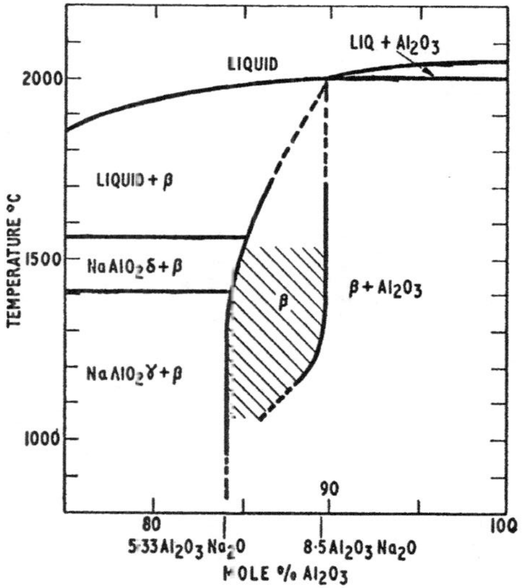

Figure 4

Phase diagram of Na_2O–Al_2O_3

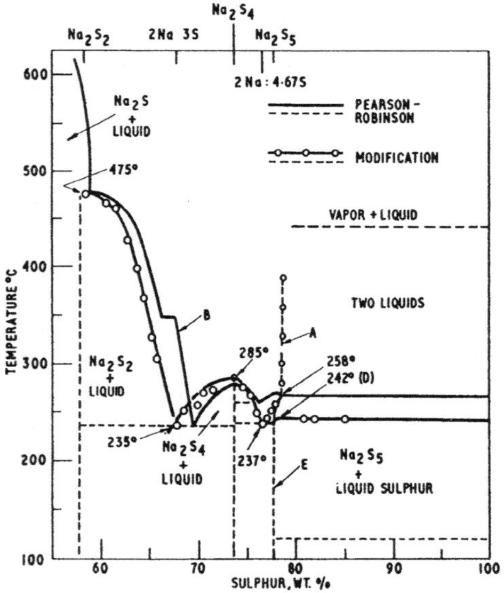

Figure 5

Phase diagram for the system Na_2S–S

diagram, Figure 4)[5]. By the addition of suitable oxide dopants the pure β'' phase may be formed, although more generally the electrolyte consists of the two phases, with individual crystallites showing ordered intergrowths parallel to the basal planes[6].

The β'' phase is a significantly better ionic conductor than the β phase:

Phase	Specific Resistance at 360°C
β	16 ohm-cm
β''	2.5 ohm-cm

Mixed compositions show intermediate values between these extremes[7].

In general, then, an electrolyte with a high proportion of β'' phase is desirable to give a low internal resistance to the cell. However, in selecting the preferred composition there are other important factors to be taken into consideration, including ease of ceramic fabrication, ceramic strength, resistance to atmospheric attack and stability to charge/discharge cycling. There is scope for more research before the optimum electrolyte composition is evolved.

Ceramic Fabrication Problems

Two general approaches have been adopted to the fabrication of electrolyte tubes: (1) isostatic pressing (2) electrophoretic deposition. In either case the "green" tube must be sintered at high temperature in air in order to form and densify the beta alumina phase.

Beta alumina tubes are not easy to fabricate satisfactorily, partly because of the stringent specification they must meet and partly on account of the intrinsic properties of the material itself. The specification includes:

- imperviousness at operating temperature
- high strength tubes
- straightness and close dimensional tolerances over ~30 cm length
- good surface finish
- stability to atmospheric exposure
- freedom from residual stresses

This is certainly a more stringent specification than is normally applied to other ceramic tubes. Moreover, there are certain fabrication problems which are peculiar to β Al_2O_3. One of these is the difficulty of reacting fully and densifying without allowing grain growth to take place; if any $NaAlO_2$ is formed at the grain boundaries, this rapidly promotes grain growth through liquid phase sintering. Another problem area is the rapid loss of sodium oxide, by decomposition and vaporisation at the sintering temperature.

Various practical expedients have been adopted by different laboratories to overcome these problems and the art of producing satisfactory tubes for use in experimental cells is now reasonably well developed. Nevertheless, the ceramic fabrication problem is highly interactive with both the composition chosen and the electrolyte failure problem and there is further technology to be perfected before the economic mass production of electrolyte tubes with optimum properties becomes a routine matter.

Electrolyte Failure Mechanism

Electrolyte tubes have been known to undergo cracking or fracture in use and also to fail through electronic short circuiting. Examination of fracture surfaces often reveals a microscopic inclusion or inhomogeneity in the ceramic. Short-circuiting is a more fundamental defect and occurs when sodium metal penetrates the tube wall along grain boundaries[7]. This is termed "dendritic growth". There is evidence that this dendritic growth is associated with the charging process and the dendrite starts to grow from the sodium side[8]. As may be expected, the effect is worse for coarse-grained electrolyte or high charge rates. There is thus a marked dependence of the electrolyte life, in terms of Ah/cm^2 passed, on the ceramic specification and the current density during recharge. Until better electrolyte is developed, this factor may limit the maximum practical recharge rate, a key factor in battery design.

Physical Chemistry of the Cathode

The phase diagram of the system Na_2S–S is shown in Fig. 5. At the cell operating temperature (350 ± 50°C) sulphur is completely immiscible with sodium pentasulphide and two liquid phases co-exist in the composition range S to Na_2S_5. The density difference between the two liquids is not large (1.66 g cm^{-3} for S and 1.85 g cm^{-3} for Na_2S_5 at 360°C) but their viscosities differ enormously. The presence of convective effects in the liquid cathode, as a result of resistive heating, will tend to oppose phase segregation and, on a simple model, one might expect discharge to begin with the formation of droplets of Na_2S_5 in sulphur as the continuous medium and proceed via a phase inversion to droplets of undischarged sulphur in a liquid Na_2S_5 phase. In practice the situation is complicated by the presence of the graphite felt which impedes the flow of liquid and may be preferentially wetted by sulphur. A simpler system exists in the single phase Na_2S_5-Na_2S_3 region and it is well known that Na/S cells are charged more easily in this region than in the two phase region.

Electrochemistry of the Sulphur Electrode

Since sulphur is a non-conductor of Na^+ ions and of electrons, the initial discharge can occur only at the 3 phase contact of the electrolyte, the melt and the graphite:

$$\underset{\text{(electrolyte)}}{2Na^+} + \underset{\text{(carbon)}}{2\varepsilon} + \underset{\text{(sulphur melt)}}{5S} \longrightarrow \underset{\text{(polysulphide melt)}}{Na_2S_5}$$

The polysulphide liquid so formed serves to extend the 3 phase contact since it is a reasonable Na^+ ion conductor. Further sulphur is brought into contact with carbon fibres as a result of the volume change associated with discharge. Discharge beyond Na_2S_5 to Na_2S_3 is straightforward.

During recharge the reverse procedure takes place. The situation is more critical in that care must be taken not to form insulating layers of sulphur around the graphite fibres, as these will inhibit electron exchange, or around the ceramic tube, as this would prevent Na^+ diffusion back to the electrolyte surface. Either effect is detrimental to efficient recharge. In general, for rapid recharge, the sulphur electrode thickness must not exceed 10 mm and the detailed design of the sulphur electrode is important.

Corrosion Phenomena

The thermal and electrochemical corrosion of stainless steel in the S/polysulphide melt has been reported previously (9). Thermal corrosion gives rise to an adherent, black sulphide deposit on the surface of the steel, but electrolytic corrosion is essentially a stripping process with the stainless steel cell body being progressively dissolved in the sulphide melt. Electrochemical experiments have revealed that this occurs predominantly during the final stages of discharge and is associated with polarisation of the melt and the formation of Na_2S_2 close to the steel surface.

The question of a suitable container for the cathode melt is one of the key problem areas in developing the Na/S cell; research at present is concentrating on the development of protective coatings for use on steel containers.

4.2 Engineering Problem Areas

Sealing of Cells

The satisfactory sealing of Na/S cells is another difficult technical problem. It is necessary to provide three separate seals - to seal both the sodium and the sulphur electrodes from the atmosphere and also from each other. In addition to being leak-tight, these seals must be chemically and electrochemically compatible with the reactants they contain (sodium and S/Na_2S_x, respectively) at temperatures up to 400^oC and for periods of several years. Ideally, the seals will have some flexibility to compensate for the rigidity and brittle nature of the ceramic tube.

A further sealing requirement is that the anode and cathode shall be electrically insulated from each other. According to the design either insulating seals may be used, or electrical insulators must be incorporated in addition to the seals. Again, such insulators must withstand electrical breakdown over the expected life of the cell.

Our experience at Harwell is that this is one of the most difficult problem areas and the life of test cells is generally terminated by seal failure. Various possible solutions are under investigation.

Optimisation of Cell Design

Before it is possible to design a vehicle traction battery it is necessary to settle on a cell design. The optimisation of cell design is a central issue which must take into account not only the technical constraints within the cell, listed above, but also constraints imposed by the vehicle design and intended duty cycle as well as considerations of safety and production engineering. Particularly important parameters are:

Sulphur annulus thickness: this is determined by the energy/power output required, the desired cell resistance and the time available for recharge. Safety considerations are also important.

Cell diameter: this is determined by the requirement to match the molar quantities of sodium and sulphur, making due allowance for incomplete reaction and the volume changes during discharge.

Cell length: large cells are desirable in order to minimise battery production costs, but practical lengths are set by ceramic fabrication problems and space considerations in the vehicle.

Materials and methods of construction: the technical problems of corrosion and sealing are mentioned earlier. These technical problems must be solved within a framework which is economically viable and the solution must lead to a cell design which is suitable for assembly by standard production engineering techniques. The development of an optimum cell design is a progressive and iterative process involving both experimentation and design calculations.

Battery Design

Having settled upon an optimum cell design for a given vehicle and duty cycle, it is now necessary to assemble the cells into a traction battery. Using the 50 Wh cells developed by the U.K. Electricity Council, the delivery van in Table 1 would require 1000 cells and the city bus 9000 cells. Clearly, this is unlikely to be an economic solution and larger cells are necessary, especially for the bus. Even so, it is likely that 500-1000 cells will be involved.

The assembly of these cells into a battery must take account of electrical engineering considerations, thermal control problems, questions of accessibility and maintenance and safety factors. With as many as 500-1000 cells involved it is evident that a series/parallel network will be required; the design of this network poses particular problems for Na/S cells because of two characteristics they possess:

(1) a fully charged cell will not pass current; it is therefore important that cells in a series chain shall all be well balanced or the first cell to be fully charged will prevent the others charging further.

(2) spontaneous cell failure, e.g. through electrolyte or seal breakdown, can lead to either open-circuit or short-circuit conditions. An open-circuited cell will isolate those cells in series with it, while a short-circuited cell will discharge those in parallel with it.

Another consideration in the series/parallel network design exercise is the wish of the vehicle operator to have at least 3 or 4 separate networks of cells, each providing full motor voltage, so that in the event of one network failing completely it will still be possible to drive the vehicle to its garage. This is important if the battery is to be kept hot. This requirement will impose a limitation on the maximum motor voltage or on the minimum number of cells which can be employed.

The internal resistance of a Na/S cell is higher than that of a comparable lead/acid cell. This higher resistance leads to reduced voltage at high current drains and limits the voltage efficiency of the Na/S battery to, typically, 65-85% depending upon the power output. This is not entirely wasteful as some heat output is required, in any event, to maintain the battery at operating temperature.

Heat transfer and insulation

The cell operating temperature range (300-400°C) is determined by the freezing point of Na_2S_4 (285°C) and by accelerated corrosion and seal degradation above 400°C. The battery must be thermally insulated so as to reduce the rate of heat loss on standing and also to assist in equilibration of temperature between cells. The van battery tested by the Electricity Council was contained in a double skinned box, well insulated with microtherm. Practical experience with this battery confirmed thermal analysis which showed that it is quite practicable to leave the battery standing idle for several hours without excessive cooling occurring. Similarly, reasonable runs at full power are possible without overheating. The amount of thermal insulation used is a compromise between these two factors.

In operation it is envisaged that the vehicle will not normally be left for more than a few hours out of use without connecting to a battery charger. Trim heaters will be built into the battery box and these can be energised either from the mains, when connected, or else from the battery itself.

Safety criteria

The question of safety is a complex issue and depends very much upon the battery application. We may distinguish between the consequences of spontaneous cell failure, caused by electrolyte fracture, internal short circuiting etc, and externally imposed cell failure as a result of impact damage. For spontaneous cell failure, the principle being adopted in the U.K. is that a burn-out in one cell shall not propagate to adjacent cells. There is good evidence that this can be achieved; the features of cell design which minimise the effects of spontaneous failure have been discussed in a recent paper[4].

Externally imposed impact damage is more difficult to

counter, but by suitable design of the battery housing and its central placement within the vehicle structure it should withstand all but the most severe collisions. One suggested criterion is that the battery should be capable of withstanding the same deceleratory forces as the human body. With thoughtful engineering this appears to be a practical goal.

Production Engineering

The final, and major constraint of the designer is that the cell and battery designs which evolve shall be amenable to mass production techniques at realistic costs. This is too extensive a topic to discuss here; suffice it to say that each proposed material of construction and component design must be reviewed critically in the light of this criterion. The size of the operation can be gauged by considering a modest market for 60,000 urban delivery vans with a battery life of 3 years. The annual requirement would then be for 20,000 traction batteries containing, say, 15-20 million cells. If a production line assembled cells at the rate of two per minute and operated three shifts, all the year, it would still require ~20 such production lines to manufacture this number of cells. The need for simplicity and automation is evident.

5. Conceptual Design Studies

As an exercise, conceptual design studies were undertaken for batteries to power a city bus and a light delivery van, with the performance targets listed in Table 1. Here we outline the exercise for the Leyland National bus.

The highly interactive nature of the performance targets and the various scientific and engineering constraints have been emphasised. It follows that the design process is likely to be an iterative procedure, leading to progressively better solutions. This preliminary exercise is a first step along this path.

A practical battery design must evidently be a compromise of performance, economics, servicability and safety. For this exercise an attempt was made to meet the requirements of servicability and safety by (a) subdividing the battery into a number of units - battery boxes (b) assembling the cells into modules contained in welded cartons. With the design selected, the 450 kWh bus battery was composed of 1320 cells, each of 336 Wh. The vehicle had five battery boxes, all mounted centrally in the chassis, between the wheels (Figure 6). Each battery box was thermally self-contained, with insulation, trim heaters etc, and enclosed six cartons of cells (Figure 7). Each carton was an all-welded structure containing 44 cells in series (Figure 8).

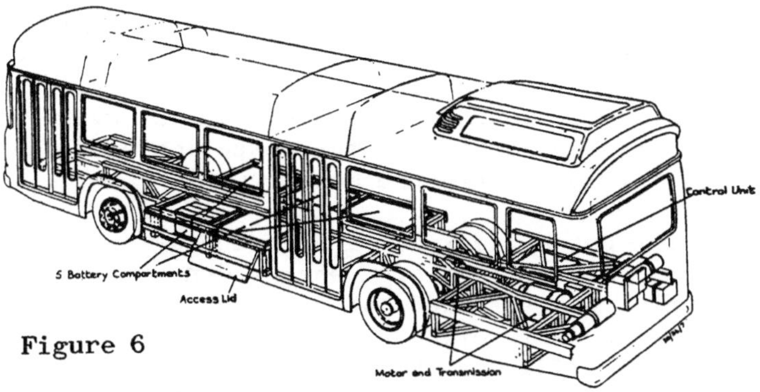

Figure 6

Battery layout in bus (conceptual design study)

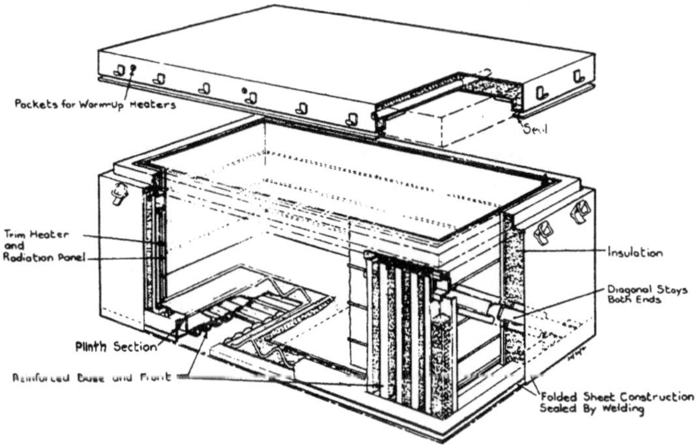

Figure 7 Battery compartment for bus

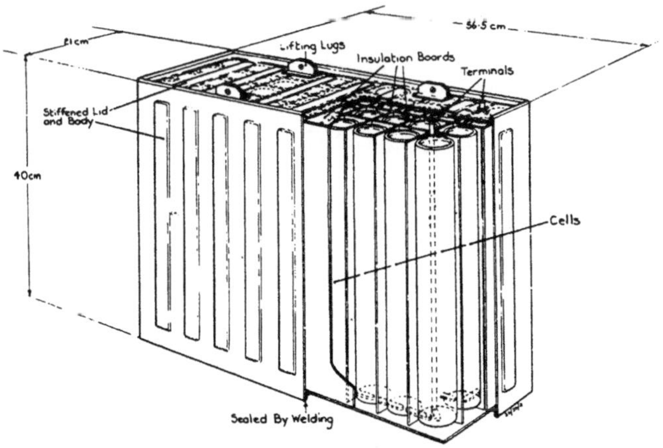

Figure 8 Single battery module for bus

The principle of tertiary containment was adopted to provide additional protection in the event of cell failure or accident damage. The carton of 44 cells is regarded, in this design, as a disposable unit which would be discarded in the event of cell failure. This design simplifies the detection of failed cells and the electrical wiring within the battery box.

Because the bus battery has a high energy/power ratio it is possible to design a cell of high energy density. The particular cell design which was selected had a calculated energy density of 310 Wh/kg and each cell weighed 1.08 kg. Thus the total mass of the cells would be 1425 kg. With a permissible battery mass of 2200 kg, this leaves 775 kg mass for the cartons, battery boxes, insulation etc. Detailed analysis showed this to be a feasible proposition. The importance of design for safety may be appreciated when it is realised that this bus battery contains 260 kg of sodium and 490 kg of sulphur at 350°C. The wisdom of subdividing this into a large number of small cells is apparent.

In conclusion, it should be emphasised that this was a preliminary exercise, undertaken to show that it was feasible to attain the performance targets. Further developments will undoubtedly change the design and it may well be necessary to accept a certain lowering of energy density in order to achieve acceptable safety standards. However, the broad feasibility of designing a Na/S traction battery meeting the performance targets for urban road vehicles is now established, provided the technical problems which determine the reliability and life of cells can be solved.

References

1. "Fast Ion Transport in Solids" Proceedings of NATO sponsored Advanced Study Institute, Belgirate, Italy, 1972 (Ed. W. van Gool) North Holland (1973)

2. N. Weber and J. T. Kummer. Adv. Energy Conv. Eng. 913-916 ASME Conference, Florida (1967)

3. I. Wynn Jones. International Society of Electrochemistry Biennial meeting, Marcoussis, France, May 1975 "A Prototype Sodium/Sulphur Motive Power Battery"

4. M. D. Hames and A. R. Tilley. International Society of Electrochemistry Biennial meeting, Marcoussis, France, May 1975. "Safety in Sodium/Sulphur Batteries"

5. Y. Le Cars, J. Thery and R. Collongues. Rev. Int. Hautes. Temper et Réfract 9, 153-60 (1972) "Domaine d'existence et stabilite des Alumines β et β" dans le systeme Al_2O_3-Na_2O"
J. Fally, C. Lasne, Y. Lazennec, Y. Le Cars and P. Margotin J. Electrochem. Soc. 120, 1296 (1973) "Study of a Beta Alumina Electrolyte for Sodium/Sulphur Battery"

6. D. J. M. Bevan, B. Hudson and P. T. Moseley. Mat. Res. Bull. 9, 1073-1084 (1974) "Intergrowth of Crystal Structures in Beta Alumina"

7. J. L. Sudworth, M. D. Hames, M. A. Storey, M. F. Azim and A. R. Tilley. Power Sources 4 (Editor D. H. Collins, Oriel Press) 1, (1973) "An Analysis and Laboratory Assessment of Two Sodium/Sulphur Cell Designs"

8. J. Fally, C. Lasne, Y. Lazennec and P. Margotin J. Electrochem. Soc. 120, 1292 (1973) "Some Aspects of Sodium/Sulphur Cell Operation"

9. R. J. Bones, R. J. Brook and T. L. Markin. Power Sources Symposium (1974) "Polarisation and Corrosion Experiments on Sodium/Sulphur Cells"

DISCUSSION

COLLINS : In designing a battery we must consider that it will be used by humans, so engineering for human use must be included. This tends to reduce specific energy figures. A review (by Butherus of Bell Telephone Labs) a few years ago of all commercially available systems indicated that you were fortunate if you obtained 20 % of the theoretical specific energies. Of the possible high power liquid electrolyte/solid electrode or liquid electrode/solid electrolyte systems for operation at 0.5 A/cm^{-1}, sodium-sulphur may not be the only solution, considerable progress has been recently made with the lithium (aluminium alloy)/eutectic-chloride-molten salt-electrolyte/sulphur (non sulphide) system.

DELL : Thank you for these comments. I agree entirely that operational factors including the requirement to engineer the system to cater for human foibles, are extremely important considerations. So far relatively little thought has been given to these operational problems and the abuses to which Na/S batteries may be subjected through human failings. I am reasonably confident that batteries with long life and acceptable performance will be built in the laboratory, but whether they will ever find widespread use is another matter.

Regarding the lithium molten salt FeS_2 battery, I am well aware of this work. But I did not discuss it as it falls outside the scope of the Summer School. However, I believe that many of the technical engineering and operational problems are common to the two systems.

ETSELL : Have you any data related to the charging and discharging characteristics of a series of cells rather than a single cell. The necessity of suitably matching a series of cells containing ceramic tubes strikes me as a formidable problem.

DELL : This is certainly true. As mentioned in the paper, if the cells are not matched, the first in a series chain to charge fully will cut-off others in the chain. In practice, this may not be a serious constraint provided the match is not too extensive and there are adequate cross connections between chains of cells. However, if one cell deteriorated prematurely, this could give rise to problems.

Most scientific investigations to date have been concerned with single cells. We are just now beginning a systematic study of the behaviour of series/parallel-connected arrays of cells.

VARIOUS UTILISATIONS OF SOLID ELECTROLYTES

M. Voïnov

BATTELLE, Geneva Research Centre, 1227 Carouge-Geneva (Switzerland)

I. INTRODUCTION

Although the high temperature fuel cell and the sodium-sulphur battery covered in other chapters have generated most of the interest in solid electrolytes, various maybe less spectacular uses of them have been proposed. These applications will be reviewed briefly here trying to point out advantages and disadvantages from a practical point of view.

For some applications the fact that the electrolyte is a solid is just used to slow down diffusion of uncharged species. For other applications solid state implies ruggedness and resistance to acceleration. For still others the key feature of interest is the fact that only one ion migrates in some solid electrolytes of interest as opposed to at least two in liquids. Although their properties do not really allow fabrication of devices that could not be conceived otherwise, the above features sometimes give a slight edge to those featuring solid electrolytes over competitive systems.

For various applications reviewed here the actual magnitude of the ionic conductivity, although of importance, is not as determining a factor as it is in fuel cells and batteries.

On the other hand properties of the electrode - electrolyte interface are often the main concern as far as they determine useful life of devices. It is easy to see that an interface solid - solid will not accommodate readily volume changes whether due to

M. Kleitz and J. Dupuy (eds.), Electrode Processes in Solid State Ionics, 431–466.

metal deposition or to gas evolution when a current flows through that interface. These volume changes produce stresses in the interface and eventually loss of contact between the electrode and the electrolyte.

The use of a liquid electrode like sodium as in the case of the Na-S battery is of course an elegant way to solve the problem but it is probably a unique possibility. Another way has been to use a liquid electrolyte separating the electrode from the electrolyte. This is the approach used in most but not all electrochemical sensors operating near room temperature. In any case, most so far developed commercial applications of solid electrolytes require very small current flow through the electrode-electrolyte interface. The limitation there is probably not in the magnitude of the current but in the total charge passed which determines the total stress on the interface. Investigation of the way interface stresses built up as well as the nucleation and growth of voids on their interfaces should yield interesting results.

Proposed and actually developed utilisations of solid electrolytes are very varied and range from simple refractory heating elements as demonstrated by Nernst at the end of the last century (1) to advanced analogue memory components (2) and high temperature water electrolyser.

Solid electrolytes are used in a traditionally distinct field of technology like metallurgy, chemical engineering and the electronics industry.

II. USE OF STABILISED ZIRCONIA AS A RADIATION SOURCE

As realised by Nernst, oxides having a high melting point and a good conductivity can be used at high temperature without a protective atmosphere either to heat or, as originally envisaged, to produce light. In fact Nernst himself was very much aware of the difficulties encountered in his days with filament lamps. Carbon or metallic filaments had limited life because of poor sealing technique of the glass bulb. A filament stable in air seemed indeed an elegant solution. Nernst having realised the adequate conductivity of mixtures of oxides at high temperature and amongst them probably stabilised zirconia, demonstrated also that the conductivities were ionic.

This he recognised as a problem as because of it he expected rapid degradation of performance due to electrolysis. To explain the long life observed although no data were given, Nernst claimed that at even higher temperature, the conductivity became electronic and consequently electrolysis did not take place any more. Whether this was an "ad-hoc" claim or not is difficult to assess. Indeed we now know that transport number of O^{--}, electrons and holes in stabilised zirconia vary with temperature and oxygen pressure (3). However, in the case for instance of zirconia stabilised with 10 Mole per cent yttria electronic conductivity becames comparable with ionic conductivity at 1200°C when the oxygen pressure is reduced to 10^{-20} atm. For the case of $Zr_{0.88} Ca_{0.12} O_{1.88}$ the oxygen transport number is still higher than 0.99 at 1700°C and oxygen pressure of about 10^{-10} atmosphere (4). To take advantage of the gradual switch to electronic conductivity as temperature increases would thus require an envelope around the zirconia and a very low oxygen partial pressure in it which is what one wanted to avoid to begin with.

In any case Nernst's glower was not a success as a light bulb, either because of electrolyte decomposition or electrode loss of contact. However, for one particular application, stabilised zirconia is still used as a radiation source and this is Infra-Red spectroscopy (5).

Infra-Red Source

A glass adsorbs strongly radiation having wave length higher than 3 µm, normal lamps cannot be used as infra-red sources. Up to 6 µm wave length window can be used. But beyond 6 µm the Nernst glower is still widely used because of its stability in air. It generally consists of a rod of mixed rare-earth oxides which must be heated with, for example, a gas flame to start it.

A typical composition is 90 % ZrO_2 by weight with 7 % yttrium oxide and 3 % erbium oxide. With platinum contact it can be operated at 1800°C and typical power dissipation is 1.2 Watt per square millimeter (6). Its relative emissivity is close to 0.8 and constant above 10 µm (fig. 1).

Tension on the glower is above 100 V and can be A.C. or D.C. but the former is preferable since then net electrolysis is minimum. Given the dimension of the glower (1 mm^2 section, 25 to 30 mm long) it is easy to check that most of the tension drop is ohmic. However, current density as high as 100 A per square centimeter of zirconia are used.

The life of the Nernst's glower is somewhat impredictable and can be from six months to only six hours. Loss of contact between electrode and electrolyte is the most common cause of failure. Because of the low specific heat of stabilised zirconia, the glower must be protected from draughts.

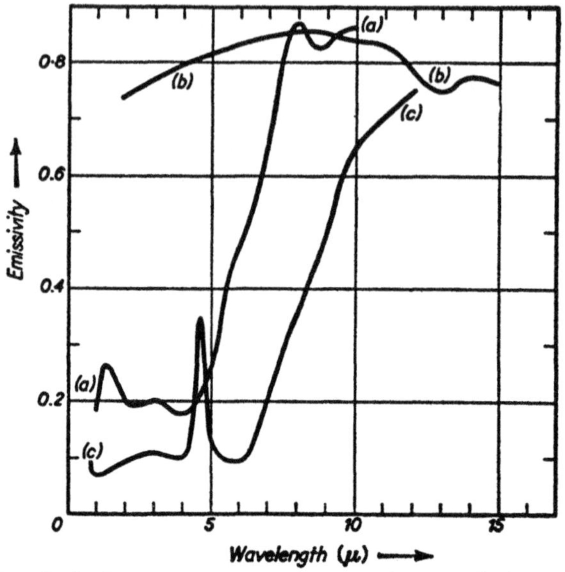

Fig. 1 Emissivities of source materials. (a) Nernst glower; (b) Silicon carbide (globar); (c) Mixture of thoria and ceria.

Heating Elements

The use of oxygen conducting stabilised zirconia as heating elements in high temperature furnaces is of interest both because of its high melting temperature (> 2500°C) and of its stability in oxidising atmosphere (fig. 2).

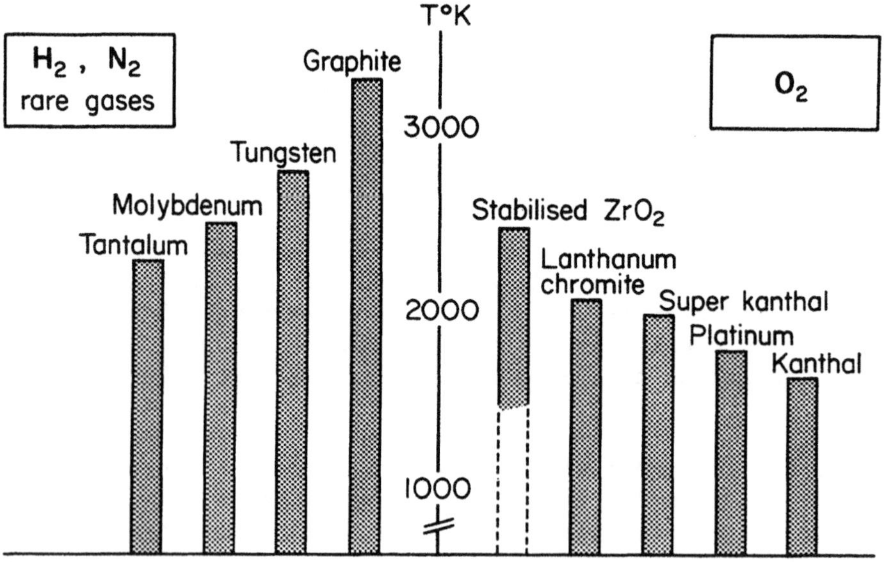

Fig. 2 Temperature limit of utilisation of heating elements.

Because of the high resistivity of the material at low to medium temperature it is not practicle to heat it by passage of current below around 1200°C. It is preferable to preheat it to that temperature where the heat dissipated by conduction overcomes radiation and conduction losses in the furnace. On the other hand, the conductivity of the elements increases rapidly with temperature: for ZrO_2 4 % CaO conductivity increases by five order of magnitudes from 500 to 2000° K. So without precaution one can get thermal run-away and melting of the elements. Typical variation of resistance-current-tension and power with temperature are given in figure 3.

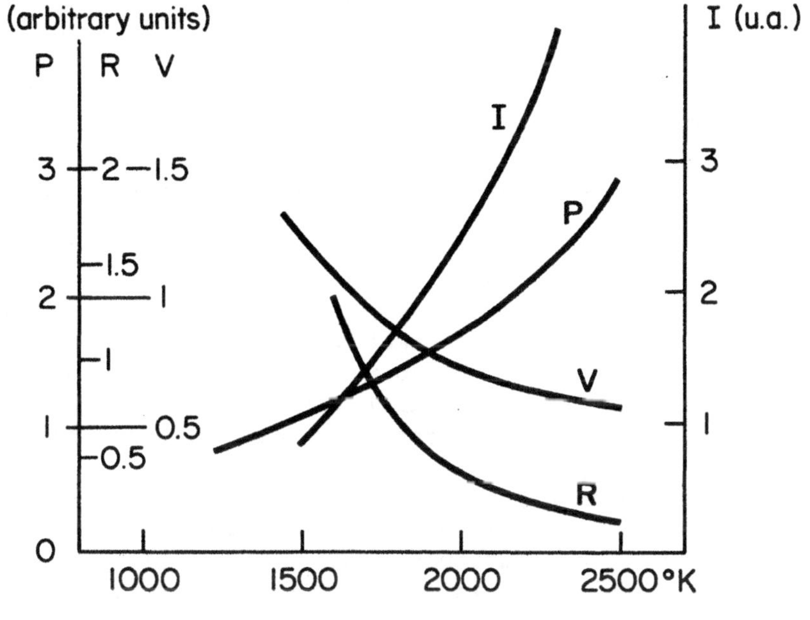

Fig. 3 Typical variation with temperature of resistance current tension and power of a stabilised zirconia heating element.

In practice other factors than the conductivity of the material are important for the construction of furnaces. Thus zirconia is sensitive to mechanical and thermal shock. For instance its elastic modulus is only 15.000 kg/mm^2 at 1500° K and its fracture strength 5 kg/mm^2 at the same temperature.

The current leads which are of course of a different material have to be maintained at a lower temperature because they would not resist the high temperature one wants to reach with stabilised zirconia. This can be done by giving the zirconia heating element a suitable shape but then it has to sustain a thermal gradient and thus a mechanical stress proportional to the difference of temperature. From figure 2 it is seen for instance that Platinum contact are limited to about 1700° C. So with this contact the stabilised zirconia will have to sustain something like a 400° to 500°C temperature gradient if maximum temperature of the elements is sought. Even with lanthanum chromite which can operate to 2000° K a 200° C temperature gradient will exist.

With industrial furnace one thousand of hours of operation at 2100° K were obtained in 1973 (7) with specially designed heating elements withstanding 150° C cm^{-1} thermal gradient and having platinum contact.

In another design (8) contact to the stabilised zirconia is via lanthanum chromite itself connected at 300° C with classical aluminium contact. Other lanthanum chromite elements are used to bring the zirconia to a temperature at which it can operate.

Although the main problem encountered has been stated (7) as creeping of the zirconia at high temperature, it seems also that the behaviour of lanthanum chromite as an oxygen electrode is not as well documented as for instance Platinum.

III. ELECTROCHEMICAL SENSORS

Various electrochemical sensors are used to measure concentrations of different species in mixtures. Only sensors using crystalline solid electrolytes will be considered here. Solid electrolytes, specially those having only one mobile ion, have been used first in ion specific electrodes operating near room temperature in wet chemicals. However, new sensors utilising stabilised zirconia and working at high temperature are also developed for various applications all involving O_2 measurement and chemicals in equilibrium with it.

III.1. Ion Specific Electrodes

In principle any solid electrolyte having only one moving ion can be used to build an ion specific electrode. For this application the solid electrolyte is used as a separator between two reference electrodes (fig. 4). One of these electrodes is in contact with a reference solution on one side of the solid membrane while the other reference electrode sees the sample solution in which one wants to know the ion activity through a salt bridge. Excellent account of the theory of membrane potentials have been published (9).

In the case when two ions are mobile in the solid electrolyte membrane the membrane potential will be :

$$E = \frac{RT}{F} \log \frac{a^{z}_{x_1} + \frac{U_y}{U_x} K_{xy} \; a_{y_1}}{a_{x_2} + \frac{U_y}{U_x} K_{xy} \; a_{y_2}} \qquad (1)$$

where x and y are the two monovalent moving ions and a^*_1, a_2 their activities on the two sides of the membrane, U their mobilities in the membrane and K is the equilibrium constant characterising the exchange between species in solution and in the membrane.

In practice one will use membranes where only one specie can migrate and the membrane potential will then be :

$$E = \frac{RT}{F} \log \frac{a_1}{a_2} \qquad (2)$$

where a_1 and a_2 are activities on both sides of the considered ion specific membrane.

For a given choice of reference electrodes and reference solution in which for example a_2 is known the response of the ion specific electrode to a change of concentration in the sample solution will be :

$$E \text{ (Volt)} = A + 0.059 \log a \qquad (3)$$

for a monovalent cation at 25° C.

The factor A depends on the nature of the reference electrodes, on the concentration of the reference solution and includes too a liquid-liquid junction potential at the salt-bridge connection.

A great advantage of crystalline membranes is generally their high specificity compared to liquid membranes for instance. They are in fact selected so that only one ion specie can penetrate in the lattice of the membrane which simplifies the response of the electrode as shown above.

It can be foreseen that one limit of utilisation of these membranes will theoretically be fixed by the solubility product K of its constituent. Thus if the solubility product of membrane MX specific for M^+ is K it will inject M^+ in solution up to a concentration of :

$$a_{M^+} = \sqrt{K}$$

at least when a true equilibrium is established.

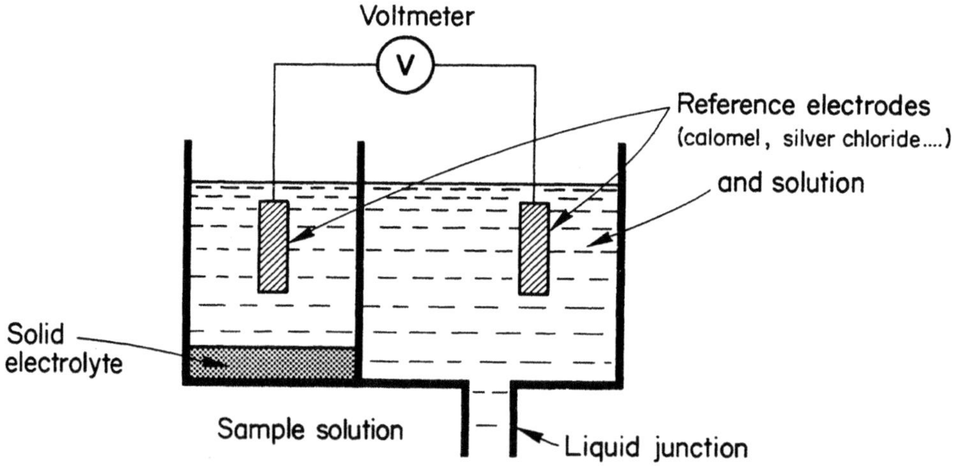

Fig. 4 Schematic diagram of the apparatus used in ion-selective electrode measurement.

Another type of specific ion electrode uses a crystalline solid electrolyte in conjunction with a little soluble salt of the ion to be measured. For instance with MX above can be used MY to measure Y or M'X to measure M'. In the first case one has a membrane potential :

$$E = A + B \log [M^+]$$

and $$[M^+][X^-] = K$$

thus $$E = A + B \log \frac{K}{[Y^-]}$$

In the second case one has similarly :

$$E = A + B \log [M^+]$$

and $$[M^+][X^-] = K_1$$

$$[M'^+][X^-] = K_2$$

Thus $$[M^+] = \frac{K_1}{K_2} [M'^+]$$

and $$E = A' + B' \log [M'^+]$$

Operation of these electrodes depends on the establishment of solubility equilibrium and interference will occur when less soluble compounds than MY or M'X can be formed.

In the following table I are listed electrodes based on two well-known ionic conductors at room temperature namely doped LaF_3 and Ag_2S.

TABLE I

reproduced from reference 9

Ion determined	Membrane	Principal interferences
F^-, La^{+++}	LaF_3	OH^-
CI^-	$AgCI/Ag_2S$	Br^-, I^-, $S^=$, NH_3, CN^-
Br^-	$AgBr/Ag_2S$	I^-, $S^=$, NH_3, CN^-
I^-	AgI/Ag_2S	$S^=$, CN^-
SCN^-	$AgSCN/Ag_2S$	Br^-, I^-, $S^=$, NH_3, CN^-
$S^=$, Ag^+	Ag_2S	Hg^{++}
CN^-	AgI/Ag_2S	I^-, $S^=$
Cu^{++}	CuS/Ag_2S	Hg^{++}, Ag^+
Pb^{++}	PbS/Ag_2S	Hg^{++}, Ag^+, Cu^{++}
Cd^{++}	CdS/Ag_2S	Hg^{++}, Ag^+, Cu^{++}

Lanthanum Fluoride membrane

The conductivity of Lanthanum Fluoride has been reported (10) to be of the order of $10^{-7}\Omega^{-1}cm^{-1}$ at room temperature and the moving specie is F^-. This measurement was made with evaporated gold electrodes or silvered epoxy contact. A fast rising step voltage was applied and the "instantaneous" current read. This procedure of course leads to a very dubious result. A full analysis of the frequency dependence of the impedance is needed if one wants to eliminate electrode polarisation effects. More recent results seem to indicate a conductivity close to $10^{-4}\Omega^{-1}cm^{-1}$ at room temperature (11).

Apparently the conductivity can be further increased by doping with Eu^{++} (12) but no actual data are available.

Because of the low solubility product of LaF_3 (approximately 10^{-23}), it can be used over a wide pH range to measure F^- activity.

The only significant interference comes from hydroxide ions and is probably due to the fact that the solubility product of La $(OH)_3$ ($10^{-21.7}$(13)) is comparable to the one of LaF_3. In acid solution, say below pH 4 to 5, protonation of F^- occurs as HF has a pK of 3.14 and there is a positive shift in potential (fig. 5).

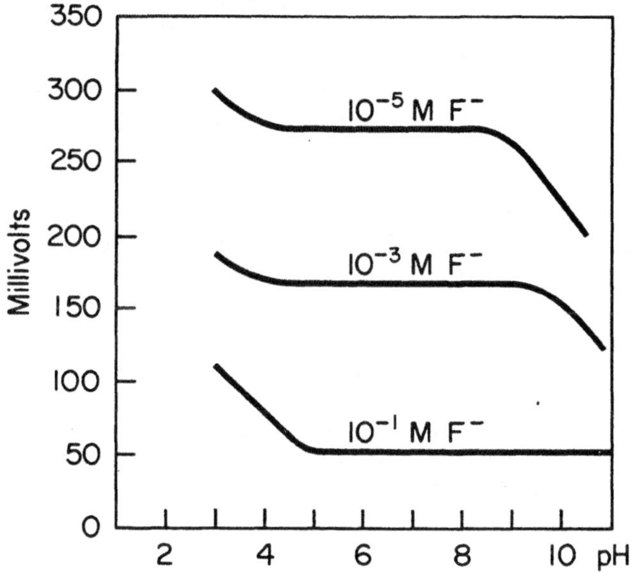

Fig. 5 Electrode response as a function of pH and fluoride concentration. Hydroxyl ion interference depends on the fluoride level. Decreased F^- activity in acid solutions is attributed to formation of associated species.

A lanthanum fluoride membrane can also be used to measure La^{+3} activity by equilibrating sample solution with LaF_3. To reach equilibrium faster, finely dispersed powder is stirred in the sample solution. The amount of F^- actually measured will be a function of the original La^{+3} activity.

It has been claimed too (14) that the conductivity of ultra high vacuum evaporated thin (2000 to 12000 A°) film of LaF_3 between a Bi, Al, La, Pr or Au "anode" and a grid cathode made of silver or gold varied linearly with O_2, CO_2, SO_2 or NO partial pressure. For NO_2 the variation is logarithmic and for CO there is no variation. To explain these rather surprising results the authors speculate that other species than F^- might be able to migrate through LaF_3 (15).

Silver Sulfide membrane

Although by far not the best silver ion conductor known with a published resistivity of 10^3 Ωcm (16) at room temperature, beta silver sulfide Ag_2S has one of the lowest solubility products in water reported: 1.6 10^{-49} at 18° C. Furthermore, it resists attack by reducing and oxidising agent very well and it equilibrates rapidly with sample solution. The fact that a certain degree of electronic conductivity appears depending on the exact stoichiometry should not be a problem as long as there are no metallic electrodes in contact with it. We will come back to this problem with oxygen probes using stabilised zirconia.

Both free S^{--} and Ag^+ can be determined directly to the level of respectively 10^{-19} and 10^{-20} M provided there is an excess of complexed Ag^+ or S^{--}. Main interference is Hg^{++} because HgS solubility product is even lower ($\triangleq 10^{-53}$ to 10^{-49}) than Ag_2S.

As shown in Table I, silver sulfide can be used in conjunction with either silver halides or insoluble sulfides to have a whole series of ion specific electrodes. Notably absent in Table I are carbonate and chromate both having solubility product close to 10^{-12}. They probably should be added in Table I as interfering with anions measurements.

For cations, interferences are easily determined again in view of solubility products :

$$\text{CuS} : 8.5.10^{-45}$$

$$\text{PbS} : 3.4.10^{-28}$$

Lastly, deviation from Nernstian behaviour is observed in acid solution (fig. 6) because of the increased solubility of metal sulfides resulting from HS^- and H_2S formation.

III.2. Oxygen concentration cells and stabilised zirconia

Because of the conduction of electricity by O^{--} in stabilised zirconia and other refractory materials these can be used to measure oxygen pressure provided reference electrode can be applied on them. In this case, "reference electrode" refers to a metal which can exchange electrons with O_2 and O^{--} at a fast enough rate that its potential does not vary appreciably during measurement.

Theoretically, an oxygen concentration cell can be represented schematically as :

$O_2(f_1)$	porous or permeable electrode	Solid electrolyte O^{--} mobile (stabilised zirconia)	porous or permeable electrode	$O_2(f_2)$

where f_1 and f_2 are fugacities.

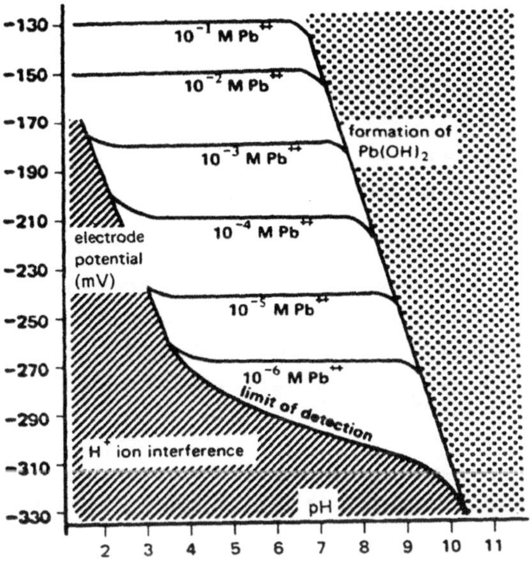

Fig. 6 Effect of pH on a lead-selective electrode. pH varied with $HClO_4$ and NaOH at the indicated background level of $Pb(ClO_4)_2$. (ref. 9).

The electric tension between the two electrodes is :

$$E = \frac{1}{4F} \int_{\mu_2}^{\mu_1} t_{O^{--}} \, d\mu \qquad \text{(A)}$$

where μ_i are chemical potentials of O_2 and $t_{O^{--}}$ the oxygen transport number. At equilibrium, this transport number depends on the oxygen chemical potential (3), that is to say on the oxygen partial pressure. Thus it is known that at high oxygen pressure a contribution to conductivity by positive holes appear while at low oxygen partial pressure an electronic conductivity becomes important.

Because of the variation of t_{O--} with μ_{O_2} integration of (A) is not straightforward but can be found in reference 17.

In practice, one will always try to work in a temperature partial pressure region where t_{O--} is higher than 0.98. Under these conditions (A) is approximated by :

$$E = \frac{RT}{4F} \log \frac{P_1}{P_2} \qquad (B)$$

where P_1 and P_2 are the oxygen partial pressure on both sides of the membrane and are supposed to be the same as fugacities. Knowledge of E and P_1 allows determination of an unknown P_2.

Operation of the device is limited to temperature above 600° C either by the conductivity of stabilised zirconia or by electrode polarisation. Extending down the useful temperature range of operation would probably require voltmeters having higher input impedance than presently known.

Apart from purely thermodynamic investigations, many uses of high temperature oxygen probes using zirconia have been proposed (18).

For instance high temperature oxygen measurement can be used to monitor combustion processes. By knowing the oxygen content of exhaust gas, adjustments can be made to improve combustion efficiency for example in central heating installations or in steam boilers. An oxygen probe based on stabilised zirconia has also been used to control sintering atmosphere in the manufacture of ferrites. Measurement of oxygen partial pressure in reactive cathodic sputtering is also an interesting possibility. Knowledge of the amount of dissolved oxygen in molten glass and molten metals are industrially important and it has been claimed that, under certain conditions, oxygen measurement allows control of carburising furnace atmosphere. As another example, the use of stabilised zirconia in chemical oxygen demand (COD) measurement of water has sometimes been advertised (19).

Measurement of the amount of oxygen dissolved in molten metals

Low oxygen copper has to contain less than 0.001 per cent oxygen and stabilised zirconia can be used to monitor oxygen content of the molten copper before casting.

Corrosivity of molten liquid sodium used as a coolant in modern fast breeder reactors is determined in part by its oxygen content. Rapid and precise knowledge of this content is thus an important safety precaution.

In steel making too, quick knowledge of the amount of dissolved oxygen is required specially in modern basic oxygen furnaces where 100 to 200 ton batches are treated in about 45 minutes. There is no time to take many samples to the laboratory.

For direct oxygen measurement in steel, various probes featuring stabilised zirconia are commercially available. In these probes the liquid metal itself is used as one of the electrodes. To protect the solid electrolyte when it is inserted through the slag layer into the molten steel, a thin copper or steel cap covers it. This cap dissolves rapidly in the steel and measurement is performed after about 15 seconds when thermal equilibrium is reached. Probes have very limited life at their temperature of utilisation which is around 1600° C and have to be discarded after each measurement.

Given the logarithmic response of the probe, its best precision is obtained when the partial pressure to be measured is equal to the reference pressure. The oxygen pressure used as a reference in the electrochemical cell can be fixed in various ways but two reference media are utilised in practice : air and metal-oxide couples (19).

Air of course is very convenient to use as its oxygen content is very constant at 20.9 % and independent of temperature. However, oxygen partial pressure in it is much higher than for oxygen dissolved in steel and consequently measurement precision is not optimised.

Utilisation of other gases like CO, CO_2 and H_2 have been proposed to reduce the reference partial pressure. However, their use create further complication as their flow and temperature have to be controlled precisely during actual measurement.

Provided they reach equilibrium fast enough at measuring temperature ($\hat{=}$ 1600° C) and do not react with the solid electrolyte, metal-oxide couples allow fixation of oxygen pressure to desired value. The equilibrium can be written :

$$MO \rightleftarrows M + 1/2\ O_2 \qquad \Delta G^{o}$$

Knowledge of the standard free enthalpy of this reaction allows calculation of oxygen partial pressure to be put in equation (B).

$$2\ \Delta G^{o} = RT \log P_1$$

$$E = \frac{2\ \Delta G^{o}}{4F} - \frac{RT}{4F} \log P_2$$

Of course as stated above this formula is valid as long as conductivity in the stabilised zirconia membrane is purely ionic. We know this to be the case within a certain temperature, oxygen partial pressure domain (21). Consequently oxygen partial pressure measurement can be done using the above formula in the same domain. This is illustrated in fig. 7 where standard free energies of formation of some oxides are plotted in a familiar way as a function of temperature. There is also on this diagram a logarithmic scale allowing to plot RT log P_{O_2} and thus to determine equilibrium partial pressure for oxides at various temperatures from the relation:

$$\Delta G^{o}_{T} = RT \log P_{O_2}$$

On the same diagram the electrolytic domains for calcia stabilised zirconia and yttria doped thoria is shown. There is a certain uncertainty on the exact limit of these domains. One can see for instance that for measurement in molten copper at say 1200° C, no problem with electronic conductivity of zirconia should be met down to an O_2 partial pressure of 10^{-16} atmosphere. On the other hand measurement in steel at 1600° C appears to be limited to between 10^{-8} and 10^{-13} atm.

Given the relationship between O_2 partial pressure and concentration in liquid iron (table) one wants to measure, it is desirable that the electrolytic domain does extend to 10^{-13} atm.

TABLE II

Relationship between equilibrium partial pressure and concentration of O_2 dissolved in molten iron at 1600° C

O_2 pressure	% O_2
10^{-10}	0.02
10^{-12}	0.0025
3.10^{-13}	0.001

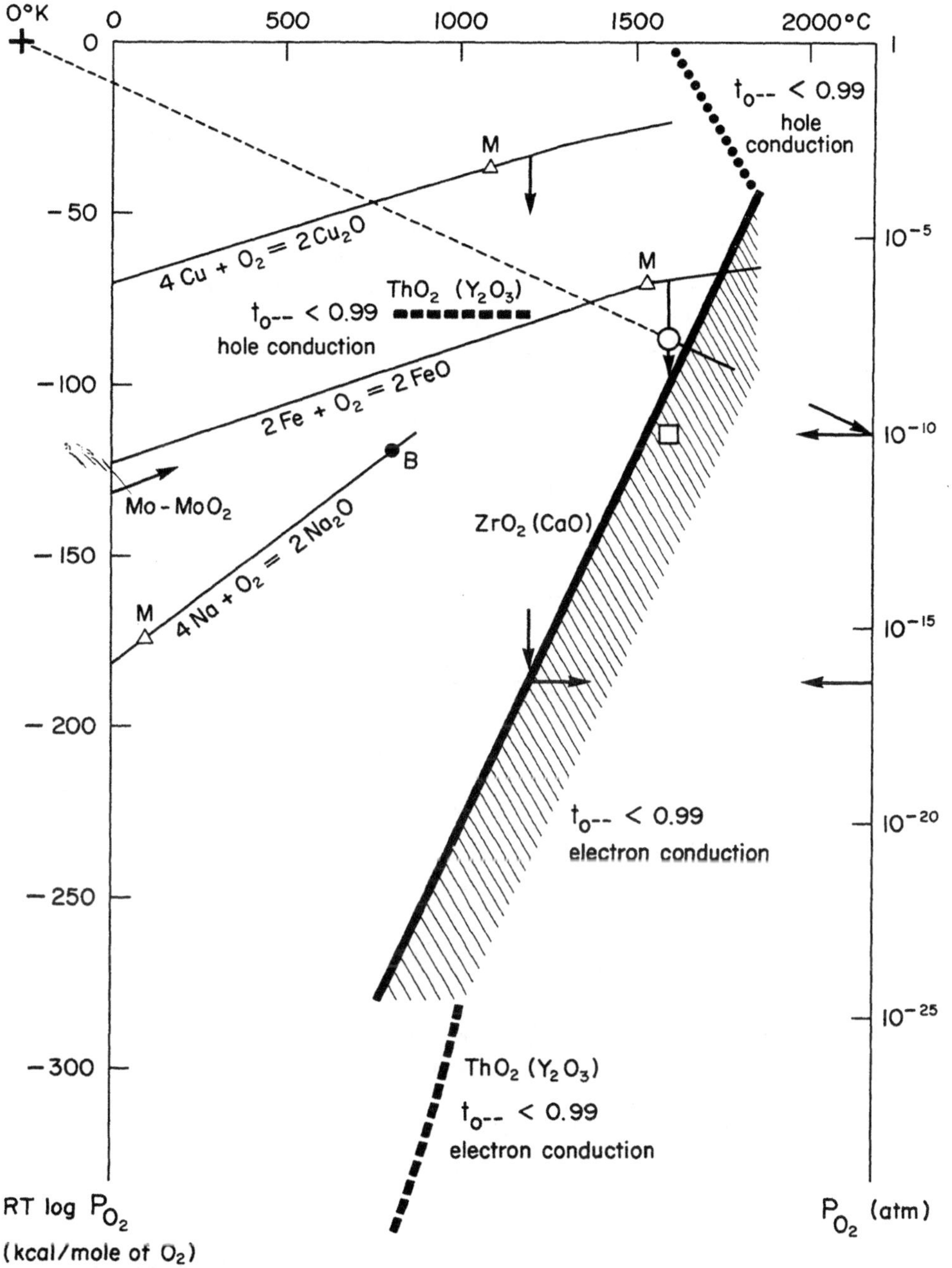

Fig. 7 Diagram free energy. Temperature Oxygen Partial Pressure and electrolytic domain of solid electrolytes. In molten iron at 1600°C and oxygen partial pressure of 10^{-10}atm(O), electronic conduction in stabilised zirconia might appear ⓧ. In molten copper at 1200°C, oxygen partial pressure can probably be determined safely down to 10^{-16}atm. with stabilised zirconia.

As far as oxygen measurements in liquid sodium are concerned, they will be done under conditions where there is electronic conduction in stabilised zirconia. Other materials having better characteristics will have to be used.

Electronic conductivity is responsible for deviation of the solid electrolyte oxygen probe from Nernstian behaviour.

Knowing the pressure p - at which ionic conductivity is equal to the electronic conductivity at a given temperature one can estimate the error made when neglecting electronic conductivity. If P_o is the reference oxygen partial pressure and P is the pressure to be measured, the tension in the case when there is only electronic conduction is :

$$E = \frac{RT}{F} \text{ lin } \frac{P_-^{1/4} + P_o^{1/4}}{P_-^{1/4} + P^{1/4}}$$

By neglecting electronic conductivity one determines a pressure P' from :

$$E = \frac{RT}{F} \text{ lin } \frac{P_o}{P'^{1/4}}$$

The relationship between the true value and the apparent one is:

$$P^{1/4} = P'^{1/4} \left(1 + \frac{P_-^{1/4}}{P_o^{1/4}} \right) - P_-^{1/4}$$

One can see that the true value will be closer to the measured value using the Nernst relation when P- is very small and the reference pressure much bigger than P-. Taking P- as 10^{-15} atm., real activity and measured activity for O in liquid steel have been calculated at 1600° C (20) (Table) with air as reference :

Real Activity %	0.005	0.010	0.020	0.040	0.060	0.08	0.1
Measured (Nernstian) %	0.006	0.012	0.022	0.044	0.064	0.085	0.106

Apart from being responsable for deviation from Nernstian behaviour, electronic conductivity can produce variation in O_2 content of the sample. If the reference pressure is higher than the pressure to be measured, the internal short-circuit due to electronic conductivity results in injection of O_2 in the sample.

This oxygen is removed from the surface by diffusion in the liquid but the interface concentration and consequently the f.e.m. read will bear little relationship with the true value and will be also time dependent. If the reference pressure is smaller than the pressure to be measured, then O_2 is extracted and as it is available to the zirconia interface by diffusion there will again be a time dependent response.

In actual practice, the most often used reference media in commercial probes are air with or without circulation and metal-oxide couples Mo-MoO_2 or Cr-Cr_2O_3 for which equivalent oxygen partial pressure at 1600° C are respectively 10^{-7} and 10^{-13} atm. (22). Internal electrodes are either platinum or molybdenum.

When Mo-MoO_2 or Cr-Cr_2O_3 are used relative error in oxygen activity in the steel are expressed respectively by (20) :

$$\frac{\Delta a}{a} = 1.23\ \Delta E + \left[0.48 - 0.12 \log a\right] \Delta T$$

$$\frac{\Delta a}{a} = 1.23\ \Delta E + \left[0.57 - 0.12 \log a\right] \Delta T$$

where ΔE and ΔT are the precision on tension and temperature readings and are expressed in millivolt and degree centigrade.

Because of the temperature dependence of the oxygen partial pressure of these systems, it is particularly important to know it with precision. As a rule of thumb 10 % precision in O_2 determination in steel requires knowledge of the temperature within 10° C and the tension within 2 mV.

Control of atmospheres

Measurements of O_2 in atmospheres have been used or are proposed for the control of combustion processes or atmosphere composition in metal heat treatment and carburising. Basically the method consists in measuring O_2 partial pressure which is related to other compounds in the atmosphere by chemical equilibria :

$CO + 1/2\ O_2 \rightleftarrows CO_2$ in so-called exothermic gases

$H_2 + 1/2\ O_2 \leftrightarrows H_2O$ in cracked ammonia

and $2\ CO \rightleftarrows C + CO_2$ in so-called endothermic atmosphere

Exothermic gas and cracked ammonia

These atmospheres are used for annealing stainless steel and other metals and alloys and for bright annealing of stainless steel. Composition of these atmospheres vary depending on the nature of the metal or alloy treated. For so-called exothermic gases, this variation is done by changing the air fuel ratio prior to combustion and an oxygen probe can be used to do that automatically.

As shown in fig. 8 where electrolytic stability domain and oxygen potential of various atmospheres (23) is plotted, if the annealing temperature is 1200° C the O_2 partial pressure has to be higher than 10^{-16} or 10^{-20} depending on the actual limit of the electrolytic domain for the measurement to be reasonably feasible with stabilised zirconia. If the higher value is retained, then the oxygen probe will not be very useful for cracked ammonia monitoring which is considerably outside the electrolytic domain of stabilised zirconia. If the lowest value is taken then at 1200° C cracked ammonia atmosphere having dew-point of -70° C cannot be monitored without due correction for electronic conductivity of the electrolyte.

In that respect precise determination of the limits of the electrolytic domain would be very useful.

Carburising atmospheres

In so-called endothermic gases the carbon potential in a metal can be related to the CO/CO_2 ratio provided there is the equilibrium :

$$2\ CO \rightleftarrows [C]_M + CO_2$$

When the CO content of the atmosphere is constant and when the equilibrium :

$$2\ CO + O_2 \rightleftarrows 2\ CO_2$$

is established, there is a relation between the oxygen partial pressure in the endothermic gas and the carbon potential in iron. Again one should be careful not to enter a domain where electronic conductivity becomes non-negligible.

Thus at 1000° C and for zirconia the oxygen pressure has to be bigger than 10^{-17} atm. which corresponds to an oxygen free energy of about - 100 kcal. For an endothermic gas prepared from propane and containing 31 % ($H_2 + H_2O$) and 23 % ($CO + CO_2$) (24) (fig. 9) this means that at that temperature carbon concentration higher than 0.15 % cannot be determined without errors due to electronic conductivity. On the other hand, if the limiting oxygen pressure can be as low as 10^{-22} as is the case for yttria doped

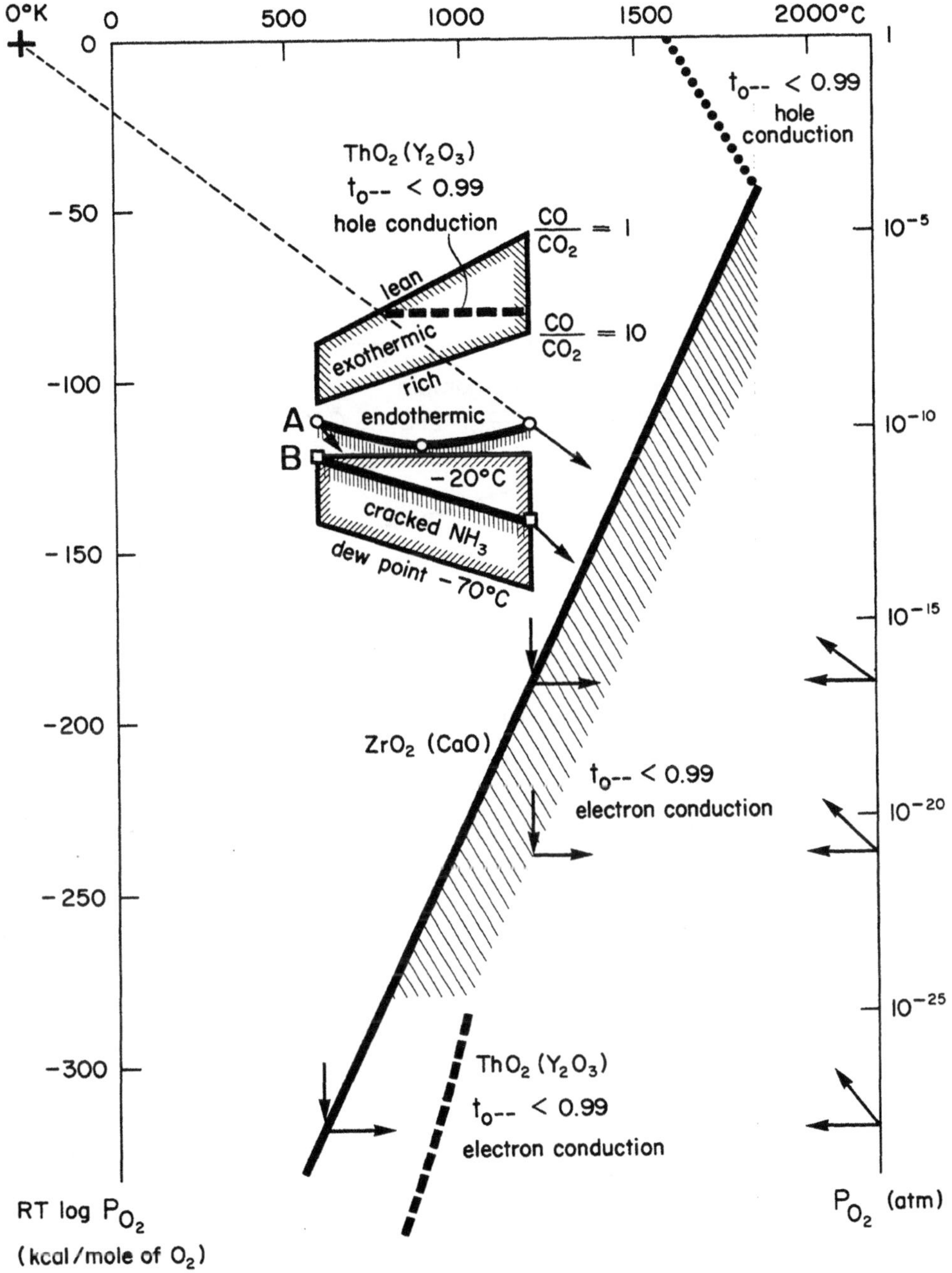

Fig. 8 Representation of typical exo and endothermic atmospheres on a diagram Free Energy-Oxygen Pressure-Temperature and electrolytic domain of high temperature solid electrolytes. With stabilised zirconia oxygen measurements are possible above line A if one takes the upper limit of the electrolytic domain and above line B if one takes the lower value.

thoria (21), oxygen free energy has to be lower than - 130 and then no problem will be met with electronic conductivity in the above atmospheres.

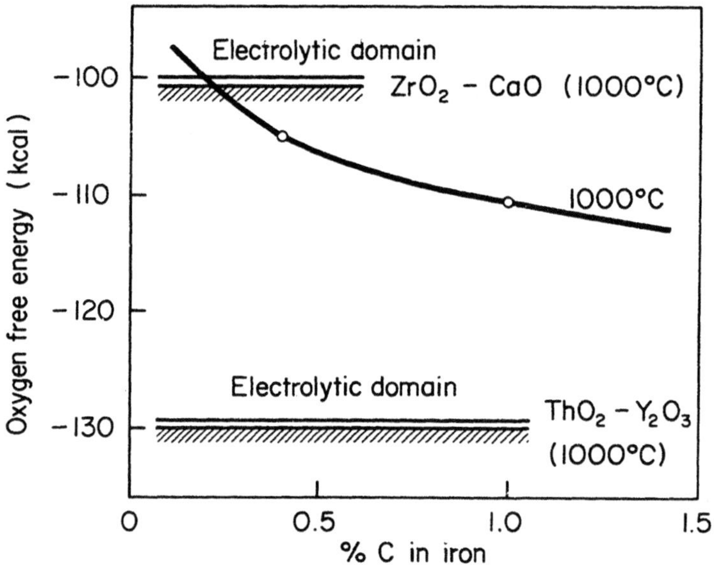

Fig. 9 Electrolytic domain and carbon concentration determination in iron by P_{O_2} determination in a carburising atmosphere (31 % $H_2 + H_2O$, 23 % $CO + CO_2$) (24).

Chemical Oxygen Demand (COD) measurement

To measure rapidly COD ranging from 1 to 5000 mg/l of stabilised zirconia (ZrO_2) a rather ingenious system was advertised by Philips (19) (fig. 10). A sample of water (10 µl) is vapourised and organic matters in it are burned on Pt at 900° C. The combustion gas is oxygen which is injected by electrolysis through stabilised zirconia and controlled also by an oxygen probe using zirconia. After combustion, oxygen pressure is sensed by another stabilised zirconia probe and still another electrolysis reestablishes the original O_2 pressure. Integration over time of current passed to keep the oxygen pressure constant after combustion gives a direct measure of the amount of oxygen answered thus COD. It was claimed that COD values can be obtained in two minutes which represent a vast saving of time over wet chemical methods.

As a clean surface water has a COD of 20 - 30 mg O_2/l and as the solubility of O_2 in fresh water at 20° C is around 10 mg/l, samples must be degassed before combustion. At the highest sensitivity reported (10^{-3}g/l O_2) and with 10 µl samples the charge to be measured is 125 µC.

Gas flow and temperature controls should be quite difficult to realise.

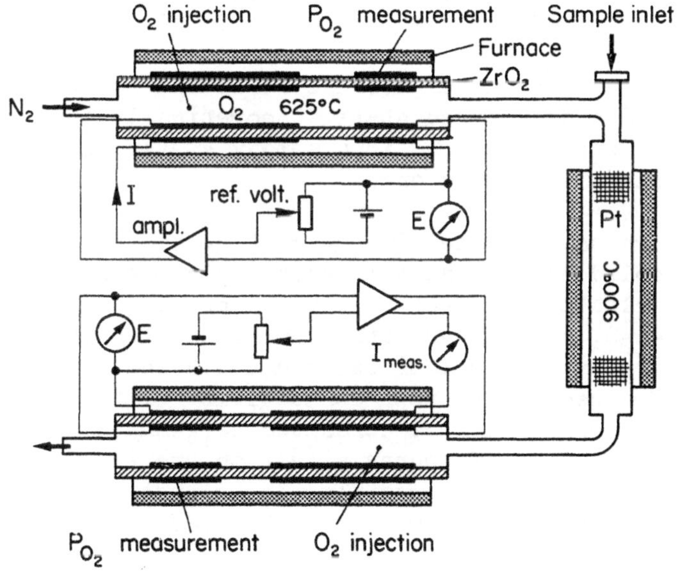

Fig. 10 Chemical Oxygen Demand measurement (Advertised in Scientific American 1972).

III.3. Other probes

Probes using solid electrolytes have been used in various investigations of thermodynamic properties but have not been yet followed by commercialisation (25). Most spectacular of these is the report (26) that Al N is a conductor presumably for triply charged nitrogen.

IV. UTILISATION OF SOLID ELECTROLYTES IN ELECTRONIC CIRCUITS

As early as 1954 Letaw and Bardeen proposed (27) some electrolytic analogue transistor using at that time liquid electrolytes. As they found that the response time of the device was long they thought of using it only to study more readily phenomena occuring in ordinary transistors. Although they attributed long response times to the low mobilities of ions in solutions, they were not in fact depending on ionic conduction for operation but on ion diffusion. Response times observed were times to establish steady-state diffusion conditions. Establishing steady-state conduction conditions takes much less time and interface phenomena, not ion mobilities, are then controlling response times.

Over the years various devices featuring solid electrolytes have been described for use in electronic circuits.

Binary Memory

Since electrochemical storage cells can be switched between a charged and a discharged state, theoretically they should be usable as bi-stable devices in a memory system,with the two states corresponding to zero and one.

To increase response time by getting rid at least of diffusion problems, solid electrolytes in which only one specie is moving should be selected. In fact in preliminary experiments conducted with AgI, Ag_2S, Ag_2HgI_4 and other non specified mixtures (28) response time as low as 1 ms using a 200 V addressing signal were measured. Because of the poor conductivity of these known silver compounds, experiments were conducted between 50 and 140° C. Electrodes used were of course silver on one side and Pt, Pd, Au, C on the other side. Response time is in part limited by double layer capacity of about 200 μF/cm^2 (29, 30).

In the "write" mode the decomposition of the electrolyte occured which must have led to degradation of performances although no cycling performances are available.

Coulometers and Timers

The coulometer dates back to the official birth of Electrochemistry with Faraday's experiments. Utilisation of solid electrolytes offers advantages in resistance to acceleration but more importantly it allows miniaturisation readily by using evaporation and masking technique such as developed in the electronic industry. Complex arrays of thin film coulometers have been thus produced (31). Another feature of solid electrolyte coulometers is that they can be operated over wider temperature ranges than is possible when one uses water. Some organic polar solvents also have this advantage.

Solid electrolyte coulometers developed so far use silver conducting compounds Ag_3SBr (32) Ag_3 SI (33) and Rb Ag_4 I_5 (34) between a silver and a metallic electrode. In operation a known amount of silver is transported and plated on the other electrode. After storage, this amount is stripped away by a current in the reverse direction and transported back to the silver electrode. Under ideal circumstances the tension on the coulometer during stripping should remain constant until the stripping is completed and then, at constant current, it should jump brusquely to a value where another electrochemical reaction can support the current. In practice the transition is not as abrupt as one would like it to be (fig. 11) because of double layer changing, varying activity of silver atom at low coverage and interaction between silver and the electrode.

The amount of stored charge is limited by loss of contact between Ag and the electrolyte which leads to polarisation at constant current. During discharge it is the auxiliary electrode which becomes polarised (fig. 12).

When stripping current is varied, stripping curves are shifted because ohmic drop and electrode polarisation vary with current density. Thus if the cut-off voltage is fixed, errors will be introduced depending on the steepness of the transition. Electrode polarisation can indeed become important as exchange current between 0.6 and 7.8 mA/cm^2 with a 200 mV tension variation per decade of current has been measured (35) for the Ag-Ag_4 RbI_5 interface.

To eliminate ohmic drop and electrode polarisation the use of pulse read-out has met with some success (36).

Charge storage over long periods of time will result in loss by silver diffusion in the electrode. This phenomena is relevant for small charges stored over long times. A 13 % loss after 30 h has been observed with AgBr (37) and about the same value after 12 days.

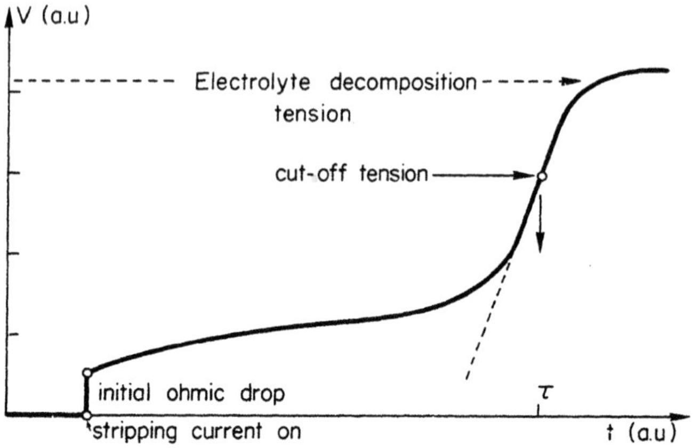

Fig. 11 Features of a stripping curve illustrating initial ohmic drop, cut-off tension, transition time τ and electrolyte decomposition tension.

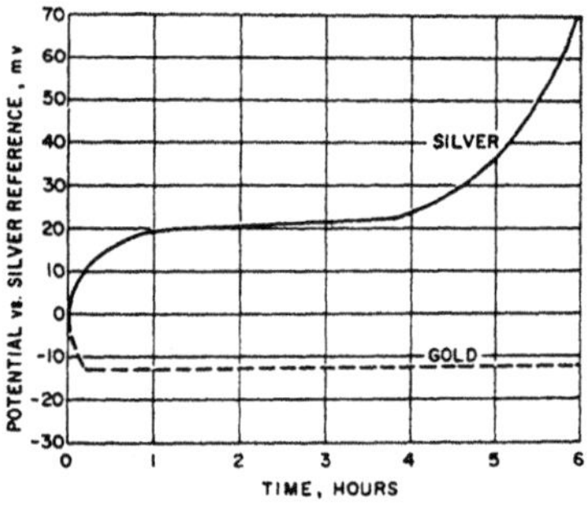

Fig. 12 a Polarisation during charge cycle for Ag_3SBr. Charge: 21,600 sec at 25 μA.

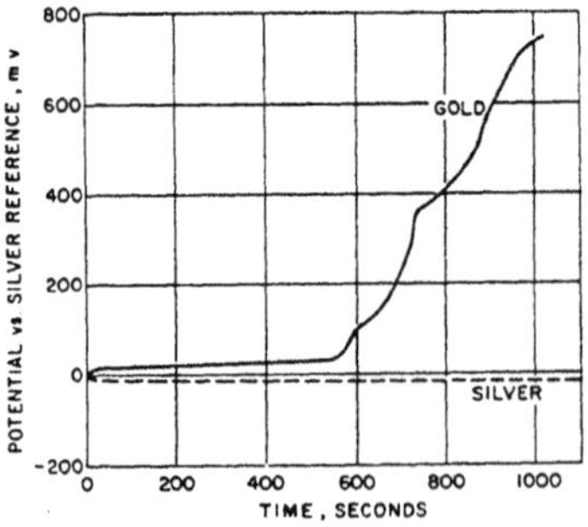

Fig. 12 b Polarisation during stripping cycle for Ag_3SBr. Charge: 800 sec at 25 μA; stripping current: 25 μA.

(12a and b from reference 33)

On the other hand, the fact that during storage there is not tension between the electrodes is a very favourable feature for long storage.

Main features of solid electrolyte coulometers have been reviewed (38). As a rough rule coulometric accuracy is better than 5 % for current density from 5 to over a 100 μA/cm^2 and charging time from 50 up to 250 s. In general, these accuracies are obtained within 1 hour of the end of the charge.

One can imagine that silver dendrite growth and internal short circuit is a major problem when cycling these coulometers. However, cycling (2000 μC/cm^2 or roughly 10 monolayers) over 20.000 cycles have led to no apparent damage (31) with thin film evaporated AgBr. In practice, solid electrolyte coulometers having storage capacity of 6000 μAh have been developed (39).

Solid electrolyte coulometers like other coulometers can be used in circuits indicating operating times larger than 1000 h or in delay line having a range from 1 s to 1 month or in cyclic delay line (fig. 13) (40).

Analogue memory

Of course a coulometer as above with or without a solid electrolyte can be used as an analogue memory. A known amount of charge is passed through it, plating a known amount of metal on an indicator electrode which can be read by current reversing. In order not to erase the information when reading it, it has been proposed (36) to use two back-to-back coulometers to transfer the information to another one when reading it (fig. 14).

Another type of electrochemical analogue memory has been investigated (2, 41, 42). In this type of device a mixed electronic and Ag^+ conductor is sandwiched between two layers of $RbAg_4I_5$.

Mixed conductor

$$Ag \;\Big|\; Rb\,Ag_4\,I_5 \;\Big\|\; \begin{matrix} Ag^+ \\ e^- \end{matrix} \;\Big\|\; Rb\,Ag_4\,I_5 \;\Big|\; Ag$$

Because of the electronic conductivity of the compound in the centre of the device, its potential versus the silver electrodes can be measured directly without further reference electrodes. On the other hand, this potential is determined by the Ag^+ activity in the mixed conductor. If the stoichiometry of the mixed conductor varies this potential will change.

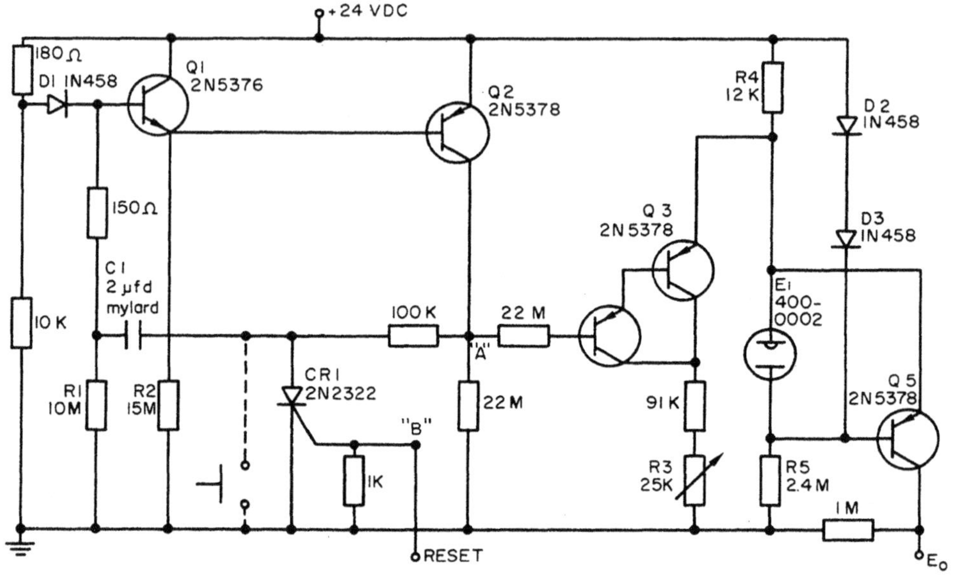

Fig. 13 Repetitive delayed signal circuit featuring a coulometer (E_1) (reference 40).

Charge time and charge current are determined respectively by R_1 C_1 and R_3. When resetting, C_1 is discharged rapidly either with a manual switch (---) or by applying an appropriate signal in "B". E_1 is then charged through D_2 D_3 and R_3. When C_1 is charged, Cl_3 Cl_4 switch and E_1 is discharged through R_4 R_5. End of discharge triggers Cl_5 producing the wanted timed variation of E_o. From E_o electrical resetting can be done to obtain a repetitive signal. (E_1 400-0002: identification code from the Berman Corp. Bissett).

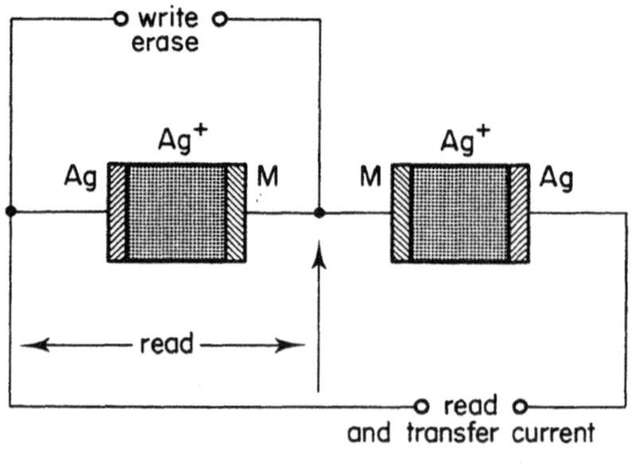

Fig. 14 Circuit to read an analogue electrochemical memory without losing its content.

The principle of the operation of the analogue memory is thus: A writing or erase current injects or pumps Ag^+ out of the mixed conductor. One silver electrode is used for that purpose. The other one is used to read the information stored as silver ions in the mixed conductor.

In practice, mixed conductors like β-Ag_2Xe and β-Ag_2 Te have been tried (44) but the compound: $(Ag_2Se)_{0.925}$ $(Ag_3\ PO_4)_{0.075}$ is preferred (43). Electronic conductivity of this particular compound is $2.10^3\ \Omega^{-1}cm^{-1}$ while the ionic conductivity is as good as the one Rb $Ag_4\ I_5$ ($0.26\ \Omega^{-1}cm^{-1}$) at room temperature. The potential versus a silver reference is changed by up to 150 mV for apparent current density from 100 µA to 10 mA/cm^2 and 0.2 g of mixed conductor. This corresponded to a charge of 0.318 ± 0.002 C.

This maximum potential change of 150 mV did not decrease by more than 2 % after 4 hours of storage. After 360 h a 5.8 mV potential variation was reported (2).

The fact that the time necessary to read such a device is much faster than with analogue coulometers is of course an advantage. But addressing and erasing time would be just as long as compared to electronic solid state devices.

Energy Storage Device

The double layer between an inert electrode like carbon and a solid electrolyte like Rb $Ag_4\ I_5$ can be as high as 20 $\mu F/cm^2$ (35) in potential region where there is no decomposition of the electrolyte.

It has been shown (43) that a rather high overall capacitance (Table III) could be obtained by optimising the total interface area in a given volume. Such a capacitance is constituted by a carbon electrode mixed with Rb $Ag_4\ I_5$ and another silver electrode. Tension on it is limited to about 0.6 V where electrolysis occurs.

TABLE III	Volt	C (Farad/cm^3)	CV (coulomb/cm^3)	1/2 CV^2 (Joules/cm^3)
ESD (1 cell)	0.5	9.9	4.95	1.23
ESD (10 cells in series)	5	0.136	0.68	1.73
Aluminum	5	$0.55.10^{-3}$	$2.75.10^{-3}$	$6.8.10^{-3}$
Tantalum	6	$0.34.10^{-3}$	$2.04.10^{-3}$	$6.3.10^{-3}$
Ceramic	1000	10^{-7}	10^{-4}	5.10^{-2}

V. DISPLAY DEVICES

Although apparently still in very early stages of development both active and bistable passive display devices featuring solid electrolytes have been patented.

It has been claimed (44) for instance that sodium beta-alumina can be used to inject holes in a semiconducting material like cadmium fluoride doped with some rare earths. Electron-hole recombination is accompanied by emission of light. However, the fact that more light was obtained when the interface CdF_2-β-Al_2O_3 was wetted with an aqueous solution of sodium hydroxide casts some doubt on the actual usefulness of β-Al_2O_3 in such light emitting devices or on their mode of operation.

As far as bistable passive display devices are concerned it is known that light adsorption characteristics of substances like WO_3 or MoO_3 are modified by electric field (45). The phenomena consist in the appearance of an adsorption band and devices featuring it are unproperly called "electrochromic". It was also found that instead of metallic contact (46) liquid electrolyte contact to the substances could be used. The next step was then to replace the liquid electrolyte with a solid one and the use of sodium β-alumina has been patented (47). By application of a proper polarity tension, sodium is supposed to be injected in WO_3 apparently changing its light adsorption characteristics. It is doubtful that sodium tungsten bronzes Na_xWO_3 are formed in this way. On the other hand, as it is known that these bronzes change colour with the amount of sodium in them, they could conceivably be used instead of pure WO_3 provided the rate of exchange between them and β-Al_2O_3 is fast enough. Their advantage over WO_3 is that they are better electronic conductors. Apart from the difficulty of making a good solid to solid contact, response time of such devices in the write or erase mode will be limited by room temperature solid state diffusion of sodium in tungsten trioxide.

In any case, energy to write or erase devices based on such "electrochromes" will be typical of the necessary energy used in wet electrochemical display devices such as those using metal plating or polyviologen (48). Operation of these devices require from 1 to 10 mJ/cm^2 with response times in the tens of millisecond range.

The use of known solid silver and copper ion conductors in display devices has also been proposed (49).

A difficulty common to all display featuring electrolytes is the detection of the end of erase. Beyond that time, if the erasing current is still applied there will be decomposition of the electrolyte. Electrolyte decomposition will also happen if one applies an erasing current on a non-written part. Detection of end of erase can be done with a reference electrode or probably with an electronic circuit such as the one in figure 14. In fact, during erasing the potential behaves like the tension on a Coulometer during stripping.

Solid electrolytes offer an advantage over liquid electrolytes because slow processes associated with ionic diffusion in solutions are eliminated.

VI. THERMOGALVANIC CELLS

Thermogalvanic cells featuring solid electrolytes have received little attention from scientists although they have many interesting features as compared with those using liquid electrolytes.

Solid electrolytes can withstand higher temperatures than liquids which is important for the Carnot efficiency of a thermogalvanic cell. In the case of devices featuring a liquid electrolyte and a redox couple, the electroactive species diffuse from one electrode to the other. To have high steady state current, diffusion gradient should be as steep as possible which means bringing the electrodes close to each other. There results an increase in thermal loss by conduction. On the other hand, with a solid electrolyte such as β-$A\ell_2O_3$ the electroactive species migrate in the electrolyte where it is the only possible current carrier. Consequently the current will not be limited by mass diffusion but by heat diffusion in metallic electrodes or by the electrical resistance of the solid electrolyte.

With a thermogalvanic cell featuring β-$A\ell_2O_3$ and liquid sodium, between 0.3 and 0.4 volt open-circuit tension has been measured for overall temperature gradient of about 250°C (50). The particular design used did not allow precise knowledge of this gradient across the β-$A\ell_2O_3$.

The figure of merit for a thermogalvanic cell I_2 - AgI - I_2 has been calculated (51) to be around 10^{-3} between 350° and 500°C when the electrical resistivity of AgI is close to 0.4Ωcm. This figure compares favourably with values of $2.5.10^{-3}$ for n-type $Bi_2(Te\ Se)_3$ at 100°C and 2.10^{-3} at 400°C for p-type $(Ag\ Sb\ Te_2)._{15}(Ge\ Te)_{.85}$.

On the other hand, the sensitivity of solid electrolyte ceramics such as β-$A\ell_2O_3$ or zirconia to thermal stresses can be considered to be a drawback.

REFERENCES

(1) W. Nernst, Z. für Elektrochemie No. 2, p.41 (1899).

(2) T. Takahashi and O. Yamamoto, J. of Applied Electrochem., 3, 129-135 (1973).

(3) L.D. Burke, H. Rickert and R. Steiner, Z. für Phys. Chem. Neue Folge, Bd. 74 S, 146-167 (1971).

(4) W.A. Fischer and D. Janke, Archiv für das Eisenhüttenwesen, 39, No. 2, p.89-99 (1968).

(5) R.G. White, Handbook of Industrial Infrared Analysis, Plenum Press (1964).

(6) A.E. Martin, Infrared Instrumentation and Techniques, Elsevier (1966).

(7) A.M. Anthony, K. Dembinski and L. Dupont, "Sciences of Ceramics", Vol. 7, p.137 (1973).

(8) M. Moise, CENS, SEEN, BP No. 2, Gif-sur-Yvette 91190 (F), Private Communication.

(9) G. Eisenman in Ion-Selective Electrodes, National Bureau of Standards, Special Publication, 314 (1969).

(10) A.Sher, R. Solomon, K. Lee and M.W. Muller, Phys.Rev., Vol. 114, No. 2, p.593-604 (1966).

(11) L.E. Nagel and M. O'Keeffe, Fast Ion Transport in Solids, North-Holland pub. co, p.165-172 (1973).

(12) M.S. Frant and J.W. Ross, Science, 154, 1553-1554 (1966).

(13) L.G. Sillen and A.E. Martell, Stability Constants of Metal Ion Complexes, Spec. Pub. No. 17, The Chem.Soc.London (1964).

(14) B.C. LaRoy, A.C. Lilly and C.O. Tiller, Abstract No. 167 and 168, Electrochem. Spring Meeting, Houston (1972).

(15) B.C. LaRoy, A.C. Lilly and C.O. Tiller, J. of the Electrochem. Soc., Vol. 120, No. 12, p.1668-1673 (1973).

(16) R.G. Cope and H.J. Goldsmid, Brit.J.Appl.Phys., 16, 1501, (1965).

(17) Schmalzried H., Z.für Phys.Chemie, B 38, p.87-102 (1963).

(18) R.T. Sprouie, Materials Engineering, Vol.79, No 1 (1974)

(19) Advertisement in Scientific American, Vol. 226, No. 6 p.85 (1972).

(20) M. Chastant, C. Gatellier, Mme M. Jon et M. Olette, Rapport IRSID REI38, p.951-994 (1974.

(21) J.W. Patterson, J. of the Electrochem.Soc., 118, 7, p.1033-1039 (1971).

(22) M. Chastant, C. Gatellier, M. Olette, Proceedings of SIMAC 1972, p.11/1 - 11/6.

(23) R.G. H. Record, Control and Instrumentation, p.31-34, (Feb. 1971).

(24) L.H. Fairbank and L.G.W. Palethorpe "Controlled Atmospheres for the Heat Treatment of Metals", The Iron and Steel Inst., Special Report 95.

(25) W.L. Worrel, Ceramic Bulletin, Vol. 53, No. 5 (1974).

(26) W.A. Fischer and B. Schuh, Arch. Eisenhüttenwes. 46. No. 2 p.101 (1975).

(27) H. Letaw Jr. and J. Bardeen, J. of Appl. Phys. Vol. 25, No5, p.600-606 (1954).

(28) R.V. Andes and D.M. Manikowski, Abstract No. 157, Electrochem. Soc., Pittsburgh Meeting (1963).

(29) D.O. Raleigh, J. of Phys. Chem., 70, 3, p.689-698 (1966).

(30) R.D. Armstrong, T. Dickinson, W.P. Race, R. Whitfield, Electroanal. Chem. and Interf. Electrochem., 27, 158-160 (1970).

(31) J.H. Kennedy and F. Chen, J. of the Electrochem. Soc., 118, 7, p.1043-1047 (1971).

(32) J.H. Kennedy and F. Chen, J. of the Electrochem. Soc., 116, 2, p.207-211 (1969).

(33) O. Yamamoto and T. Takahashi, Denki Kagaku 36, 894 (1968).

(34) Gould Ionics Inc., Bull. No.118, 119, 120.

(35) R.D. Armstrong, T. Dickinson and P.M. Willis, Electroanal. Chem. and Interf. Electrochem., 57, 231-240 (1974).

(36) J.H. Kennedy, F. Chen and J. Willis, J. of the Electrochem. Soc., 117, 263 (1970).

(37) J.H. Kennedy, F. Chen and Andy Clifton, J. of the Electrochem. Soc., 115, 9, p. 918-924 (1968).

(38) J.H. Kennedy in Physics of Electrolytes, Vol.2, p.931-956, Academic Press (1972).

(39) Gould Ionics, model U-6000.

(40) Neue Züricher Zeitung, Mittagsausgabe Nr. 447, 25 Sept.1972.

(41) T. Takahashi and O. Yamamoto, J. of the Electrochem. Soc., 118, 7, p.1051-1057.

(42) T. Takahashi and O. Yamamoto, Abstract No. 176, Spring Meeting of the Electrochem. Soc., Houston (1972).

(43) J.E. Oxley, Abstract No. 175, Spring Meeting of the Electrochem. Soc., Houston (1972).

(44) U.S. Pat. 3.445.724.

(45) S.K. Deb., Appl. Optics Suppl. No. 3, p.192-195 (1969).

(46) U.S. Pat. 3.807.832.

(47) U.S. Pat. 3.712.710.

(48) H.T. Van Dam and J.J. Ponjée, J. of the Electrochem. Soc., 121, 12, p. 1555-1558 (1974).

(49) M. Voïnov, 3rd Meeting, European Solid State Electrochem. Group, Bressanone, Italy (1974).

(50) U.S. Pat. 3.458.356.

(51) J.L. Weininger, J. of the Electrochem. Soc., 111, 7, p.769-774 (1964).

DISCUSSION

HEYNE : You did not mention the oldest and mostly used ion selective electrode, that for pH, the glass electrode. But this raises the interesting point that ions can be measured that are not the charge carriers in the solid. The same applies to β-alumina which works as a reasonable pH electrode. Can you say something about the mechanism ?

VOÏNOV : If β-Al_2O_3 can be used in a pH electrode,it is because it can exchange Na^+ with H^+ or OH^- and then there is a membrane potential, like in the case of the glass electrode, which depends not only on concentrations but also on the mobilities of Na^+ or H^+/OH^- in β-Al_2O_3 and on the exchange equilibrium constant.

HEYNE : The figure you mentioned for the lower limit of detection for the C.O.D. measuring instrument is of the right order of magnitude. Background current is of the order of 1 μA, time of complete cycle 2 or 3 minutes. So about 150 μC is the equivalent of the lower limit of C.O.D.

BONANOS : With reference to the electronic conductivity of Y.S.Z. : if we use Y.S.Z. to compare two O_2 partial pressures which differ significantly, any electronic conductivity in the solid electrolyte will lower the E.M.F. measured and give erroneous results. If however, the electrolyte is used mainly to detect a difference in O_2 partial pressures, one of which can be varied at will so as to make this difference zero (i.e. an O_2-partial-pressure-null-detection-system). The electronic conductivity would not affect the accuracy, but only the sensitivity of the system. This set-up could possibly extend the range of usability of Y.S.Z. to condition where the electronic conductivity is significant.

VOÏNOV : It sounds like a very good idea but I have a feeling that null detection will be very sluggish precisely because of electronic conduction.

MESCHTER : Voïnov is too pessimistic about the uses of solid electrolytes ; attaining the thermodynamic potential is not the important requirement. In fact, one device in cars uses pure ZrO_2 as the "electrolyte" as long as the measurements are reproducible the device will be acceptable. We now have more options for measurement of oxygen in liquid steel. Dr. Jancke's group at the Max-Planck-Institut für Eisenforschung (Düsseldorf) has developed $CaZrO_3$ for this purpose. It allows accurate measurement of oxygen contents in steel at lower concentrations than does $ThO_2(Y_2O_3)$ and is claimed to have better thermal shock resistance than either $ZrO_2(CaO)$ or $ThO_2(Y_2O_3)$.

ETSELL : N-type conductivity in zirconia does not interfere with oxygen measurements in steel at oxygen concentrations generally present in steelmaking operations (> 25 ppm). Furthermore, corrections can be successfully applied to small electronic contributions. Interference from electrode polarization arising from oxygen transfer across the electrolyte does not constitute a serious problem at high temperatures where diffusion processes are rapid providing the reference electrode is properly chosen.

Transport numbers for ZrO_2-CaO and ZrO_2-Y_2O_3 from 600°-1600°C are summarized and evaluated in :

- J. Electrochem. Soc. <u>119</u>, 1-7, (1972)

The ThO_2-Y_2O_3 reference should have been :

- Z. Naturforsch, <u>27a</u>, 1138, (1972)

VOÏNOV : At 1600°C, 25 ppm corresponds roughly to an oxygen partial pressure of 10^{-12} atm. Most authors have reported that electronic conduction becomes apparent in calcia stabilized zirconia (t_{ion} < 0.99) at 10^{-8} to 10^{-10} atm. If the limit is 10^{-10} atm this corresponds roughly to 200 ppm. Pushing the limit to 10^{-12} atm would be very nice.

The fact that diffusion is rapid does not mean you will have a linear gradient of chemical potential thus a steady state. To have that, you must, one way or another, establish a diffusion boundary layer, for instance by stirring.

Den ENGELSEN : We made some experiments using AgI based electrolytes in order to study the possibility of display applications. Anodically generated iodine gives deeply colored charge-transfer complexes with perylene and related aromatic compounds. We have found that the visibility of these complexes was higher than metallic silver deposited at the cathode. The reason is that one monolayer of such a charge transfer complex absorbs already a substantial amount of light. For 1 cm^2 of a perylene monolayer to become visible, we need about 10^{-4}C. There are two difficulties connected with this type of display: the high resistance of perylene without iodine, the low diffusion of iodine at erasing.

GPSR Compliance

The European Union's (EU) General Product Safety Regulation (GPSR) is a set of rules that requires consumer products to be safe and our obligations to ensure this.

If you have any concerns about our products, you can contact us on ProductSafety@springernature.com

In case Publisher is established outside the EU, the EU authorized representative is:

Springer Nature Customer Service Center GmbH
Europaplatz 3
69115 Heidelberg, Germany

Batch number: 10816188

Printed by Printforce, the Netherlands